PERGAMON INTERNATION
of Science, Technology, Engineering

The 1000-volume original paperback librar
industrial training and the enjoyment of leisure

Publisher: Robert Maxwell, M.C.

APPLIED ELECTROMAGNETISM

APPLIED ELECTRICITY AND ELECTRONICS
General Editor: P. HAMMOND

Other titles of interest in the
PERGAMON INTERNATIONAL LIBRARY

ABRAHAMS & PRIDHAM
Semiconductor Circuits: Theory Design and Experiment
Semiconductor Circuits: Worked Examples

BADEN FULLER
Engineering Field Theory

BADEN FULLER
Worked Examples in Engineering Field Theory

BINNS & LAWRENSON
Analysis and Computation of Electric and Magnetic Field Problems, 2nd Edition

BROOKES
Basic Electric Circuits, 2nd Edition

CHEN
Theory and Design of Broadband Matching Networks

COEKIN
High Speed Pulse Techniques

CRANE
Electronics for Technicians

CRANE
Worked Examples in Basic Electronics

DUMMER
Electronic Inventions 1745–1976

FISHER & GATLAND
Electronics—From Theory into Practice, 2nd Edition

GARLAND & STAINER
Modern Electronic Maintenance

GATLAND
Electronic Engineering Application of Two-Port Networks

GUILE
Electrical Power Systems, Vols 1 & 2

HAMMOND
Electromagnetism for Engineers, 2nd Edition in SI/Metric Units

HANCOCK
Matrix Analysis of Electrical Machinery, 2nd Edition

HARRIS & ROBSON
The Physical Basis of Electronics

HINDMARSH
Electrical Machines and their Applications, 3rd Edition

MURPHY
Thyristor Control of AC Motors

Applied Electromagnetism

BY

P. HAMMOND

Head of the Department of Electrical Engineering
Southampton University

PERGAMON PRESS

OXFORD · NEW YORK · TORONTO · SYDNEY
PARIS · FRANKFURT

U.K.	Pergamon Press Ltd., Headington Hill Hall, Oxford OX3 0BW, England
U.S.A.	Pergamon Press Inc., Maxwell House, Fairview Park, Elmsford, New York 10523, U.S.A.
CANADA	Pergamon of Canada Ltd., 75 The East Mall, Toronto, Ontario, Canada
AUSTRALIA	Pergamon Press (Aust.) Pty. Ltd., 19a Boundary Street, Rushcutters Bay, N.S.W. 2011, Australia
FRANCE	Pergamon Press SARL, 24 rue des Ecoles, 75240 Paris, Cedex 05, France
FEDERAL REPUBLIC OF GERMANY	Pergamon Press GmbH, 6242 Kronberg-Taunus, Pferdstrasse 1, Federal Republic of Germany

---

First edition 1971

Reprinted with corrections 1978

Library of Congress Catalog Card No. 79–149027

Printed in Great Britain by A. Wheaton & Co. Ltd., Exeter

ISBN No. 0 08 016382 3 flexicover
0 08 016381 5 hardcover

TO DAVID AND LOIS

On a huge hill,
Cragged, and steep, Truth stands, and hee that will,
Reach her, about must, and about must goe.

(JOHN DONNE, Satyre 3)

CONTENTS

PREFACE

ELECTROMAGNETISM provides the logical framework of electrical engineering. For this reason it is becoming generally recognized that a course in electromagnetism is an essential part of a broadly based curriculum in electrical engineering and applied science. It would be idle to deny that this is a difficult subject, but the difficulties can be surmounted. The substance of this book has been taught for several years in an undergraduate course and a substantial proportion of the students have shown by their interest that they have enjoyed the experience. Not only have they found the subject relevant and useful, but it has also given shape to the rest of their studies.

In choosing the title *Applied Electromagnetism* I am stressing that the end-product of electromagnetic devices and systems has determined both the subject-matter and, more importantly, the approach to the subject. The mathematical treatment is subservient to the physical content. Engineering students have a disconcerting habit of asking for a physical explanation, and this is what I have sought to provide in the book. On the other hand, I have avoided detailed description of apparatus. This would have made the book very bulky and it would have interrupted the argument. Many applications of the theory are given in the exercises at the ends of the chapters, and the student should test his grasp of the subject by working these exercises.

This book has been in the making for many years. Now that it is almost finished I am very conscious of my debt to many people. There are those who taught me, there are the authors of the books I have consulted, and there are the students who by their clarity of mind and inquisitiveness have forced me to come to grips with difficulties I had overlooked. I cannot list their names, but I hope they will accept this acknowledgement.

Three people, however, I do want to thank by name: Mr. C. J. Carpenter and Dr. R. L. Stoll read the typescript of the book and made many helpful comments; Dr. K. K. Lim gave valuable help with Chapter 13. Finally, I want to thank the publishers for their encouragement and skill.

The international system (SI) of units is used throughout the book.

Southampton 1970 P. HAMMOND

CHAPTER 1

ELECTROMAGNETISM AND ELECTRICAL ENGINEERING

1.1. Engineering and Mathematics

Authors of scientific textbooks generally give no reason for their work except possibly in a preface. But few people read the preface, especially if it is a long one. Therefore, since it is important that reader and author should establish some mutual understanding, I am putting my reasons for this book in the first chapter. This should have two advantages. First, you may use this part of the book as a map to survey the country through which we are to pass. Secondly, you will find the first chapter of the book easy and so will be spared the annoyance of having to deal with a book constructed like a medieval castle with a raised drawbridge, which forces the reader to swim across the moat of a difficult first chapter to test his powers of endurance. A third possibility, which is hardly an advantage, is to treat this chapter as a preface and to avoid reading it.

An explanation is required in any case in a book addressed to students of electrical engineering who want to *use* the subject, be it in constructing power systems or radio transmitters or in serving mankind through medical electronics or any of the multitudinous ways open to electrical engineers. It is not sufficient to draw the attention of such readers merely to the beauty of the mental constructs which are used in electromagnetism, the subject must be useful in its application. Happily there is no conflict between usefulness and beauty. The ideas underlying electrical science are beautiful because they contain in a few simple laws or relationships a virtually inexhaustible number of useful applications.

It is not surprising, though it may be thought unfortunate, that these fundamental ideas are expressed in a language which is not that of everyday speech. Some of the words are indeed the same, but they carry a carefully restricted definition. Often words are replaced by mathematical symbols in order to make use of the conciseness and precision of mathematical formulation. This has the immense additional advantage of enabling the engineer to use the marvellous logical framework built by the labours of mathematicians for their own amusement and delight.

Without mathematics as an aid to thought, the development of electromagnetism would have been almost impossible. Only very few people have made progress in the subject without mathematics. The exception to prove the rule was Michael Faraday, one of the greatest thinkers of all time, who used no mathematics at all. He invented such pictorial terms as tubes of magnetic flux and described his law of electromagnetic induction purely in words. The advances he made were without precedent. Nevertheless, it was only when Clerk Maxwell clothed Faraday's words in mathematics that their full power became apparent. To many electrical engineers, especially in their student days, mathematics appears an unnecessary theoretical burden imposed on a practical subject. A study of Faraday's researches shows that mathematics is, however, the lesser evil. Faraday himself suffered from repeated attacks of complete mental exhaustion. Anybody who has tried to formulate such a relationship as that between magnetic and electric fields in words only will understand and share in this exhaustion.

Although mathematics is essential, there is a marked difference in attitude between engineers and mathematicians. The rules which a mathematician has to observe are very strict. His system is complete in itself and no non-mathematical evidence may be admitted in his arguments. The engineer or scientist appeals to measurement as well as to mathematical logic. Indeed, he is forced to test his mathematics by measurement, and often he can shorten an argument by direct appeal to experiment.

This implies that the engineer regards statements about measurable quantities as more interesting and valuable than statements that cannot easily be put to the test of measurement. Moreover, a statement about a quantity which can never be tested is meaningless in engineering although it may form the basis of meaningful mathematics. To cite but one example: "electric current" is easily measured and is of the greatest value in engineering calculations; "electric field strength" cannot be directly measured but is used to predict measurable effects and provides a useful mental stepping stone in the theory; but "internal dielectric field stress" cannot be associated with any measurement and its value to engineers is questionable. Thus while the rules of mathematics are stricter than the rules of engineering evidence, the subject-matter of engineering science is more restricted than that of mathematics because every symbol must be associated, however remotely, with experiment. The value of mathematics lies, for an engineer, chiefly in organizing his knowledge. It should relate measurable phenomena in such a way as to suggest new experiments and undisclosed properties of nature. The value of a theory depends on its fruitfulness rather than on elegance, although the two are not incompatible.

Throughout this book the sequence of thought will lead from experiment to mathematical formulation and on to further experiment. It is very important to keep this in mind, because it is fatally easy to lose sight of the experimental starting and finishing points in a book which necessarily must spend most of its time with the stages in between. If it were possible to argue from one experiment to another without using as a bridge the formal language of electromagnetism, then that language should be avoided by engineers. But since this procedure proved an intolerable burden to the mind of a Faraday, most readers of this book will agree that it would be wise, in the words of Maxwell, to "avail ourselves of the labours of the mathematicians". We shall use these labours to the best advantage if we ask of every instrument reading in an experiment: How can this be formulated in electromagnetic language, and ask of every equation and mathematical relationship: How can this be tested by experiment, and if it withstands the test, how can it be used in designing and building useful devices?

1.2. Electricity and Mechanics

Many readers of this book will be students of a course in electrical engineering and will have scant sympathy for the efforts of some of their lecturers to introduce other related subjects. This is understandable, because nobody can learn everything, and it is essential to specialize if one wants to make progress. This book is a specialist book. But no successful engineer can afford to be too narrow a specialist. At the very least he must be able to discuss his work with other kinds of engineers, and to do this he should master the language of other branches of physical science. It is better still to know how to use the ideas of other branches of the subject and to acquire a common vocabulary for a group of sciences. This can be done very successfully in the physical sciences by making *mechanics* the central subject. The powerful notions of mechanics such as force, momentum, and energy can be readily applied to electrical problems just as well as to the study of mechanics or fluid flow or thermodynamics. It would be folly to ignore this underlying unity, and in the book we shall try to keep it constantly in mind.

The subject of mechanics seeks to describe the way in which different parts of a system act on each other. The word mechanism immediately calls to mind a system of wheels and rods through which forces and motions can be transmitted from one part of a system to another. But mechanics can deal equally well with interactions which would not normally be recognized as mechanisms. The solar system has no linkages between its various members, but celestial mechanics is one of the oldest branches of the subject. It uses the interaction between the sun and the planets to predict the motion of the system. Mechanics is thus able to deal with action at a distance just as with action by means of physical links. Nor is mechanics limited to large systems. Through the endeavours of scientists in the first half of this century, mechanical terminology can be applied to the smallest constituents of matter by means of wave mechanics and quantum mechanics. It is true that the rules of mechanics have to be modified in these branches of the subject, but the concepts of mechanics are deliberately retained.

The position is similar in electricity. The great "electricians" of the past like Cavendish, Franklin, Ampère, and Maxwell sought to establish the subject as a branch of mechanics. They wrote about electro-"statics" and electro-"dynamics", and were at pains to point out the similarities and differences between ordinary mechanics and the mechanics of electricity. It would be foolish to cultivate so narrow an outlook as to turn one's back on all this careful work. Few people would do so deliberately; yet it can happen almost by accident that in the pursuit of a specialist interest contact is lost with other parts of the subject. An engineer has to co-operate with other people, and he cuts himself off at his peril. In this book we shall therefore be very careful to highlight the close relationship between electricity and mechanics.

In any case, this relationship was not arbitrarily devised by the early pioneers. It would be more correct to say that it is a natural relationship which was discovered by them, or perhaps that the concepts of mechanics are particularly well fitted to the structure of human thinking about the physical world. As we make our way through the subject in this book, we shall frequently observe how a mechanical analogy lights up something which first appeared obscure and bewildering. Unfamiliar concepts are set in a well-known framework, and we begin to be able to predict the behaviour of complicated apparatus by means of a few general notions.

1.3. Electricity and the Concept of Electric Charge

Newton established his system of mechanics on the three concepts of length, time, and mass. With these three basic "dimensions" it is possible to build a complete consistent mechanical science. To express the electrical properties of a system, an additional concept is required, and the natural choice is to use as a fourth dimension—the quantity of electricity or electric charge.† The choice is natural because this quantity is found to be "conserved" in electrical systems. Conservation laws, such as conservation of energy, feature prominently in mechanics. It is clearly very convenient to focus attention on quantities which are invariant during a process and do not suddenly appear or disappear.

The choice of electric charge has another enormous usefulness. It is the astounding fact that the force between stationary electric charges varies inversely as the square of the distance between the charges. This inverse square law is, of course, also observed in gravitational attraction. Thus the theory of gravitation and the theory of electricity are bound to exhibit close similarity. We have here the first example of the relationship between electricity and mechanics, which we discussed in the previous

† This does not conflict with the SI system which uses current as the fourth unit quantity. The SI system is used to define the quantity, whereas we are defining a concept. It is easier to *think* about charge and to *measure* current.

section. The whole mathematical apparatus of the theory of gravitation is at our disposal and can be translated term by term and equation by equation into the language of electrostatics.

On the other hand, there is an additional effect in electrostatics which so far has not been observed in gravitation. Experiment discloses two sorts of charge, which are labelled positive and negative because their effects cancel. Like charges repel and unlike charges attract. Another difference is that of magnitude. The electrostatic force exerted by an electron, for instance, is enormously bigger than its gravitational force. The ratio of the forces is of the order of 10^{42}. This makes it difficult to separate the two sorts of electric charge. When engineers speak of charged bodies they usually mean bodies with a slight excess of positive or negative charge. It is thought that in the universe there is as much positive as negative charge.

The strength of the electric interaction makes it possible to transmit appreciable force and energy by means of small charges. It also makes it possible in many cases to neglect the effect of mass when considering the motion of electric charges. In ordinary mechanics, of course, the existence of charge is ignored because bodies are assumed to have negligible net electric charge. In some applications, however, both mass and charge must be considered. This is true in vacuum and gas discharges and in electrical machines, where the mechanical inertia of the moving masses interacts with the electrical inertia of the moving charges.

The inverse square law makes use of the notion of point mass or charge and is therefore a mathematical abstraction. Finite mass or charge cannot be compressed into a point of zero dimensions. There is, however, no real difficulty if by a point we mean not a mathematical point but a region the dimensions of which are small compared with the dimensions of the apparatus, and also small compared with such other significant lengths as wavelength or skin-depth, which will be discussed later in the book. If, in addition, we want to avoid the discussion of the complicated forces inside atoms, we must choose our point to be a region of the order of 10^{-10} m which is the order of magnitude of an atomic diameter. This means that if we restrict the theory to distances at least a thousand times as large, and hence of the order of 10^{-7} m, from an electric charge ot atomic size, we shall be able to regard such a charge as a point source.

Our theory is, therefore, connected with *macroscopic* phenomena. The discrete microscopic nature of matter, charge, and energy plays no part in it. Electricity is treated as a continuous fluid made up of arbitrarily small droplets. This theory is often called the "classical" theory of electromagnetism. It covers all large-scale effects but is insufficient to explain such things as conduction processes in matter, which depend on interatomic structure and need quantum mechanics for their elucidation.

1.4. Electricity and Magnetism

The subjects of electricity and magnetism grew up more or less independently, although the investigators of electrical phenomena were generally interested in magnetism and vice versa. Many people felt intuitively that there was some connection between the subjects, and their belief was reinforced by the observation that electric storms sometimes affected the strength of magnets used in magnetic compasses. Nevertheless, there was no clear understanding of the connection between the two phenomena of electricity and magnetism. It became general to postulate two fluids—electric charge and magnetic pole strength.

The usefulness of this division became obvious when it was discovered in 1750 that the forces between magnets could be explained in terms of an inverse square law between droplets of the magnetic fluid or magnetic *poles*. Rather strangely this was 17 years before the discovery of the inverse square law of electricity. To mathematical minds this meant that a complete theory of magnetostatics could be developed on the basis of magnetic pole strength, and this was done. It is quite fantastic that gravitation, electricity, and magnetism should all share this simple law of the inverse square of the distance. To turn one's back on such a powerful simplification in our understanding of these phenomena would indeed be foolish.

Unfortunately, however, there are difficulties in treating magnetism entirely in terms of magnetic poles. It had been known for centuries that the magnetic fluid could not be isolated. Every positive pole has a negative one attached to it. In the eighteenth century, when experimenters learned to *pipe* electricity from one place to another by means of conductors and when Franklin even drew the electric fluid from a thundercloud, nobody was able to do the same with magnetism. It is true that magnetism could be induced in bodies near to magnets and it could also be transferred to some extent by rubbing, but clearly the phenomena were vastly different.

It took well over half a century to solve the problem of magnetism. The solution, when it came, diminished the importance of magnetic poles as a concept in the theory because it was discovered that magnetism was not an independent phenomenon but an effect associated with moving electric charges. Oersted in 1820 observed magnetic effects near an electric current; Ampère extended this work and formulated a complete theory based on four famous experiments; Faraday found the complementary effect that a changing magnetic field produces an electric current; and Maxwell completed the work by extending Ampère's theory to embrace changing currents as well as constant ones.

These investigations showed that the two phenomena of electricity and magnetism could not be separated but were aspects of the one phenomenon of electromagnetism. They also showed that magnetism was always associated with moving electric charges. Even in permanent magnets the magnetic effects arise from moving charges within the material, as was brilliantly postulated by Ampère long before the advent of atomic theory.† Magnetic energy was thus found to be *kinetic* in contrast with the *potential* energy of electrostatics. Magnetism could be replaced by electrodynamics, and the need for a separate magnetic fluid in the theory disappeared. It also became clear why no individual poles had been found. Every moving charge gives rise to an equivalent dipole.

But this argument can be pressed too far. Granted that kinetic and potential energy are different, they can yet be transformed into each other. The complete Maxwell theory exhibits beautiful symmetry between the two types of energy or between magnetism and electricity. This symmetry is not accidental. There is an inverse square law of magnetism as well as electricity. Granted that there are only dipoles, why should one not close one eye and look at each pole one at a time? The labours of the mathematicians should not lightly be discarded. Even Ampère himself, after saying that all magnets consist of moving charges, worked out his

† However, Ampère considered only orbital motion and did not know about the spin motion which is so important in ferromagnetism.

problems on the opposite assumption that all moving charges are collections of poles. He found the mathematics of the inverse square law more convenient. We shall use the same criteria. To understand the phenomena we shall generally think in terms of electric charges, but where it saves labour we shall feel free to use magnetic poles. Other wiser men have done the same, and that is why the subject is called by the curious hybrid name of electromagnetism.

Summary

Figure 1.1 summarizes the chapter, although real life is, of course, more complicated. The double arrows between *Mathematics* and *Experiment* show that there is a two-way process, and sometimes new mathematics arises in this way. Often this mathematics has already been developed by mathematicians, but its field of application is new. It is also possible that new concepts may arise, but this is the work of genius and has in the past happened very infrequently, perhaps two or three times in a century.

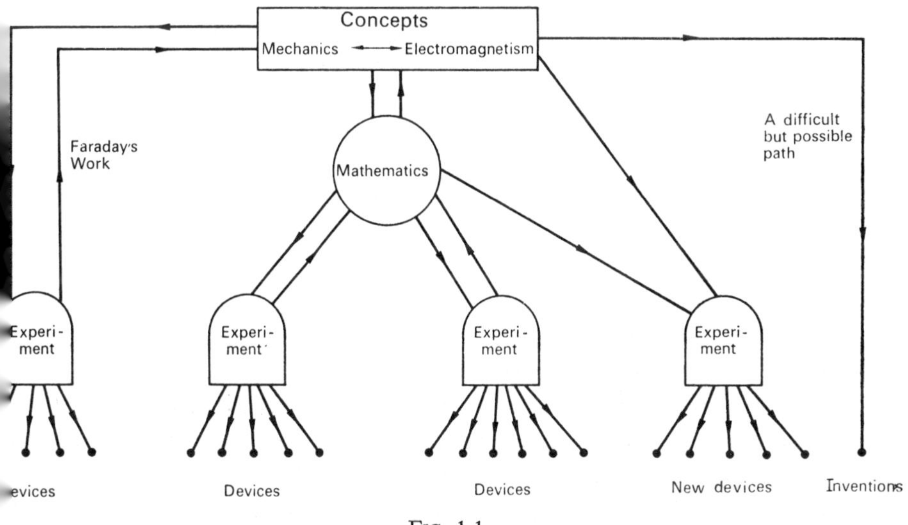

FIG. 1.1

Exercises

1.1. Discuss the advantages and disadvantages of storing energy by means of (a) electric capacitors, (b) super-conducting magnets, (c) high-pressure steam boilers, (d) electric batteries, (e) water reservoirs, (f) nuclear reactors, (g) fuel cells. Pay particular attention to the cost per unit of energy, the energy per unit volume and the ease or difficulty of making the energy available to the user.

1.2. What is the purpose of a pumped-storage hydro-electric scheme?
Prepare a financial report in support of such a scheme for an industrial country with negligible hydro-electric energy.

1.3. Discuss the relative advantages of distributing natural gas in pipes or of burning it in electric power stations and distributing the energy through transmission lines.

1.4. List the social benefits which would result from abolishing petrol and diesel engines in vehicles and replacing them by electric drives.
What technical problems stand in the way of such a development?

1.5. Discuss the relative advantages and disadvantages of communication systems using radio links or telephone cables.

1.6. Compare the tractive effort obtained in a diesel locomotive and a train driven by a linear induction motor.

1.7. Outline the experiments undertaken by Ampère to investigate the relationship between electricity and magnetism. What conclusions did Ampère reach?

1.8. Outline the experiments undertaken by Hertz to show that electromagnetic waves can be transmitted through space.

1.9. Outline the experiments undertaken by Faraday to investigate the generation of electricity from magnetism.

1.10. "Wilhelm Gilbert is the father of modern field theory." Discuss this statement with reference to Gilbert's book *De Magnete*. What differences did Gilbert see between magnetic and electric fields?

1.11. Why did Franklin postulate one electric "fluid" rather than two?

Note. These exercises are not examination questions. They require the collection and assessment of information from libraries and other sources. It is hoped that some readers will find their curiosity aroused sufficiently to set out on a tour of exploration, which will make the details of the rest of the book more meaningful.

CHAPTER 2

A REVIEW OF ELEMENTARY ELECTROMAGNETISM

THIS book is addressed chiefly to readers who have already studied the subject at an elementary level and are familiar with the physical phenomena but who lack the mathematical language which is essential for detailed discussion and understanding. However, before we embark on the full mathematical formulation we shall in this chapter briefly review the elementary theory. The purpose of the review is to define the starting point of this book and to refresh the reader's memory. Readers who find the contents of the chapter too condensed may like to refer to some convenient elementary text.†

2.1. Electrostatics

In order to make the development of the subject as simple as possible we shall deal first with the interaction of electric charges at rest. Of course these charges will have had to be assembled, but in electrostatics it is assumed that the observer has not arrived on the scene until all the charges are in their final static positions.

Electric charge is defined by the experimental relationship of the inverse square law, which states that the force between two point charges Q_1 and Q_2 which are separated by the distance r is directed along r and is given by

$$\mathbf{F} = \frac{Q_1 Q_2}{4\pi\varepsilon_0 r^2}\,\hat{\mathbf{r}}. \tag{2.1}$$

The symbol $\hat{\mathbf{r}}$ is a unit vector in the r-direction and $4\pi\varepsilon_0$ is a constant of proportionality which is determined by fixing the units of force,

† See, for instance, the author's *Electromagnetism for Engineers, An Introductory Course*, Pergamon Press, 1965.

charge and distance. In the SI (MKS) system ε_0 is approximately $8{\cdot}854 \times 10^{-12}$ F/m. The constant 4π is introduced explicitly into eqn. (2.1) to ensure that in general 4π occurs in problems of spherical geometry and 2π in cylindrical problems, and that π is absent from rectangular problems. The electric constant ε_0 is often called the "permittivity of free space", which suggests that it defines a measurable property. However, since the inverse square law is already used to define the charge, it cannot also define another quantity. In fact, ε_0 is nothing more than a dimensional constant. Its dimensions are (coulomb second)^2/(kilogram) (metre)^3 or, more simply, farad/metre.

It is often convenient to consider eqn. (2.1) in two stages. Let the charge Q_1 produce an electric field strength $\mathbf{E}_1$ at the position where Q_2 is placed. Then

$$\mathbf{F} = \mathbf{E}_1 Q_2 \tag{2.2}$$

and

$$\mathbf{E}_1 = \frac{Q_1}{4\pi\varepsilon_0 r^2}\,\hat{\mathbf{r}}. \tag{2.3}$$

Generally we work with the notion of electric field strength **E** rather than force **F**, and this enables us to consider the charges as sources which produce a *field* at a distance from themselves. The introduction of the concept of electric field does not add any measurable data, but it provides a useful mental stepping-stone or concept. In terms of a possible experiment, the electric field strength would have to be measured by the force on a charge of unit strength.

Often it is preferable to formulate a problem in terms of energy rather than force because energy can be added arithmetically, whereas forces have to be added as vectors. The work done in assembling the two charges of eqn. (2.1) is

$$W = -\int_{\infty}^{r} \mathbf{F}\,.\,d\mathbf{r}\dagger = \frac{Q_1 Q_2}{4\pi\varepsilon_0 r}. \tag{2.4}$$

The negative sign shows that work is done against the force **F**. This ex-

† The scalar product $\mathbf{A}\cdot\mathbf{B}$ of two vectors **A** and **B** is to be understood as $AB\cos\alpha$, where α is the angle between **A** and **B**. See Fig. 2.1. Physically **Force** · **Distance** = Work done.

pression can again be split into two statements

$$W = V_1 Q_2 \tag{2.5}$$

and

$$V_1 = -\int_{\infty}^{r} \mathbf{E} \,.\, d\mathbf{r} = \frac{Q_1}{4\pi\varepsilon_0 r}. \tag{2.6}$$

Thus V_1 is the work done in bringing a unit charge to within a distance r of Q_1. Because this work is recoverable and depends only on position, the energy is called *potential* energy and V is called the *electrostatic potential.* The unit of potential difference is the volt. If no work has to be done in moving a charge from one place to another, then those places are at the same potential—and vice versa.

It follows that any conductor in an electrostatic problem is at constant potential. If it were not so, then the conduction charges would move to positions of lower energy. But they do not, because we have said that it is a problem in electro*statics*. For the same reason the net charge in conductors is located only on the outside surface of the conductors, which is the position of minimum energy. The charges, since they are of the same kind, seek to get away from each other as far as possible.

As well as using the electric field strength we shall often want a quantity which is directly related to charge without using the electric constant. We define the vector **D** by

$$\mathbf{D} = \varepsilon_0 \mathbf{E} \tag{2.7}$$

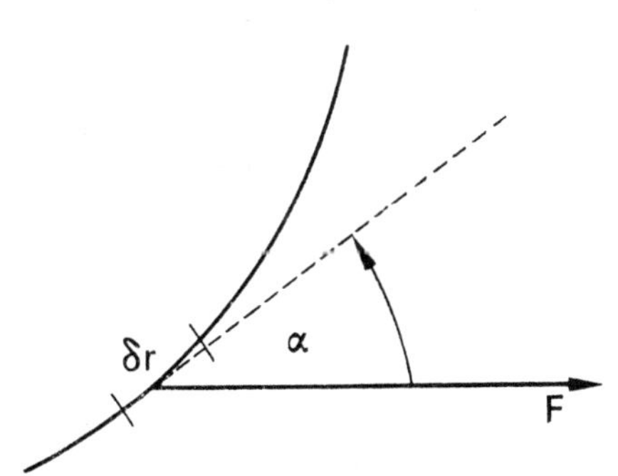

$\mathbf{F}.\delta\mathbf{r} = F\delta r \cos\alpha$

FIG. 2.1. The scalar or dot product of two vectors.

or

$$\mathbf{D}_1 = \frac{Q_1}{4\pi r^2}\hat{\mathbf{r}}. \tag{2.8}$$

If we take a surface integral† of **D** over a sphere of radius r which has Q_1 at its centre,

$$\oiint \mathbf{D} \,.\, d\mathbf{s} = \oiint \frac{Q_1}{4\pi r^2}\, ds = Q_1 \,. \tag{2.9}$$

(The direction of the vector $d\mathbf{s}$ is along the outward normal to the area ds.) Thus the surface integral of **D** is equal to the enclosed charge. This result is generalized by Gauss's theorem, which shows that the surface need not

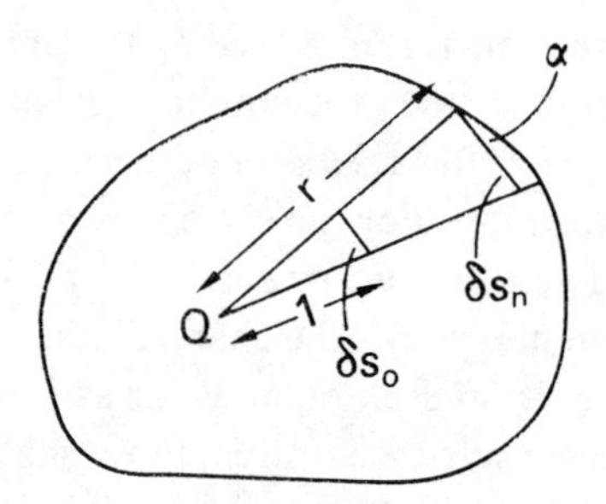

FIG. 2.2. Gauss's theorem.

be spherical and the charge does not need to be at any particular point within the enclosed volume. Consider this with reference to Fig. 2.2.

$$\oiint \mathbf{D} \,.\, d\mathbf{s} = \oiint \frac{Q}{4\pi r^2}\, ds \cos\alpha = \frac{Q}{4\pi} \oiint \frac{ds_n}{r^2} \tag{2.10}$$

also

$$\frac{\delta s_0}{1^2} = \frac{\delta s_n}{r^2} \tag{2.11}$$

and

$$\oiint ds_0 = 4\pi \,, \tag{2.12}$$

† The reader should ensure that he understands this important concept. The mathematics is explained in eqn. (2.10) and Fig. 2.2. Physically the surface integral is the outflow across a surface. At each element of surface we multiply the velocity (or equivalent vector) normal to the surface by the area of the surface element.

so that

$$\oint\!\!\oint \mathbf{D} \cdot d\mathbf{s} = Q\,. \tag{2.13}$$

The surface integral over any open or closed surface is defined by

$$\Psi = \iint \mathbf{D} \cdot d\mathbf{s}\,, \tag{2.14}$$

where Ψ is called the *electric flux* by analogy with fluid mechanics. The charge Q is the source of the flux, and Gauss's theorem states that the flux over any arbitrary *closed* surface is always equal to the strength of the enclosed sources. This is a direct consequence of the inverse square law. The vector **D** is called the *electric flux density*.

The fluid analogy suggests that it may be convenient to divide the electric field into tubes of flux. These tubes start, and end, on electric charge, and by definition contain a constant amount of electric flux. They have in general a varying cross-section (see Fig. 2.3). Since the flux in them is

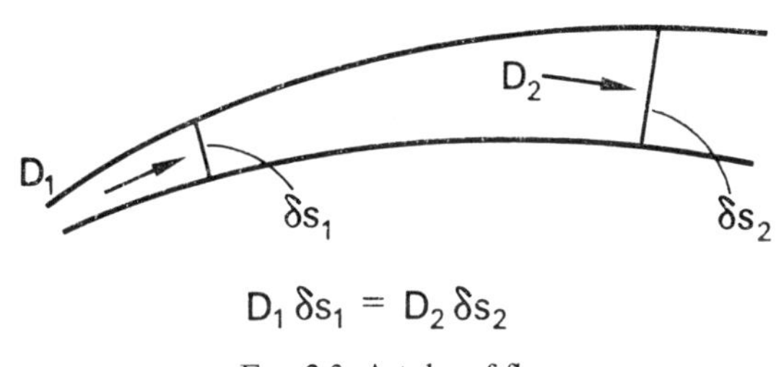

FIG. 2.3. A tube of flux.

constant, the flux density is inversely proportional to the cross-sectional area. Since electric field strength is proportional to flux density, this field strength is weak where the tubes are widely spaced and strong where they are crowded together. Although the analogy with fluid flow is helpful, it is important to remember that in electrostatics there is no flow of anything, and the term electric flux can be rather misleading.

Our consideration of electrostatics started with force and energy. Electrostatic devices are used either to produce force or to store potential energy. Often we shall have to deal with devices in which equal amounts of positive and negative charge have been separated by some mechanical, chemical, or thermal process. Since the potential difference is proportional to the charge, it is convenient to use the ratio of *charge/potential differ-*

ence which is independent of charge. This ratio is called the *capacitance*. In the idealized case of Fig. 2.4 of two parallel charged plates of area s separated by the distance t, the potential difference is $Qt/\varepsilon_0 s$ and the capacitance is ε_0/t per unit area of the surface of one plate. This assumes

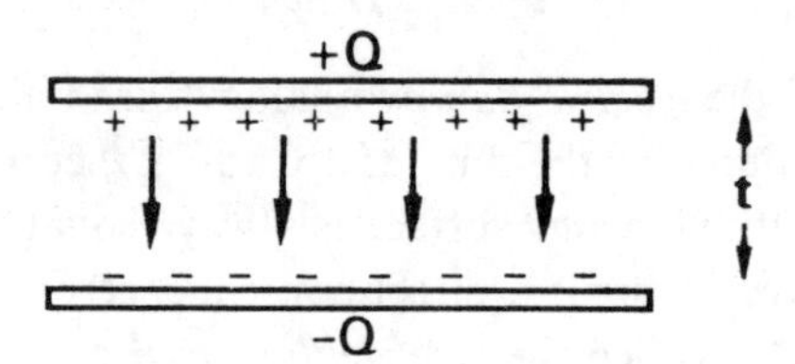

FIG. 2.4. Parallel-plate capacitor.

that the charges are uniformly distributed over the plates and that there is no fringing flux at the edges. The capacitance is independent of the amount of the charge and depends only on the geometry of the "capacitor", as the device is called. It is important to note that capacitance is associated with every assemblage of charges.

The capacitance can be increased by inserting a polarizable material between the plates. Such a material is called a dielectric. Induced charges appear on the surface of a dielectric when it is inserted in an electric field. Figure 2.5 shows a simple arrangement. If we neglect the space between

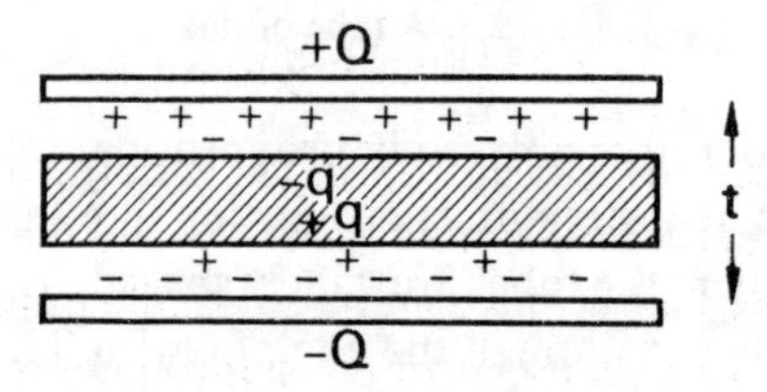

FIG. 2.5. Parallel-plate capacitor with dielectric.

the plates and the dielectric, it is clear that the potential difference now depends on $(Q - q)$ and not on Q only. The capacitance Q/V is therefore increased in the ratio $Q/(Q - q)$. If the potential difference V is fixed, a larger charge can be stored on the plates.

The induced charge q is generally proportional to the electric field strength $\mathbf{E}$, and hence in our simple case to V or $Q - q$. The proportionality constant is called the *electric susceptibility* χ_e. Thus $q \propto (Q - q)$ and $q = \chi_e (Q - q)$. The increase in the capacitance is in the ratio $Q/(Q - q) = 1 + \chi_e$. This ratio is called the *relative permittivity* and is given the symbol ε_r, where $\varepsilon_r = 1 + \chi_e$. The increased capacitance is therefore $\varepsilon_r \varepsilon_0 / t$ per unit area. Sometimes $\varepsilon_r \varepsilon_0$ is written simply ε and is called the absolute permittivity. It is now clear that the misleading term "permittivity of free space" arose from the juxtaposition of ε_0 and ε_r. It should be noted that the relative permittivity ε_r is non-dimensional and measurable, being the ratio of two capacitances. It is a property of the dielectric material.

It is often convenient to carry the distinction between the *free* charge Q on the metal plates and the induced *bound* charge q on the dielectric into the description of the electric field itself. This can be done by deciding to associate **D** with Q only, whereas **E** is associated with both Q and q. Thus the sources of **E** are all the charges Q and q, and **E** is the *total* field, whereas **D** has only the free charges Q for its sources and is a *partial* field vector. This means that the relationship $\mathbf{D} = \varepsilon_0 \mathbf{E}$ of eqn. (2.9) can only be true in the absence of bound charges, i.e. outside a dielectric. Within the dielectric we must now write

$$\mathbf{D} = \varepsilon_0 \varepsilon_r \mathbf{E}. \tag{2.15}$$

2.2. Magnetostatics

In the previous section we were concerned with the potential energy of static electric charges. We must now review briefly the effects associated with moving charges, but in this section we shall confine our discussion to charges moving with velocities of constant magnitude though not of constant direction. This will enable us to defer the discussion of changes in kinetic energy.

Before we deal with the kinetic energy which forms the main subject of this section, it is necessary to deal briefly with the dissipation of energy which accompanies the motion of charges. In dielectrics some dissipation of energy accompanies the process of polarization, and in the conduction of charge through metals there is also appreciable loss. The dissipated electrical energy reappears as heat. The power loss in the conduction process is particularly important and we shall discuss it briefly. It is given by

$$P = RI^2, \tag{2.16}$$

where the resistance R is defined by Ohm's law

$$V = RI. \tag{2.17}$$

This resistance is observed to vary directly as the length and inversely as the cross-section of the conductor

$$R = \varrho \frac{l}{s}, \tag{2.18}$$

where ϱ is the resistivity† of the material and depends largely on temperature. The power loss per unit volume is

$$p = \varrho |\mathbf{J}|^2, \tag{2.19}$$

where $\mathbf{J}$ is the current density. The electric field strength required to drive the current through the resistive material is given by

$$\mathbf{E} = \varrho \mathbf{J}. \tag{2.20}$$

After this digression we come now to the discussion of the interaction of charges in steady motion. This interaction is called *magnetostatics*, because steady currents act exactly like permanent magnets and the simplest description of the phenomenon is in terms of magnetic fields. It is, therefore, convenient to define first of all the various magnetic quantities.

The fundamental magnetic quantity is the magnetic dipole consisting of magnetic poles $+Q^*$ and $-Q^*$ separated by a short distance d. The law of force between poles is the inverse square law

$$\mathbf{F} = \frac{Q_1^* Q_2^*}{4\pi\mu_0 r^2}\,\hat{\mathbf{r}}, \tag{2.21}$$

where μ_0 is the magnetic constant, which in the SI system is equal to $4\pi \times 10^{-7}$ H/m. It is often called the *permeability of free space*, but like ε_0

† It is sometimes more convenient to use $\sigma = 1/\varrho$, where σ is called the conductivity of the material.

it does not describe any measurement. The magnetic field strength is given by

$$\mathbf{H} = \frac{Q^*}{4\pi\mu_0 r^2}\hat{\mathbf{r}} \tag{2.22}$$

and the magnetic flux density is in free space

$$\mathbf{B} = \mu_0\mathbf{H} \tag{2.23}$$

and in a polarizable material

$$\mathbf{B} = \mu_0\mu_r\mathbf{H}, \tag{2.24}$$

where μ_r is the relative permeability.

The magnetic flux is given by

$$\Phi = \iint \mathbf{B} \,.\, d\mathbf{s}\,. \tag{2.25}$$

Since all magnets consist of dipoles, there is never any net polarity, so that Gauss's theorem states that

$$\Phi = \oiint \mathbf{B} \,.\, d\mathbf{s} = 0\,. \tag{2.26}$$

Thus the flux for a closed surface is zero.

The magnetic potential difference is given by

$$V^* = -\int_1^2 \mathbf{H} \,.\, d\mathbf{r}\,. \tag{2.27}$$

So far we appear to have two independent electrostatic and magnetostatic systems whose sources are charges and dipoles. However, experiment shows that the external effect of a small current loop at a sufficient distance is exactly that of a small dipole, so that magnetostatics is linked to current electricity. Moreover, atomic theory shows that magnetism always arises from moving charges. Magnetic dipoles are retained in the theory merely for convenience (see § 1.4).

The magnetic strength of a small current loop is found to be proportional to the current and to the area of the loop, while the strength of a dipole is proportional to the pole strength and the length of the dipole. Figure 2.6 shows on the left a small loop of current I and area δs. On the right of the figure there is the equivalent dipole layer of pole strength

Q^* and thickness t. The surface facing the reader has positive pole strength and the surface at the back of the layer has negative pole strength. Experiment shows that

$$I\delta s \propto Q^*t \tag{2.28}$$

and the magnetic units are chosen to put μ_0 as the constant of proportionality, so that

$$\mu_0 I\delta s = Q^*t. \tag{2.29}$$

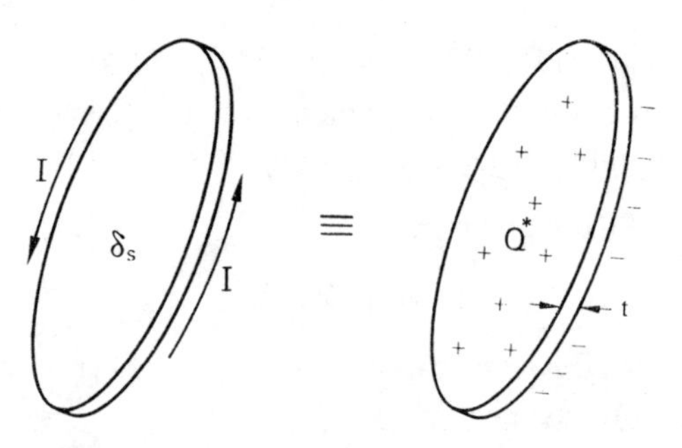

FIG. 2.6. A small current loop is equivalent to a small dipole.

A large current loop can be split up into a large number of small loops (see Fig. 2.7), and if each small loop is replaced by a magnetic dipole, the result is a double layer of magnetic pole strength, positive on one side and negative on the other. Such an arrangement is called a *magnetic shell.*

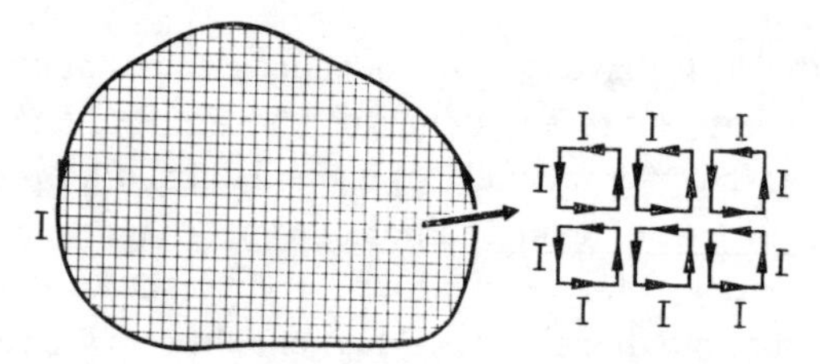

FIG. 2.7. A large current loop is equivalent to many small loops.

The sign of the pole strength is defined by saying that positive pole strength is above a counter-clockwise current loop and negative pole strength is below such a current loop.

Since the mathematics of the inverse square law are relatively simple, it is often easier to deal with the interaction of equivalent magnetic poles rather than actual electric currents. The idea of a magnetic shell is a very important one in magnetostatics. We are always at liberty to represent the action of steady electric currents by the action of an equivalent magnetic shell. This shell need not be planar. It can have any shape whatever as long as its perimeter is along the path of the electric current. Of course, the internal construction of a magnetic shell is different from that of a current loop, but the external magnetic effects are identical.

The relationship between magnetic field strength **H** and steady electric current I can be obtained by the use of a magnetic shell. Let us examine the magnetic potential difference around the shell. If the surface *density*

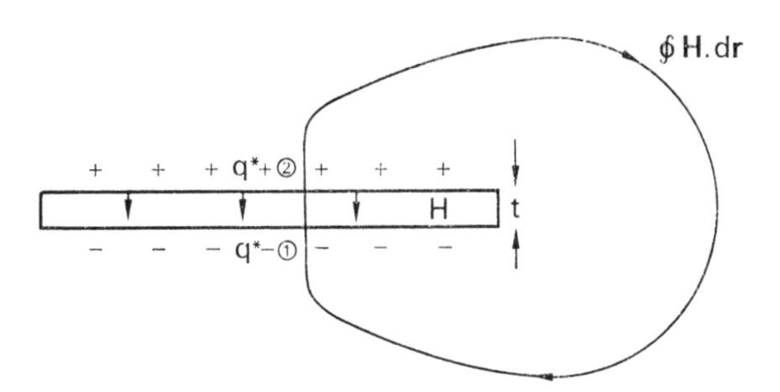

FIG. 2.8. A magnetic shell.

of pole strength is q^* (see Fig. 2.8) and the thickness of the shell is t, the potential difference between the surface s of the shell is

$$V_2^* - V_1^* = -\int_1^2 \mathbf{H} \,.\, d\mathbf{r} = Ht = \frac{q^*}{\mu_0} t. \qquad (2.30)$$

Since, moreover, in a system of inverse square law sources the potential has a unique value at every point, we have for the magnetic shell

$$\oint \mathbf{H} \,.\, d\mathbf{r} = 0. \qquad (2.31)$$

This means that the work down against the magnetic field *by* a unit pole passing through the shell is recovered by the work done *on* the unit pole by the magnetic field in passing round the outside of the shell back to the starting point.

Consider now the work done in passing around a current loop. The outside part of the work will be as before, $V_2^* - V_1^*$, and using eqn. (2.29) we can write

$$V_2^* - V_1^* = \frac{q^*t}{\mu_0} = I. \tag{2.32}$$

The work done in passing through the shell is absent because we are considering a current loop and there is no shell to pass through. The sides of the shell 1 and 2 now refer to the same place, and instead of eqn. (2.31) we must write

$$\oint \mathbf{H} \,.\, d\mathbf{r} = I, \tag{2.33}$$

where $\oint \mathbf{H} \,.\, d\mathbf{r}$ is called the *magnetomotive force* of the current I.

Work is done in passing round the closed loop embracing the current, and the magnetic potential V^* does not have a unique value at every point. Magnetic energy cannot therefore be potential energy like the energy of electrostatic systems. It is in fact the *kinetic energy* associated with moving electric charges. That is why $\oint \mathbf{H} \,.\, d\mathbf{r}$ is called magnetomotive force and not potential difference. The notion of magnetic potential difference must be restricted to problems where it is not possible to encircle a current completely, i.e. where current loops can be replaced by magnetic shells.

Equation (2.33) can be used to obtain the magnetic field strength of a current element. Figures 2.9 and 2.10 show such an element and we require to find **H** at the point P. A steady current cannot flow in such an

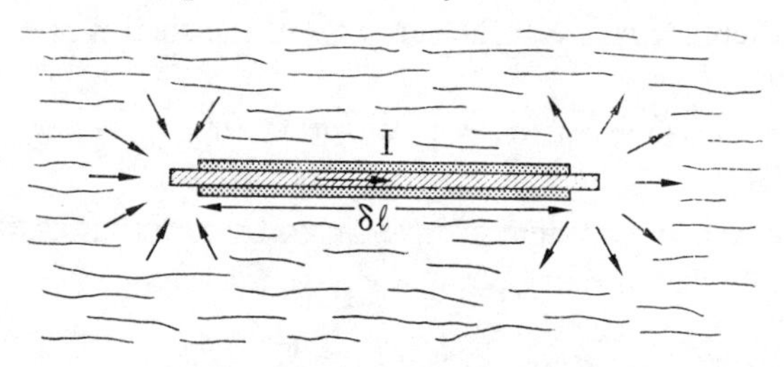

FIG. 2.9. A current element.

element unless there is a complete circuit. We can, however, imagine the element with an insulating sheath immersed in a large bath of conducting fluid and then the current could flow. Moreover, lots of such elements could be put end to end and when the last element had been put into position to form a complete circuit, the conducting fluid could be drained away. Thus this fluid would not enter into any calculations on closed circuits. Its introduction is necessary only to enable us to calculate the contribution of each element of the circuit.†

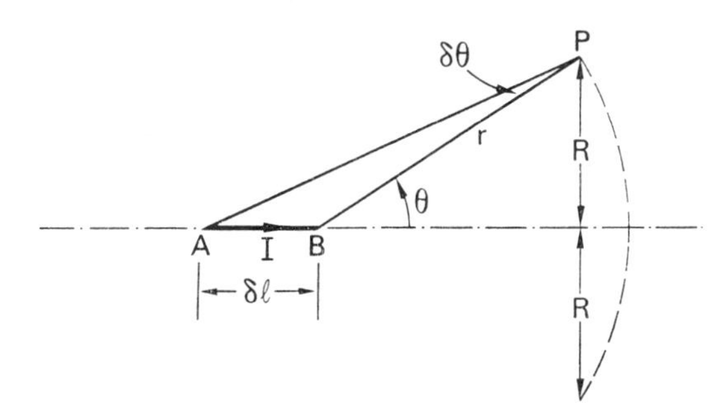

FIG. 2.10. The field of a current element.

By symmetry **H** will form circles around the axis formed by the element (Fig. 2.10). Consider the current flow through the circle passing through P. There is an outflow from the end B of the element and an inflow into the end A. The current density will be constant over a spherical cap of radius r. The area of this cap is $2\pi r^2 (1 - \cos\theta)$. Thus the outflow from B is

$$\frac{I}{4\pi r^2} 2\pi r^2 (1 - \cos\theta) = \frac{I}{2}(1 - \cos\theta) \qquad (2.34)$$

and the net outflow obtained by subtracting the inflow to A, or by differentiation, is

$$\frac{I}{2} \sin\theta\, \delta\theta .$$

Hence, writing **l** for **r** in eqn. (2.33),

$$\oint \mathbf{H} . d\mathbf{l} = H2\pi R = \frac{I}{2} \sin\theta\, \delta\theta , \qquad (2.35)$$

whence by simple geometry

$$H = \frac{I\,\delta l}{4\pi r^2} \sin\theta \qquad (2.36)$$

† The idea of a current element in a conducting fluid is due to Heaviside.

or, in vector notation,

$$\mathbf{H} = I\frac{\delta\mathbf{l} \times \hat{\mathbf{r}}}{4\pi r^2}.\dagger \tag{2.37}$$

In electrostatics it proved useful to define the quantity called capacitance, because charge and potential difference are proportional to each other and their ratio is therefore independent of either quantity. In magnetostatics we have found that the magnetic field strength is proportional to current; hence the ratio of both is similarly useful. We define *inductance* as the ratio of magnetic flux/current. Thus the inductance is given by

$$L = \frac{\Phi}{I} = \frac{\iint \mathbf{B} \cdot d\mathbf{s}}{I}. \tag{2.38}$$

Just as capacitance is a measure of the storage of electric charge and hence of potential energy, so inductance is associated with the kinetic energy of electric current. The three quantities resistance, capacitance, and inductance can be used to describe the dissipation of electrical energy into heat energy, the potential energy and the kinetic energy of electrical systems. They do not describe the processes in detail as is done by the field vectors, but they deal with complete systems or circuits. They are therefore very helpful in giving an overall view of such systems.

The *rate* of energy dissipation by resistance is RI^2, the electrostatic energy is $\frac{1}{2}(Q^2/C)$ and the electrokinetic or magnetic energy is $\frac{1}{2}LI^2$.

2.3. Electromagnetics

In electrostatics we considered the forces on electric charges due to the position of other charges, and in magnetostatics the forces exerted by magnets or steady currents on each other. We must now give an account of the forces on charges due to magnets or current systems.

Since a set of charges in uniform motion produces a magnetic field and so exerts a force on magnets, a magnet may be expected to produce a force on a moving charge.

† This notation is explained more fully in the next chapter, § 3.2.2. Here we note that if θ is the angle between $\delta\mathbf{l}$ and $\mathbf{r}$, the cross product implies $\delta l\, r \sin\theta$, whereas the dot product uses $\cos\theta$. Physically the cross product can be illustrated by the relationship Force × Distance = Torque.

The force on a magnetic pole Q^* exerted by the field of a current element is given by

$$\mathbf{F} = Q^*\mathbf{H} = Q^*I\,\frac{\delta\mathbf{l} \times \hat{\mathbf{r}}}{4\pi r^2}. \tag{2.39}$$

The magnetic flux density due to this pole at the place where the current element is located is given by

$$\mathbf{B} = -\frac{Q^*\,\hat{\mathbf{r}}}{4\pi r^2}. \tag{2.40}$$

(The negative sign is due to the change of direction of the vector $\mathbf{r}$.) Thus the force on the current element is given by

$$\mathbf{F} = I\,\delta\mathbf{l} \times \mathbf{B}. \tag{2.41}$$

For a conductor of length l in a uniform field this becomes

$$\mathbf{F} = I\mathbf{l} \times \mathbf{B}. \tag{2.42}$$

The current element $I\,\delta l$ can be viewed as a charge Q moving with velocity u, where

$$I\,\delta\mathbf{l} = Q\mathbf{u}. \tag{2.43}$$

The force on this moving charge is therefore

$$\mathbf{F} = Q\mathbf{u} \times \mathbf{B} \tag{2.44}$$

and we can define an electric field strength

$$\mathbf{E} = \mathbf{u} \times \mathbf{B}. \tag{2.45}$$

Unlike the electric field strength of electrostatics, this field strength depends on the *velocity* and not only on the *position* of its source. It cannot be derived from a potential. In terms of energy this "motional" electric field strength is associated with *electromotive force*† instead of *potential difference*. Since electric field strength is given in volts/metre, we now have two sorts of voltage: electrostatic potential difference due to stationary charges, and electromotive force due to moving charges. An ex-

† Note that the terms magnetomotive force and electromotive force are unfortunate inasmuch as these are not forces but quantities which have the dimensions energy per pole and energy per charge respectively.

ample of electromotive force is given in Fig. 2.11. A bar moves across a constant magnetic field. The e.m.f. is given by

$$\text{e.m.f.} = \int_1^2 \mathbf{E} \,.\, d\mathbf{l} + \int_1^2 (\mathbf{u} \times \mathbf{B}) \,.\, d\mathbf{l} = uBl. \tag{2.46}$$

If, as in Fig. 2.12, the bar forms part of a circuit, eqn. (2.46) can be re-written as

$$\text{e.m.f.} = \oint \mathbf{E} \,.\, d\mathbf{l} = -\frac{d\Phi}{dt}, \tag{2.47}$$

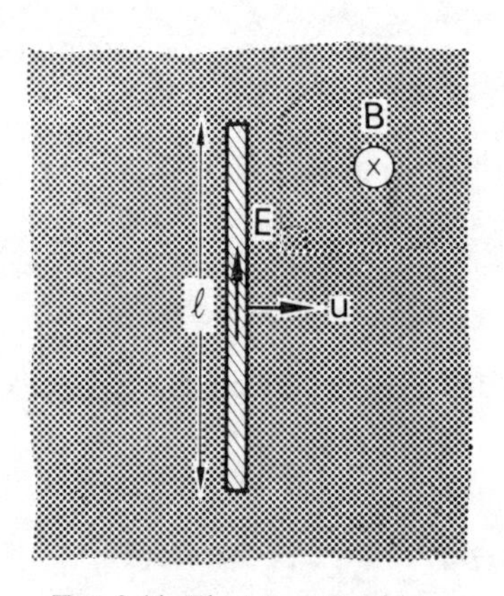

FIG. 2.11. Electromotive force.

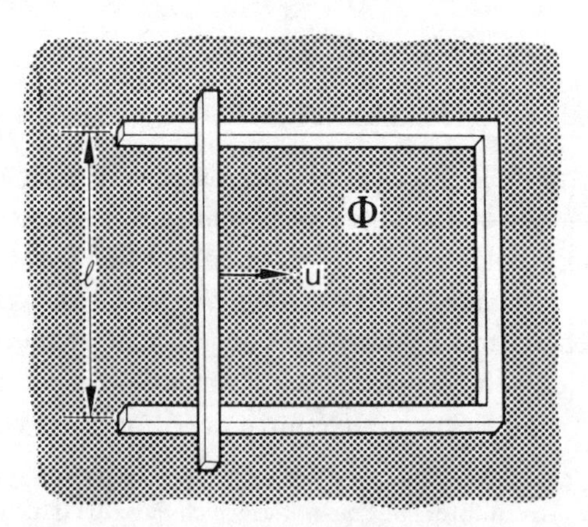

FIG. 2.12. Faraday's law.

where Φ is the magnetic flux linking the circuit. This equation is known as Faraday's law of electromagnetic induction. We have derived it from the consideration of force on a moving charge. Experiment shows, however, that the law is far wider in its application. It applies to any circuit whatever, whether stationary, moving, or undergoing deformation. The change of flux can arise not only from the motion of the circuit but also from a change in strength or position of the sources of the flux. Faraday's law is one of the most powerful and universal laws of physics.

We have seen that it is useful to distinguish between electric potential difference and electromotive force. A similar distinction is useful in magnetism. *Magnetic potential difference* can be associated with stationary magnetic *poles* as, for instance, in eqn. (2.30), and *magnetomotive force* can be associated with moving *charges*. Equation (2.33) gives the magnetomotive force of a steady current. Comparison with Faraday's law, eqn. (2.47), leads to the hypothesis that eqn. (2.33) is a special case of the more general law

$$\text{m.m.f.} = \oint H \,.\, d\mathbf{l} = I + \frac{d\Psi}{dt}, \tag{2.48}$$

where $\Psi = \iint \mathbf{D} \,.\, d\mathbf{s}$ is the *electric flux* linked by the circuit around which the m.m.f. is measured. This hypothesis led Maxwell to the idea of radiation by electromagnetic waves. To illustrate the result let us examine a very simple case in which there is an electric field in the x-direction and a magnetic field in the y-direction as shown in Fig. 2.13. There are no electric charges or currents in the region being considered and the electric and magnetic fields are assumed to be constant throughout a plane par-

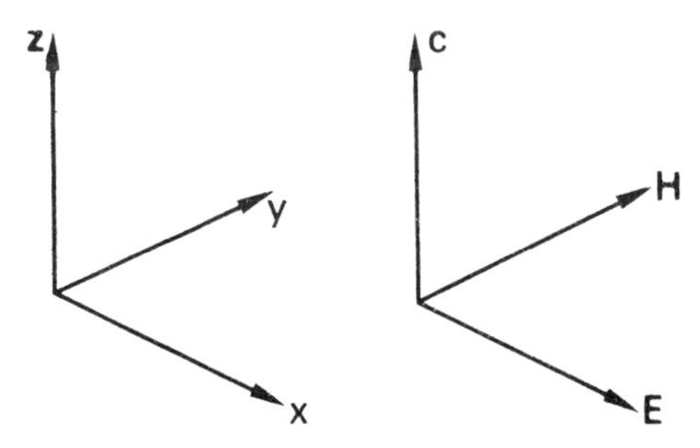

FIG. 2.13. A plane electromagnetic wave.

allel to the xy-plane. Thus $\partial E/\partial x$, $\partial E/\partial y$, $\partial H/\partial x$, and $\partial H/\partial y$ are all zero and the only variation is along the axis of z. We note in passing that such a field could arise from a large current sheet, in which the current flows in a plane parallel to the xy-plane and along a direction parallel to the x-axis.

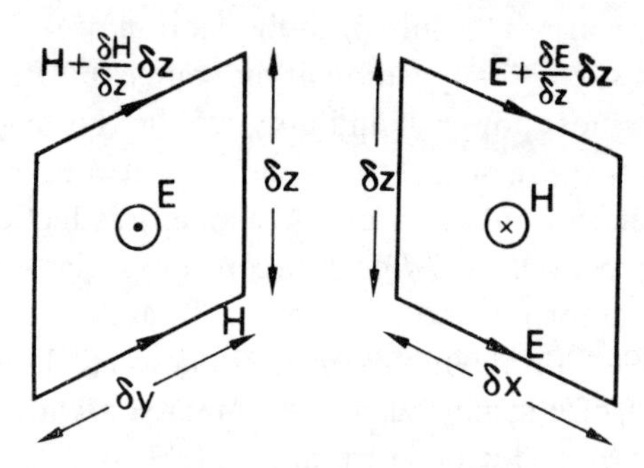

FIG. 2.14. Maxwell's equations.

Let us apply the relationships eqn. (2.47) and eqn. (2.48) to the small rectangles of Fig. 2.14.

$$\oint \mathbf{E} \, . \, d\mathbf{l} = -\frac{d\Phi}{dt} \tag{2.47}$$

becomes

$$\left(E + \frac{\partial E}{\partial z}\,\delta z\right)\delta x - E\,\delta x = -\mu_0 \frac{\partial H}{\partial t}\,\delta x\,\delta z\,,$$

whence

$$\frac{\partial E}{\partial z} = -\mu_0 \frac{\partial H}{\partial t}\,. \tag{2.49}$$

Similarly, from eqn. (2.48),

$$\oint \mathbf{H} \, . \, d\mathbf{l} = \frac{d\Psi}{dt} \tag{2.50}$$

becomes

$$\frac{\partial H}{\partial z} = -\varepsilon_0 \frac{\partial E}{\partial t}\,. \tag{2.51}$$

If we eliminate either H or E from eqns. (2.49) and (2.51) we obtain the equations

$$\frac{\partial^2 E}{\partial z^2} = \mu_0 \varepsilon_0 \frac{\partial^2 E}{\partial t^2} \tag{2.52}$$

and

$$\frac{\partial^2 H}{\partial z^2} = \mu_0 \varepsilon_0 \frac{\partial^2 H}{\partial t^2}. \tag{2.53}$$

Inspection of these equations shows that $\mu_0 \varepsilon_0$ has the dimensions $(\text{time})^2/(\text{length})^2$. We can write

$$\mu_0 \varepsilon_0 = \frac{1}{c^2}, \tag{2.54}$$

where c is a velocity. The equations (2.52) and (2.53) occur frequently in engineering problems and are known as *wave* equations. They have solutions of the type

$$E = f(z \pm ct) \tag{2.55}$$

and

$$H = F(z \pm ct), \tag{2.56}$$

where f and F represent arbitrary functions. The correctness of these solutions can be tested by substitution in the wave equation. Since E is in the x-direction and H in the y-direction, and since the waves travel in the z-direction, these waves are known as transverse waves.

Maxwell calculated c numerically and found it to be the same as the velocity of light postulated by astronomers. He inferred that light was a type of electromagnetic radiation. The fact that all electromagnetic waves irrespective of frequency travel with the same velocity c led Einstein to postulate c as a universal constant, and this led him to his theory of relativity. Thus relativity theory is in a very real sense contained in eqns. (2.47) and (2.48). These equations are often called Maxwell's equations, and they are the twin foundation stones of macroscopic electromagnetic theory.

With Maxwell's equations we have the key to the whole subject. A theoretician might say that the subject is now well understood and that there is no more to be done. But a front-door key is not the same as the house to which it gives admission. To the engineer and applied scientist a knowledge of Maxwell's equations is not the end of the story but the beginning.

One word more should perhaps be said about the velocity c before the end of this brief review. Its magnitude is found *experimentally* to be

$$c \doteqdot 2{\cdot}998 \times 10^8 \text{ m/s}.$$

In the SI system of units the ampere and hence the coulomb is *defined* by setting

$$\mu_0 = 4\pi \times 10^{-7} \text{ H/m}.$$

The electric constant ε_0 is therefore *derived* as

$$\varepsilon_0 = \frac{1}{\mu_0 c^2} \doteqdot 8{\cdot}854 \times 10^{-12} \text{ F/m}.$$

The interrelation between electricity and magnetism allows us to define only one constant. Once the magnetic constant has been chosen, the electric constant is derived from it.

Glossary

Term	*Symbol*	*Unit*	*Explanation*
Electric charge	Q	coulomb	Quantity of "electricity" defined by inverse square law
Electric constant (permittivity of free space)	ε_0	farad/metre	A dimensional constant introduced into the inverse square law to define the unit of electric charge in relation to the units of length, mass and time. In the SI system $\varepsilon_0 \simeq 8{\cdot}854 \times 10^{-12}$ F/m
Electric field strength	$\mathbf{E}$	volt/metre	Force on a unit charge at a point
Electrostatic potential and potential difference	V	volt	Work done in bringing a unit charge to a point
Electric susceptibility	χ_e	ratio of charges, i.e. pure number	A measure of polarization
Relative permittivity (dielectric constant)	ε_r	pure number	$\varepsilon_r = 1 + \chi_e$
Absolute permittivity	ε	farad/metre	$\varepsilon = \varepsilon_0 \varepsilon_r$
Electric flux density	$\mathbf{D}$	coulomb/metre^2	In free space D is just a short way of writing $\varepsilon_0 E$. In polarizable materials D is the field of free charges, whereas E is the total field. This leads to $D = \varepsilon_0 \varepsilon_r E$

Term	*Symbol*	*Unit*	*Explanation*
Electric flux	Ψ	coulomb	Surface integral of D. If the surface is closed the flux equals the enclosed charge (by Gauss's theorem)
Capacitance	C	farad	Ratio of charge to potential difference
Current	I	ampere	Rate of flow of charge through a surface
Current density	$\mathbf{J}$	ampere/metre^2	Density of current per unit area
Resistance	R	ohm (volt/ampere)	Associated with power loss due to current
Resistivity	ϱ	ohm metre	Associated with power loss per unit volume
Conductivity	σ	siemens (mho/metre)	
Magnetic pole	Q^*	weber	Quantity of "magnetic charge" defined by inverse square law
Magnetic constant (permeability of free space)	μ_0	henry/metre	A dimensional constant used to define the unit of magnetic pole, and therefore the electric current, in relation to the units of length, mass, and time. In the SI system $\mu_0 = 4\pi \times 10^{-7}$ H/m
Magnetic field strength	$\mathbf{H}$	ampere/metre	Force on a unit pole at a point
Magnetic potential and potential difference	V^*	ampere	Work done in bringing a unit pole to a point
Magnetic susceptibility	χ_m	pure number	A measure of magnetic polarization
Relative permeability	μ_r	pure number	$\mu_r = 1 + \chi_m$
Absolute permeability	μ	henry/metre	$\mu = \mu_0 \mu_r$
Magnetic flux density	$\mathbf{B}$	tesla or weber/metre^2	$B = \mu H$
Magnetic flux	Φ	weber	Surface integral of B. If the surface is closed $\Phi = 0$
Inductance	L, M	henry	Ratio of flux to current
Electromotive force	e.m.f.	volt	Work done in moving a unit charge against electric forces which are not electrostatic but are due to changing magnetic flux (Faraday's law).
Magnetomotive force	m.m.f.	ampere	Work done in moving a unit pole against magnetic forces which are not magnetostatic but are due to electric current and changing electric flux

Exercises

2.1. "The unit of electric current called the 'ampere' is that constant current which, if maintained in two parallel rectilinear conductors of infinite length, of negligible circular cross-section, and placed at a distance of one metre apart in a vacuum, would produce between these conductors a force equal to 2×10^{-7} newton per metre length." Explain how this definition of the unit of current also defines the constants μ_0 and ε_0.

2.2. Show that current density $\mathbf{J}$ (A/m²) is a vector quantity and that $\oint\!\!\oint \mathbf{J}\cdot ds = 0$ for any direct current flow. Also show that for time-varying currents

$$\oint\!\!\oint \mathbf{J}\cdot ds = -\,dQ/dt$$

at any instant, Q being the charge enclosed by the surface.

2.3. A long conducting cylinder of diameter D carries a charge Q C/m. Determine the electric field strength (a) inside the cylinder, (b) at the surface of the cylinder, (c) just outside the cylinder, (d) at a distance from the cylinder. Discuss qualitatively what is meant by a long cylinder. [Ans.: (a) 0; (b) $Q/2\pi\varepsilon_0 D$; (c) $Q/\pi\varepsilon_0 D$; (d) $Q/\pi\varepsilon_0 D$.]

2.4. Explain carefully what is meant by a tube of flux. What is the connection between this concept and the inverse square law of force between charges?

From a consideration of electric flux show that there can be no electric field inside a hollow metal container.

2.5. Distinguish clearly between relative and absolute permeability. In an experiment it is required to "screen" a region from a constant magnetic field. Two iron plates are available for this purpose. Should they be placed across the magnetic field or along it? Sketch the resulting flux distribution in either case. [Ans.: Along the field.]

2.6. Carry out a dimensional analysis in terms of the basic dimensions length, mass, time and electric charge on the following quantities; (a) magnetic flux; (b) m.m.f., (c) e.m.f., (d) electric potential, (e) magnetic potential, (f) relative permittivity, (g) magnetic pole, (h) capacitance and (i) reluctance. [Ans.: (a) ML^2/TQ; (b) Q/T; (c) ML^2/T^2Q; (d) ML^2/T^2Q; (e) Q/T; (f) $-$; (g) ML^2/TQ; (h) T^2Q^2/ML^2; (i) Q^2/ML^2.]

2.7. Show that in the electromagnetic wave described in eqs. (2.52) and (2.53) the electric field strength is in phase with the magnetic field strength and that their magnitudes are related by the factor 120π.

2.8. Just outside a flat sheet of high conductivity material there is a magnetic field parallel to the surface $H_y = H_0 \sin(\omega t - \beta x)$, where x and y lie in the plane of the surface. Assuming that the magnetic field is zero just inside the sheet determine the current flowing in the sheet and the pressure on the sheet. [Ans.: $H_0 \sin(\omega t - \beta x)$ A/m; $\frac{1}{2}\mu_0 H_0^2 \sin^2(\omega t - \beta x)$ N/m².]

2.9. A long thin tube of circular cross-section carries an electric charge Q C/m. The mean diameter of the tube is d and the wall thickness t. Determine the stress caused by the charge. [Ans.: Hoop stress $Q^2/4\pi\varepsilon_0 dt$ N/m².]

CHAPTER 3

ELECTROMAGNETIC FIELDS AND THEIR SOURCES

3.1. The Programme of Electromagnetic Theory

In the first chapter we discussed the close relationship between electromagnetism and mechanics, which enables us to describe the behaviour of electric charges in similar terms to the behaviour of gravitational mass in ordinary mechanics. In the second chapter we reviewed some of the electromagnetic relationships and saw that these were expressed in terms of force and energy. We noted that static charges are associated with potential energy which can be described in terms of electrostatic field quantities. The kinetic energy of moving charges was described in terms of magnetic field, and we also took a brief look at the dissipation of energy in resistive conductors. The rest of this book will be devoted to a closer study of these matters and particularly of the relationship between electric and magnetic fields. But before we set out we ought to consider the purpose of this study. Is our journey really necessary?

Many electrical engineers know little of electromagnetism and many textbooks on electrical technology ignore the subject or devote only a short section to it. Instead they deal at length with the principles and applications of electric circuits and especially with the three basic circuit parameters—capacitance, inductance, and resistance. The "circuit" description is very close to the language of electromagnetics. As we mentioned in the last chapter, capacitance is a measure of potential energy, inductance of kinetic energy, and resistance of the dissipation of electrical energy into heat. It is therefore not surprising that many electrical devices can be described both in circuit language and in field language. Since, moreover, the field language is more complicated, it is easy to understand why so many people avoid its use and prefer to think in terms of circuits.

If electromagnetism is to be justified it must be because it can do more than circuit theory. Is this so?

The clue to this question can be found in a close reading of the circuit books. We find that the writers simplify the subject by associating each type of energy with a physical piece of apparatus. Thus resistance is replaced by a resistor, inductance by an inductor, and capacitance by a capacitor. Whereas in electromagnetism a charge may possess potential and kinetic energy at the same time and may also be associated with the conversion of these types of energy to heat, the idealized circuit elements have only one type of energy.

Now it is, of course, true that within a limited range of operating conditions a coil may be dominantly inductive, so that the inductor can be treated as a pure inductance. Nevertheless, as the frequency is raised the capacitance effect is bound to become more marked. The circuit enthusiast then has to admit rather gloomily that the inductor has developed "stray" capacitance. Unfortunately, this stray capacitance may be far from insignificant. For instance a transformer at normal frequency is largely inductive, but when it is subject to a high frequency impulse like that of a lightning stroke, its behaviour is dominated by capacitance. At low frequencies it is an inductor, at high frequencies a capacitor. Similarly, ordinary capacitors have "stray" inductance, and at radio frequencies it becomes almost meaningless to describe an antenna as an inductor or capacitor. It will be tuned to resonance and even the resistance which it then exhibits is to do with the radiation of electrical energy and not with dissipation into heat. It is impossible to understand this action in terms of equivalent circuit elements, and this is small wonder since there is no circuit.

The source of the trouble is now clear. Circuit theory tends to confuse energy processes with pieces of hardware. This explains both its simplicity, where it can be successfully applied, and also its limitations. Of course, circuit theory can be extended to deal with infinitesimal distributed elements, but then it loses its advantages and can conveniently be replaced by field theory. In field theory we are not surprised to find that charges possess at the same time potential and kinetic energy, i.e. capacitance and inductance. Field theory, unlike circuit theory, can deal with *distributed* phenomena and forces *at a point.* Moreover, circuit theory has nothing to say about energy transfer across an air gap in a machine or through empty space in radar. Field theory finds no difficulties in such phenomena because all its laws describe the interaction of charges surrounded by empty space.

We are now beginning to see the great power of field theory and also to appreciate why it is more difficult than circuit theory. We are also able to see the programme of field theory, which is to find the forces of interaction between electric charges of arbitrary position and motion. Once we know these forces we shall be able to predict the behaviour of any electrical device whatever. This is a very ambitious programme. We shall not complete it in this book, but we shall get near its completion and shall be able to identify the unsolved difficulties.

In order to make progress with this programme we shall need to develop a mathematical language able to deal with distributed vector quantities. This language is known as vector analysis and we shall discuss it in general terms in the rest of this chapter. The results obtained in this study will then be applied to specific electromagnetic problems in the rest of the book.

3.2. Vector Analysis of the Electromagnetic Field

3.2.1. Source Coordinates and Field Coordinates

In the last section we discussed the great advantage of field theory in being able to deal with effects distributed in space. In developing the mathematical language we shall therefore generally deal with three space dimensions. Also we shall often have to deal with directed quantities such as forces. For these purposes the language of vectors is very suitable.

In the previous chapter we made use of this language in writing

$$\mathbf{E} = \frac{Q}{4\pi\varepsilon_0 r^2}\hat{\mathbf{r}}, \tag{3.1}$$

where $\mathbf{E}$ is the electrostatic field strength at a point at a distance $\mathbf{r}$ from Q. If Q is at the origin of the coordinate system there is no difficulty about this definition. But when there are many charges not all of them can be located at the origin, and we must distinguish between the coordinates of the *sources* and the coordinates of the *field point* at which **E** is to be observed.

Consider this with reference to Fig. 3.1. The *field* coordinates are $\mathbf{r} = \mathbf{i}x + \mathbf{j}y + \mathbf{k}z$, where **i**, **j**, and **k** are the three unit vectors in the x-,

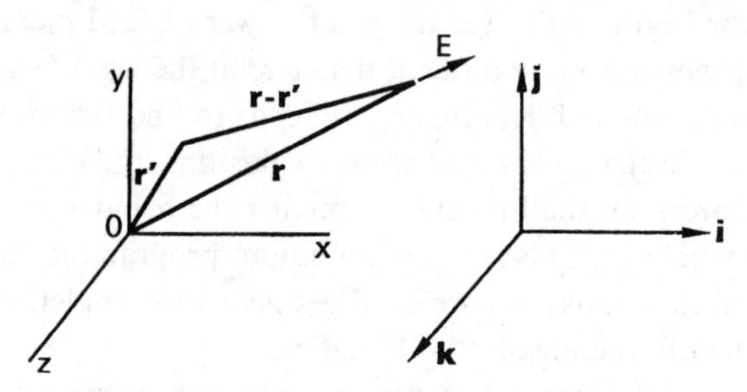

FIG. 3.1. Field coordinates and source coordinates.

y- and z-directions. The *source* coordinates are $\mathbf{r}' = \mathbf{i}x' + \mathbf{j}y' + \mathbf{k}z'$. The radius vector from source to field point is

$$\mathbf{r} - \mathbf{r}' = \mathbf{i}\,(x - x') + \mathbf{j}\,(y - y') + \mathbf{k}\,(z - z')$$

and its magnitude is

$$|\mathbf{r} - \mathbf{r}'| = \sqrt{|(x - x')^2 + (y - y')^2 + (z - z')^2|}.$$

Equation (3.1) should therefore be written more completely as

$$\mathbf{E} = \frac{Q}{4\pi\varepsilon_0\,|\mathbf{r} - \mathbf{r}'|^2}\,(\hat{\mathbf{r}} - \hat{\mathbf{r}}'). \qquad (3.2)$$

We shall often use the shortened form of eqn. (3.1), but the full implications of eqn. (3.2) should not be forgotten.

3.2.2. Scalar and Vector Products

In the previous chapter we used the scalar (or dot) product of two vectors in such relationships as $W = \int \mathbf{F} \,.\, d\mathbf{r}$. Such a product produces a scalar quantity, for instance **force . distance** = **work done**. In rectangular coordinates

$$\mathbf{A} \,.\, \mathbf{B} = A_x B_x + A_y B_y + A_z B_z \qquad (3.3)$$

where the subscripts denote the components of the vectors.

We also briefly introduced the vector (or cross) product of two vectors. This can be illustrated by the relationship **torque** = **force** × **lever arm** and is illustrated in Fig. 3.2. A torque is associated with an axis around which it acts. It is therefore called an *axial* vector and its direction is the direction of

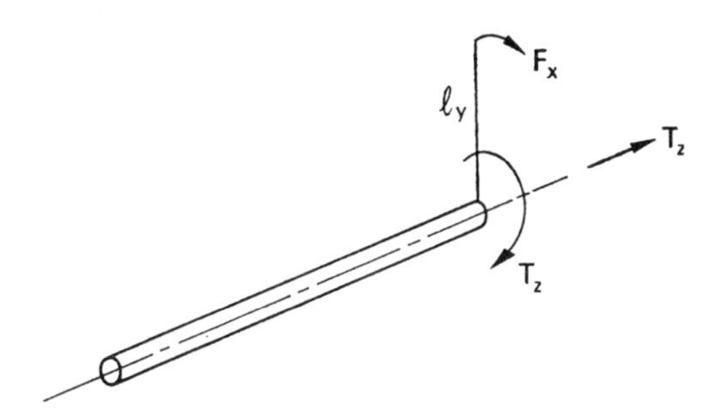

FIG. 3.2. Vector or cross product.

the axis as defined in a right-handed set of axes. Thus if the force is F_x and the lever arm l_y, then the torque is given by $T_z = F_x l_y$. Similarly, $T_x = F_y l_z$ and $T_y = F_z l_x$. In vector notation these three equations can be written $\mathbf{T} = \mathbf{F} \times \mathbf{l}$. A convenient way to remember this is to write in rectangular coordinates.

$$\mathbf{A} \times \mathbf{B} \equiv \begin{vmatrix} \mathbf{i} & \mathbf{j} & \mathbf{k} \\ A_x & A_y & A_z \\ B_x & B_y & B_z \end{vmatrix}$$

$$= \mathbf{i}(A_y B_z - A_z B_y) + \mathbf{j}(A_z B_x - A_x B_z) + \mathbf{k}(A_x B_y - A_y B_x). \tag{3.4}$$

3.2.3. The Gradient of a Scalar Field Quantity

In the previous chapter we discussed the electrostatic field $\mathbf{E}$ and its relationship to the potential V by the equation

$$V = -\int \mathbf{E} \cdot d\mathbf{r}.$$

The vector $\mathbf{E}$ can therefore be derived from a scalar V, and we must investigate how this can be done in vector notation.

We have

$$\delta V = -\mathbf{E} \cdot \delta\mathbf{r}, \tag{3.5}$$

but

$$\delta V = \frac{\partial V}{\partial x}\,\delta x + \frac{\partial V}{\partial y}\,\delta y + \frac{\partial V}{\partial z}\,\delta z. \tag{3.6}$$

This can be written

$$\delta V = \left(\mathbf{i}\,\frac{\partial V}{\partial x} + \mathbf{j}\,\frac{\partial V}{\partial y} + \mathbf{k}\,\frac{\partial V}{\partial z}\right) \cdot (\mathbf{i}\,\delta x + \mathbf{j}\,\delta y + \mathbf{k}\,\delta z). \tag{3.7}$$

We shall call the expression in the first bracket the gradient of V, hence

$$\delta V = (\text{gradient } V) \cdot \delta\mathbf{r}$$

or

$$\delta V = (\text{grad } V) \cdot \delta\mathbf{r} \tag{3.8}$$

and

$$\mathbf{E} = -\text{grad } V. \tag{3.9}$$

The gradient of V is a vector with the three components

$$\frac{\partial V}{\partial x}, \quad \frac{\partial V}{\partial y}, \quad \frac{\partial V}{\partial z}.$$

The direction of this vector can be found by considering Fig. 3.3 which shows two equipotential surfaces V and $V + \delta V$ and a radius vector $\delta\mathbf{r}$. If $\delta\mathbf{r}$ were directed along an equipotential, δV would be zero and we

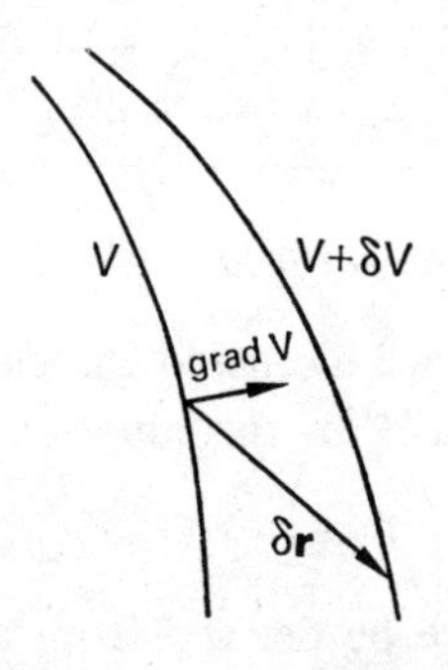

FIG. 3.3. Potential gradient.

should have, from eqn. (3.8), (grad V) . $\delta\mathbf{r} = 0$. For a dot product to be zero the two vectors must be at right-angles. Since in this case $\delta\mathbf{r}$ is along the equipotential, grad V must always be perpendicular to the equipotential. This is the direction of steepest slope, and therefore grad V is the *maximum* slope.

By the use of eqn. (3.9) the electrostatic field can therefore always be derived from a scalar potential field by finding the maximum slope (gradient) at every point of the field.

It is a great advantage to be able to describe a vector field in terms of a scalar potential because at every point only *one* scalar quantity has to be defined, whereas generally a vector needs *three* scalar quantities to define it. The question immediately arises in one's mind whether all vector fields can be expressed as gradients. The answer, unfortunately, is "No". A scalar potential field is to do with *potential* energy. Every point has a unique value of potential and no work is done by returning to the point by any path whatever. The height of a mountain is the same whether one climbs it along a gentle slope on one face or scales the sheer cliffs on another face. In a cyclic process the *potential* energy can be recovered by climbing the cliff and returning on the gentle slope or vice versa. But all the work done by the climber cannot be recovered. That work has not only to alter potential energy; there is also energy dissipation to heat. The climber knows well enough that potential energy is not the whole story. Whenever there is energy transfer or dissipation in a cyclic process it is not possible to describe the field by a single scalar potential. But, if

$$\oint \mathbf{E} \,.\, d\mathbf{l} = 0$$

we can say that $\mathbf{E} = -\text{grad } V$. Such a field is called an *irrotational* or *conservative* field.

The space derivatives

$$\mathbf{i}\,\frac{\partial}{\partial x}, \quad \mathbf{j}\,\frac{\partial}{\partial y}, \quad \mathbf{k}\,\frac{\partial}{\partial z}$$

occur very frequently and they can be thought of as a single vector operator. This vector operator is often written as ∇ and is called "del".† Thus

† Some writers call ∇ nabla.

we can write for an electrostatic field

$$\mathbf{E} = -\text{grad } V = -\nabla V. \tag{3.10}$$

3.2.4. The Sources of an Irrotational Field

To find the relation between the sources and the field quantities of an irrotational field we can make use of Gauss's theorem, which was stated in eqn. (2.13) of the previous chapter. The total "outflow" of electric flux over a closed surface is equal to the total charge surrounded by that surface.

In accordance with our programme outlined in § 3.1 we need to be able to deal with distributed quantities. Instead of Q we can write $\iiint \varrho \, dv$, where ϱ is the charge density per unit volume.†

The left-hand side of eqn. (2.13) describes the outflow of electric flux from the complete volume v. We need therefore to find the outflow per unit volume from a small element of volume.

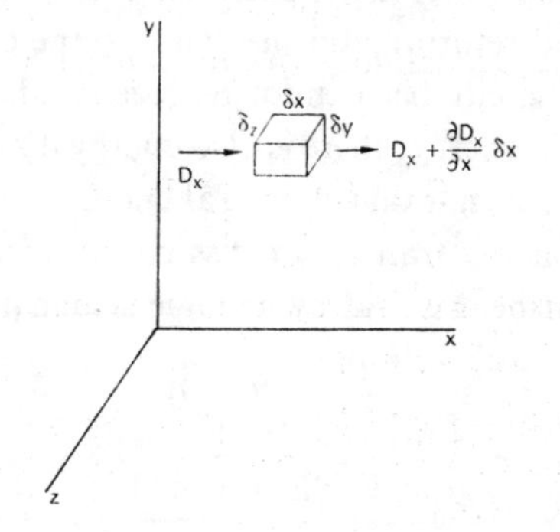

FIG. 3.4. Divergence of a vector field.

Consider Fig. 3.4 which shows the flow from a small volume $\delta x \, \delta y \, \delta z$. Consider first the flow in the x-direction. On one side the flux entering the volume is $D_x \, \delta y \, \delta z$; on the other side the flux leaving the volume is $[D_x + (\partial D_x/\partial x) \, \delta x] \, \delta y \, \delta z$. Hence the net outflow in the x-direction is given

† The symbol ϱ is used also for resistivity. The context generally makes clear which quantity is under discussion.

by

$$\left(D_x + \frac{\partial D_x}{\partial x}\,\delta x\right)\delta y\,\delta z - D_x\,\delta y\,\delta z = \frac{\partial D_x}{\partial x}\,\delta x\,\delta y\,\delta z. \quad (3.11)$$

Thus the outflow per unit volume in the x-direction is $\partial D_x/\partial x$. The *total* outflow of **D** in all directions per unit volume is called the *divergence* of **D**, or more briefly, div **D**. Thus

$$\text{div}\,\mathbf{D} = \frac{\partial D_x}{\partial x} + \frac{\partial D_y}{\partial y} + \frac{\partial D_z}{\partial z}. \quad (3.12)$$

This can be written in terms of the vector operator ∇ by noting that

$$\frac{\partial D_x}{\partial x} + \frac{\partial D_y}{\partial y} + \frac{\partial D_z}{\partial z} = \left(\mathbf{i}\,\frac{\partial}{\partial x} + \mathbf{j}\,\frac{\partial}{\partial y} + \mathbf{k}\,\frac{\partial}{\partial z}\right).\,(\mathbf{i}D_x + \mathbf{j}D_y + \mathbf{k}D_z). \quad (3.13)$$

Thus

$$\text{div}\,\mathbf{D} = \nabla\,.\,\mathbf{D}. \quad (3.14)$$

So far we have considered divergence in terms of a system of rectangular coordinates because this is the simplest geometry. However, we can easily generalize the result by defining the divergence independently of the coordinates. Then

$$\text{div}\,\mathbf{D} = \lim_{v \to 0} \frac{1}{v} \oint\!\!\oint \mathbf{D}\,.\,d\mathbf{s}. \quad (3.15)$$

An alternative form of this statement is obtained by adding the small volumes to make a large volume:

$$\iiint \text{div}\,\mathbf{D}\,dv = \oint\!\!\oint \mathbf{D}\,.\,d\mathbf{s}.\dagger \quad (3.16)$$

By use of eqn. (2.13) we have

$$\oint\!\!\oint \mathbf{D}\,.\,d\mathbf{s} = \iiint \varrho\,dv. \quad (3.17)$$

Therefore from eqns. (3.16) and (3.17)

$$\text{div}\,\mathbf{D} = \varrho. \quad (3.18)$$

Thus the *source* of the field **D** is given by the charge density ϱ in accordance with eqn. (3.18).

† This is commonly known as the divergence theorem.

The notion of divergence is illustrated by Fig. 3.5. It is helpful to think of it in terms of fluid flow. To test for divergence take a small closed surface of wire mesh. If there is a net outflow (or inflow) across the mesh, then there is divergence at that place.

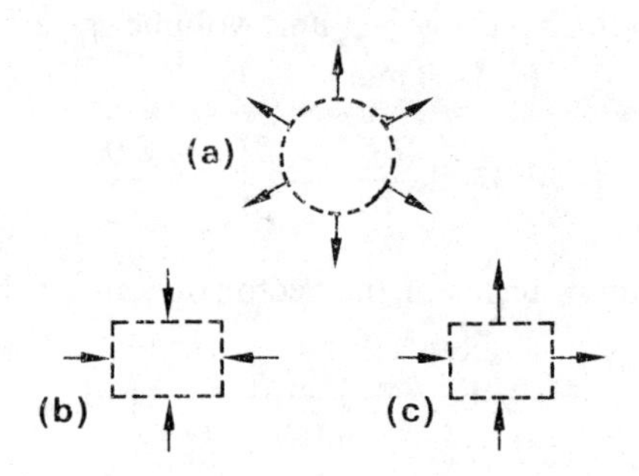

(a) divergence (positive)
(b) divergence (negative)
(c) no divergence

FIG. 3.5. Test for divergence

We can now derive the relationship between the source distribution ϱ in free space and the potential V.

$$\operatorname{div} \mathbf{D} = \varrho, \quad \mathbf{D} = \varepsilon_0 \mathbf{E} \quad \text{and} \quad \mathbf{E} = -\operatorname{grad} V.$$

Hence

$$\operatorname{div} \operatorname{grad} V = -\operatorname{div} \mathbf{E} = -\varrho/\varepsilon_0. \tag{3.19}$$

But

$$\operatorname{div} \operatorname{grad} V = \nabla \cdot \nabla V = \nabla^2 V, \tag{3.20}$$

where the operator div grad, which is called the Laplacian, is written ∇^2.†

Thus

$$\nabla^2 V = -\varrho/\varepsilon_0 \tag{3.21}$$

is the desired relationship. Equation (3.21) is called Poisson's equation and it relates the variation of the potential at a field point to the charge density at the same point. The integral form of the relationship can be

† Called "del squared".

obtained from eqn. (2.6) in the previous chapter,

$$V = \frac{Q}{4\pi\varepsilon_0 r} = \frac{1}{4\pi\varepsilon_0}\iiint \frac{\varrho}{r}\, dv'. \tag{3.22}$$

We write dv' to remind ourselves that the integration has to be carried out over the *source* coordinates. Equation (3.22) is the solution of eqn. (3.21), as can be verified by substitution in that equation.† We can therefore write **E** in terms of the sources ϱ as

$$\mathbf{E} = -\frac{1}{4\pi\varepsilon_0}\nabla \iiint \frac{\varrho}{r}\, dv' = -\frac{1}{4\pi\varepsilon_0}\iiint \varrho\nabla\left(\frac{1}{r}\right) dv'. \tag{3.23}$$

Notice that the gradient is taken with respect to the *field* coordinates and that therefore it does not affect the *source* coordinates.

Often there is no charge density at the field point. The equation then simplifies to

$$\nabla^2 V = 0, \tag{3.24}$$

and is called Laplace's equation.

It is of interest to examine the operator ∇^2 in cartesian coordinates. Written out fully it is

$$\nabla^2 = \left(\mathbf{i}\frac{\partial}{\partial x} + \mathbf{j}\frac{\partial}{\partial y} + \mathbf{k}\frac{\partial}{\partial z}\right) . \left(\mathbf{i}\frac{\partial}{\partial x} + \mathbf{j}\frac{\partial}{\partial y} + \mathbf{k}\frac{\partial}{\partial z}\right)$$

$$= \left(\frac{\partial^2}{\partial x^2} + \frac{\partial^2}{\partial y^2} + \frac{\partial^2}{\partial z^2}\right), \tag{3.25}$$

which is easily remembered.

The discussion in this section has used the electrostatic field as an example. The treatment is, however, quite general and can be applied to any irrotational field. Examples of such fields are the gravitational field, the magnetostatic field in regions free from electric current, and the velocity field of the flow of incompressible fluids free from viscosity.

3.2.5. The Sources of a Non-Conservative Field

As an example of a non-conservative field let us take the magnetic field of a steady current. We have

$$\oint \mathbf{H} . d\mathbf{l} = I. \tag{3.26}$$

† See Exercise 3.10.

In order to be able to deal with distributed quantities we can write

$$I = \iint \mathbf{J} \, . \, d\mathbf{s}, \tag{3.27}$$

where **J** is the current density per unit area. To find the magnetic field **H** at a point we can replace the line integral of eqn. (3.26) by a lot of tiny adjacent loops, as we did in the previous chapter when we considered the magnetic shell (see Fig. 2.7). The only difference is that we were then dealing with a current loop and now we have a loop of magnetic field strength.

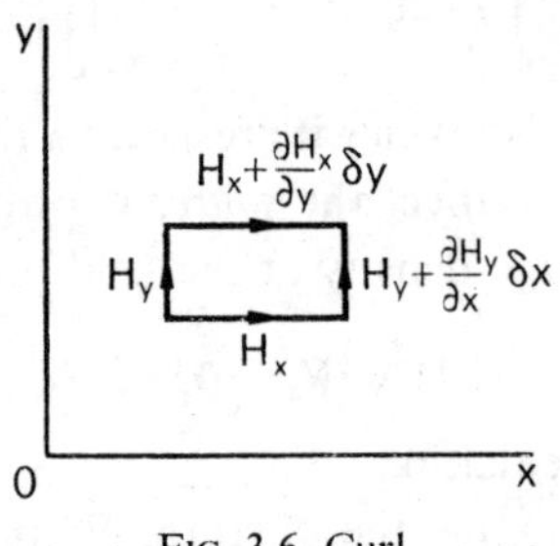

FIG. 3.6. Curl.

Consider the typical small loop in Fig. 3.6. The line integral is

$$\oint \mathbf{H} \, . \, d\mathbf{l} = \left(H_y + \frac{\partial H_y}{\partial x}\,\delta x\right)\delta y - \left(H_x + \frac{\partial H_x}{\partial y}\,\delta y\right)\delta x - H_y\delta y + H_x\delta x$$

$$= \left(\frac{\partial H_y}{\partial x} - \frac{\partial H_x}{\partial y}\right)\delta x\,\delta y. \tag{3.28}$$

The line integral has been taken in the counter-clockwise direction and with a right-handed set of axes this associates the integral with the positive z-direction. The current flow through the small loop is $J_z\,\delta x\,\delta y$.

Thus

$$\frac{\partial H_y}{\partial x} - \frac{\partial H_x}{\partial y} = J_z. \tag{3.29}$$

Similarly, we can show that

$$\frac{\partial H_z}{\partial y} - \frac{\partial H_y}{\partial z} = J_x \tag{3.30}$$

and

$$\frac{\partial H_x}{\partial z} - \frac{\partial H_z}{\partial x} = J_y. \tag{3.31}$$

Thus the small line integrals are associated with a vector which we shall call *curl* **H**, where

$$\text{curl } \mathbf{H} = \begin{vmatrix} \mathbf{i} & \mathbf{j} & \mathbf{k} \\ \frac{\partial}{\partial x} & \frac{\partial}{\partial y} & \frac{\partial}{\partial z} \\ H_x & H_y & H_z \end{vmatrix} = \mathbf{J}$$

and by comparison with eqn. (3.4) we can write this in terms of the operator ∇

$$\text{curl } \mathbf{H} = \nabla \times \mathbf{H} = \mathbf{J}. \tag{3.32}$$

We can generalize the idea of curl by obtaining in any arbitrary coordinate system the curl as the line-integral per unit area for a small loop. Thus

$$\text{curl } \mathbf{H} = \nabla \times \mathbf{H} = \hat{n} \lim_{s \to 0} \frac{1}{S} [\oint \mathbf{H} \,.\, d\mathbf{l}]. \tag{3.33}$$

An alternative form is given by adding adjacent loops

$$\oint \mathbf{H} \,.\, d\mathbf{l} = \iint \text{curl } \mathbf{H} \,.\, d\mathbf{s}, \tag{3.34}$$

which is known as Stokes's theorem. This should be compared with eqns. (3.15) and (3.16). The divergence is the outflow per unit volume and the curl is the circulation per unit area.

Consideration of Fig. 2.7 shows that the surface s need not be in one plane. It can have any shape whatever and can still be built up from meshes each small enough to be considered plane. Every small mesh can therefore be considered as a vector and can be resolved in the three directions of the axes of coordinates. The total circulation can be obtained by adding all the meshes. This is true because adjacent meshes cancel each other except at the perimeter (see Fig. 2.7).

Another example of a non-conservative rotational field is given by

Faraday's law. Equation (2.47) states that

$$\oint \mathbf{E} \cdot d\mathbf{l} = -\frac{d\Phi}{dt}. \tag{2.47}$$

Since Φ is the surface integral of the magnetic flux density $\mathbf{B}$, this can be written

$$\oint \mathbf{E} \cdot d\mathbf{l} = -\frac{d}{dt}\iint \mathbf{B} \cdot d\mathbf{s}. \tag{3.35}$$

If the surface s does not change with time, this can be written

$$\oint \mathbf{E} \cdot d\mathbf{l} = -\iint \frac{\partial \mathbf{B}}{\partial t} \cdot d\mathbf{s}. \tag{3.36}$$

Using Stokes's theorem, eqn. (3.34), we have

$$\oint \mathbf{E} \cdot d\mathbf{l} = \iint \text{curl}\, \mathbf{E} \cdot d\mathbf{s} = -\iint \frac{\partial \mathbf{B}}{\partial t} \cdot d\mathbf{s} \tag{3.37}$$

and

$$\text{curl}\, \mathbf{E} = -\frac{\partial \mathbf{B}}{\partial t}. \tag{3.38}$$

Thus the electric field has a non-conservative part due to changing magnetic flux density as well as a conservative part due to electric charge density.

To visualize the curl of a field it is again useful to think of fluid flow. To test for curl insert a small paddle wheel into the fluid. If it rotates there is curl at that point. (Test this statement with reference to Fig. 3.7.) The

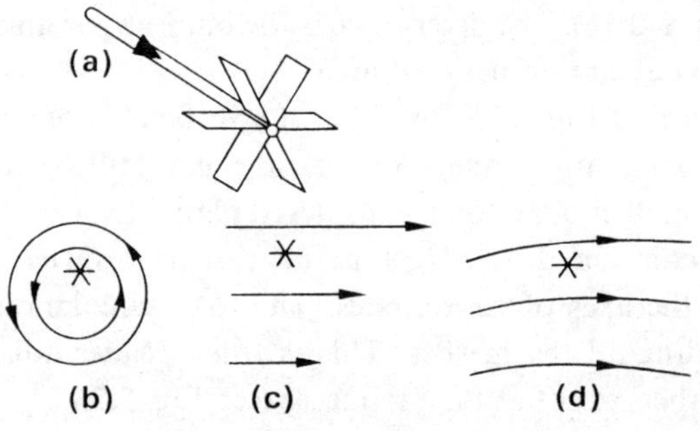

FIG. 3.7. Test for curl: (a) "curl meter", (b) curl, (c) curl, (d) no curl.

rotation will, of course, depend on the position of the axis of the paddle wheel. Curl is a *vector* and the axis of the paddle wheel defines the direction of the curl. The sign of the vector is determined by choosing a right-handed set of axes.

A field which has no divergence but only curl sources, like the magnetic field of eqn. (3.26), which arises from current sources, is said to be *solenoidal.* This name has, of course, been taken from the magnetic field of a solenoid which is of this type.

In a solenoidal field we cannot derive the field vector by taking the gradient of a scalar potential. We can, however, derive it from the curl of a vector, which by analogy is called the vector potential.

Consider eqn. (3.32),

$$\text{curl } \mathbf{H} = \mathbf{J}.$$

Let us derive **H** from a vector **A** by means of the relationship

$$\mathbf{H} = \frac{\mathbf{B}}{\mu_0} = \frac{1}{\mu_0} \text{curl } \mathbf{A}.\dagger \tag{3.39}$$

Then

$$\text{curl curl } \mathbf{A} = \mu_0 \mathbf{J}. \tag{3.40}$$

Using the vector identity (see Appendix, p. 370),

$$\text{curl curl } \mathbf{A} \equiv \text{grad div } \mathbf{A} - \nabla^2 \mathbf{A}, \tag{3.41}$$

and restricting **A** to be solenoidal vector so that div **A** = 0‡ we obtain

$$\nabla^2 \mathbf{A} = -\mu_0 \mathbf{J}. \tag{3.42}$$

This vector equation contains the three equations

$$\left.\begin{aligned} \nabla^2 A_x &= -\mu_0 J_x, \\ \nabla^2 A_y &= -\mu_0 J_y, \\ \nabla^2 A_z &= -\mu_0 J_z. \end{aligned}\right\} \tag{3.43}$$

† This discussion applies to free space. The discussion of magnetic materials is left to § 6.4.

‡ This can be done because the divergence of **A** had not been defined as yet. It is shown in § 3.2.7 that a vector is fully defined when both its divergence and curl are defined.

Comparison with eqn. (3.21) shows these equations to be Poisson's equations and their solution can therefore be written by comparison with eqn. (3.22) as

$$\mathbf{A} = \frac{\mu_0}{4\pi} \iiint \frac{\mathbf{J}}{r}\, dv', \tag{3.44}$$

and we can write **H** in terms of the source distribution **J** as

$$\mathbf{H} = \frac{1}{4\pi} \operatorname{curl} \iiint \frac{\mathbf{J}}{r}\, dv'. \tag{3 45}$$

It should be noted that only a solenoid field, i.e. a field of zero divergence, can be obtained from a vector potential, since the relationship of eqn. (3.38) implies that div $\mathbf{B} = 0$ because of the vector identity

$$\text{div curl } \mathbf{A} \equiv 0. \tag{3.46}$$

This can easily be illustrated by writing div curl **A** as $\nabla \cdot \nabla \times \mathbf{A}$. We know that the dot and cross product with the same vector is zero and this is also true for the vector operator ∇.

3.2.6. The Two Kinds of Sources of a Vector Field

In § 3.2.4 we have specified a *conservative* field in terms of its divergence sources. A conservative field cannot have curl sources because it would then cease to be conservative. There cannot be any rotation in such a field by definition.

In vector language, a conservative field can be described by the vector $\mathbf{F} = -\text{grad } V$ and the sources of the field by div $\mathbf{F} = \varrho$. That such a field has no curl can be seen from the vector identity.

$$\text{curl grad } V \equiv \nabla \times \nabla V \equiv 0. \tag{3.47}$$

Clearly the cross product of a vector with itself is zero and this is true of the vector operator ∇.

A special but important case of a conservative field arises when there are no sources within the region. Of course there will then have to be sources outside the region, otherwise the problem would be meaningless and the field would everywhere be zero.

In § 3.2.5 we specified a *non-conservative* solenoidal field by its curl sources. It is important to notice that a vector field derived solely from curl sources has no divergence, since

$$\text{div curl } \mathbf{F} \equiv \nabla \cdot \nabla \times \mathbf{F} \equiv 0. \qquad (3.48)$$

We have already discussed this in the last section in connection with the vector potential **A**.

Consideration of eqns. (3.47) and (3.48) suggests that the divergence and curl sources are independent of each other and that the *complete* specification of a general vector field should include *both* types of sources. This is true, but even more useful is the fact that such a specification is not only necessary but sufficient. *Any vector field is uniquely determined if its divergence and curl sources are given.* We shall give a proof of this statement, although it is a little cumbersome and may well be left for later reading. What is vital, however, is that the fact of the statement should be remembered. There are only *two* types of sources of a vector field—*divergence* and *curl*—and these define the field *completely.*

Before we give the proof it is worth while to list the different types of field.

(1) A vector field may have no sources in a region. It is then irrotational (conservative) in that region and can be expressed in terms of the gradient of a scalar potential (Fig. 3.8).

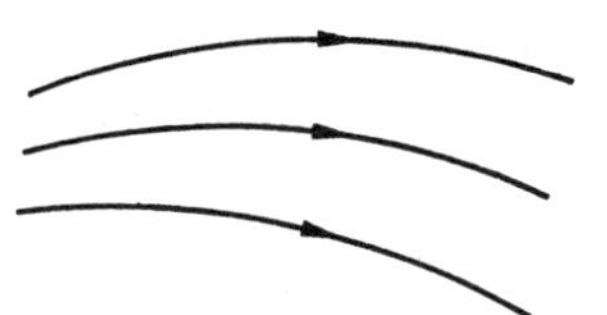

FIG. 3.8. Field without sources.

(2) A vector field may have divergence sources but no curl sources within a region. It is then irrotational (conservative) in that region and can be expressed in terms of the gradient of a scalar potential (Fig. 3.5).

(3) A vector field may have curl sources but no divergence sources within a region. It is then solenoidal in that region and can be expressed in terms of the curl of a vector potential (Fig. 3.7).

(4) A vector field may have divergence sources and curl sources within a region. It has then an irrotational and also a solenoidal part and can be expressed in terms of the gradient of a scalar potential plus the curl of a vector potential (Fig. 3.9).

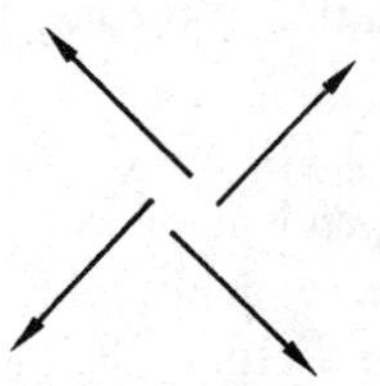

FIG. 3.9. Field with divergence and curl sources.

3.2.7.† Uniqueness Theorem of a Vector Field in Terms of its Sources

We shall now make use of the results of §§ 3.2.4–3.2.6 to obtain a formal proof of the uniqueness of a vector field in terms of its sources.

Let the sources be given by

$$\operatorname{div} \mathbf{F} = s, \tag{3.49}$$

$$\operatorname{curl} \mathbf{F} = \mathbf{c}. \tag{3.50}$$

We wish to prove that **F** can be uniquely specified in terms of these sources.

Let us first show that if

$$\mathbf{F} = -\operatorname{grad} \phi + \operatorname{curl} \mathbf{W}, \tag{3.51}$$

where

$$\phi = \frac{1}{4\pi} \iiint \frac{s}{r}\, dv' \tag{3.52}$$

and

$$\mathbf{W} = \frac{1}{4\pi} \iiint \frac{\mathbf{c}}{r}\, dv', \tag{3.53}$$

then **F** satisfies the source equations (3.49) and (3.50).

† This section could well be omitted on a first reading of the book.

Consider first eqn. (3.49):

$$\operatorname{div} \mathbf{F} = -\operatorname{div} \operatorname{grad} \phi + \operatorname{div} \operatorname{curl} \mathbf{W} = -\nabla^2 \phi = s. \qquad (3.54)$$

This is Poisson's equation and is satisfied by eqn. (3.52).

Consider next eqn. (3.50):

$$\operatorname{curl} \mathbf{F} = \operatorname{curl} \operatorname{curl} \mathbf{W} = \operatorname{grad} \operatorname{div} \mathbf{W} - \nabla^2 \mathbf{W} = \mathbf{c}. \qquad (3.55)$$

We shall be able to show that grad div $\mathbf{W}$ is zero. Assuming this result for the moment we have

$$\nabla^2 \mathbf{W} = -\mathbf{c},$$

which is Poisson's equation and is satisfied by eqn. (3.53). Hence eqns. (3.49) and (3.50) are satisfied by eqn. (3.51) which was to be proved.

To show that grad div $\mathbf{W}$ is indeed zero, as we have just assumed, consider this expression, which will be denoted by I.

$$I = \nabla \nabla . \iiint \frac{\mathbf{c}}{r}\, dv' = \iiint \nabla \nabla . \frac{\mathbf{c}}{r}\, dv',$$

since the operator ∇ is independent of the source coordinates dv'. Also, since $\mathbf{c}$ is purely in source coordinates, we can write

$$\nabla . \left(\frac{\mathbf{c}}{r}\right) = \mathbf{c} . \nabla \left(\frac{1}{r}\right) \dagger$$

and

$$\nabla \left[\mathbf{c} . \nabla \left(\frac{1}{r}\right)\right] = (\mathbf{c} . \nabla) \nabla \left(\frac{1}{r}\right). \ddagger$$

The expression therefore becomes

$$I = \iiint (\mathbf{c} . \nabla) \nabla \left(\frac{1}{r}\right) dv'.$$

Now

$$r = \sqrt{[(x - x')^2 + (y - y')^2 + (z - z')^2]},$$

and the operator ∇ has components $\partial/\partial x$, $\partial/\partial y$, $\partial/\partial z$. Because of the symmetry of the expression for r we can therefore put $\nabla = -\nabla'$, where ∇'

† See Exercise 3.15.
‡ See Exercise 3.17.

has components $\partial/\partial x'$, $\partial/\partial y'$, $\partial/\partial z'$. We thus have

$$I = \iiint (\mathbf{c} \cdot \nabla) \nabla \left(\frac{1}{r}\right) dv' = \iiint (\mathbf{c} \cdot \nabla') \nabla' \left(\frac{1}{r}\right) dv',$$

where everything is now in source coordinates. The expression I has three components, which we can consider one at a time and integrate by parts.

$$I_{x'} = \iiint (\mathbf{c} \cdot \nabla') \frac{\partial}{\partial x'} \left(\frac{1}{r}\right) dv'$$

$$= \iiint \nabla' \cdot \left[\mathbf{c} \frac{\partial}{\partial x'} \left(\frac{1}{r}\right)\right] dv' - \iiint (\nabla' \cdot \mathbf{c}) \frac{\partial}{\partial x'} \left(\frac{1}{r}\right) dv'.^{\dagger} \tag{3.56}$$

The second integral is zero because $\mathbf{c}$ has no divergence since it is equal to curl $\mathbf{F}$ by eqn. (3.50). Also, the first integral can be converted to a surface integral by eqn. (3.16). The surface of this integral can be taken out of the region in which there are sources $\mathbf{c}$, if $\mathbf{c}$ is bounded in space. Thus I_x will be zero and I will be zero, which was to be proved.

To show that the solution of eqn. (3.51) is unique we suppose the possibility of two different solutions $\mathbf{F}_1$ and $\mathbf{F}_2$ which both satisfy eqns. (3.49), (3.50), and (3.51). Let $\mathbf{G} = \mathbf{F}_1 - \mathbf{F}_2$, then

$$\text{div } \mathbf{G} = 0, \tag{3.57}$$

$$\text{curl } \mathbf{G} = 0. \tag{3.58}$$

The field $\mathbf{G}$ is therefore conservative and we can write

$$\mathbf{G} = -\nabla\psi. \tag{3.59}$$

From eqn. (3.57) we find

$$\nabla^2\psi = 0. \tag{3.60}$$

If now we apply the divergence theorem eqn. (3.16) to the vector $\psi \nabla\psi$, we obtain (see Appendix, p. 372)

$$\oiint \psi \nabla\psi \cdot d\mathbf{s} = \iiint [\psi \nabla^2\psi + (\nabla\psi)^2] \, dv. \tag{3.61}$$

If the boundary surface of the integral is taken at a sufficiently large distance from the sources, the left side of eqn. (3.61) vanishes, since ψ varies as $1/r$, $\nabla\psi$ as $1/r^2$ and s as r^2. Also the first term of the right side

† See Exercise 3.15.

vanishes by eqn. (3.60). Hence

$$\iiint (\nabla\psi)^2 \, dv = 0. \tag{3.62}$$

Since the integrand cannot be negative because it is squared, we have $\nabla\psi = \mathbf{G} = 0$ everywhere. Hence $\mathbf{F}_1 = \mathbf{F}_2$ and the solution of eqn. (3.51) is unique. The field $\mathbf{F}$ is completely specified by its divergence and curl sources.

3.3. Sources, Fields, and Potentials

In § 3.2 we were occupied with the formulation of electromagnetism in terms of the vector calculus. The power and elegance of this formulation presents a great temptation to the writer of a textbook to present the whole subject as a branch of applied mathematics. It will become easier to resist the temptation when in later chapters we deal with applications of the theory. But even at this early stage it is worth while to take a detached look at the mathematics and ask what physical information there may be in the various symbols and equations.

We remind ourselves that the subject starts with electricity as a source of mechanical force defined by the inverse square law. The primary task of an electrical engineer is always to determine the sources, be they charges or currents. Knowledge of an electrical device is incomplete until the currents and charges have been specified. The insistence of the vector calculus on a knowledge of the divergence and curl sources says the same thing mathematically. In general, electromagnetic divergence sources are static charges and curl sources moving charges or currents.

From the sources the subject proceeds to fields. The idea of a field is secondary, but nevertheless tremendously important. It provides a mental stepping stone. Instead of source A acting on source B we can postulate that source A produces a field and that this acts on source B. There is no new experimental fact in such a view. Indeed, strictly speaking, a unit charge is needed to observe the field at the field point, but this is not generally stated explicitly.

Where only two sources are under consideration, little is gained by talking about a field. But generally there will be many sources and then the field strength at a point can be used to describe the *combined* effect of these sources. Nor is this all. Often it is very difficult to determine the distribution of all the sources inside a body. In such cases one may be content to know the total source without knowing the location of its parts. Knowledge of the total source can be obtained from the field surrounding the body. Gauss's theorem, for example, tells us that we can find the total enclosed charge from the flux through the enclosing surface, and, similarly, we can find the total enclosed current from the circulation of the magnetic field strength around the perimeter [eqn. (3.26)]. Better still, we shall show in the next chapter, § 4, that we can obtain not only the magnitude of the source but also the force on it if we know the field strength at every point on an enclosing surface. This process is further extended in Chapter 10, § 3, where we find that the energy flow across a closed surface can be determined from a knowledge of the field strength of every point of an enclosing surface. It is clear that the field has many uses.

In order to make full use of the field concept we shall need a means of relating the field to the sources. Two typical relationships were given in § 3.2 as

$$\operatorname{div} \mathbf{D} = \varrho \tag{3.18}$$

and

$$\operatorname{curl} \mathbf{H} = \mathbf{J}. \tag{3.32}$$

Unfortunately these relationships are not very useful as they stand. Both of them relate the *local* variation of the field to the source density *at that point*, whereas in general we wish to find the field due to sources *elsewhere*. Often we shall wish to calculate the field at a place free of sources, for instance in the air gap of an electrical machine. Typical equations would then be

$$\operatorname{div} \mathbf{D} = 0 \tag{3.63}$$

and

$$\operatorname{curl} \mathbf{H} = 0. \tag{3.64}$$

It is evident that these equations are insufficient to tell us how to find **D** and **H** because they say nothing about the local effect of distant sources.

To overcome this difficulty the theory has recourse to what might be called a tertiary concept, which is a field quantity called potential. The idea is that the potential should be easily derivable from the sources and that the field strength can then be obtained from the potential. The sequence is source → potential → field strength.

Let us illustrate this process with reference to the electrostatic field. It was shown in § 3.2.3 that the electric field strength can be derived from a potential by

$$\mathbf{E} = -\text{grad } V. \tag{3.10}$$

The relation between this potential and the sources was discussed in § 3.2.4 and was shown to be

$$V = \frac{1}{4\pi\varepsilon_0}\iiint \frac{\varrho}{r}\, dv'. \tag{3.22}$$

Thus a knowledge of ϱ leads to knowledge of V and thence to **E**.

Similarly, in magnetostatics we found in § 3.2.5 that a *vector* potential **A** could be employed to derive the magnetic field strength from the current sources by means of the relationships

$$\mu_0\mathbf{H} = \text{curl } \mathbf{A} \tag{3.39}$$

and

$$\mathbf{A} = \frac{\mu_0}{4\pi}\iiint \frac{\mathbf{J}}{r}\, dv'. \tag{3.44}$$

These vector relationships each represent three scalar equations. For instance one component of eqn. (3.44) is

$$A_x = \frac{\mu_0}{4\pi}\iiint \frac{J_x}{r}\, dv', \tag{3.65}$$

and we notice in passing that each component of the vector **A** depends only on current in a parallel direction. We have already seen in Chapter 2 that V represents potential energy and we know that **A** is associated with the kinetic energy of the current density **J**. The fact that **A** is parallel to **J** gives us the clue that **A** has the nature of momentum, and we shall revert to this in Chapter 9. The main point here, however, is that a knowledge of **J** leads to **A**, and **A** to **H**. Equations (3.22) and (3.44), and, for the general case, (3.51)–(3.53), show how the potentials can be found by integration. This integration sums the effect of distant as well as local sources.

Before we leave the subject we must take a brief look at an alternative formulation in terms of differential equations. Let us refer again to § 3.2.4 and discuss the electrostatic field. Consider the three equations

$$\operatorname{div} \mathbf{D} = \varrho, \quad \mathbf{D} = \varepsilon_0 \mathbf{E}, \quad \text{and} \quad \mathbf{E} = -\operatorname{grad} V.$$

They can be combined into Poisson's equation

$$\nabla^2 V = -\varrho/\varepsilon_0. \tag{3.21}$$

This equation relates the variation of the potential V at a point to the charge density ϱ *at that point*. It is a differential equation and can be solved by well-established mathematical techniques. One form of the solution is the integral of eqn. (3.22). But more frequently we shall seek a solution in a finite region, where the potential (or perhaps the normal gradient of potential) is specified on the enclosing surface. In such a case we shall find a solution of the differential equation in terms of arbitrary constants and then obtain the constants from the given boundary condition. Physically the boundary conditions account for the distant sources. The uniqueness theorem of § 3.2.7 will guarantee that the solution is also unique. If there is no charge density at the place at which the variation of V is being considered, we have Laplace's equation

$$\nabla^2 V = 0. \tag{3.24}$$

The solution of this equation proceeds in the same way except that there is no "particular integral" to correspond to the right-hand side of Poisson's equation. This discussion also applies to the equations of the vector potential **A**. We shall deal in the next chapter with the process of solving these equations. The discussion in this chapter is included first to show why the potentials are introduced into the theory, and, secondly, how we can find the potentials *either* by integration of the sources *or* by the solution of a differential equation with known boundary values.

Summary

In this chapter we have examined the programme of electromagnetism. This is to determine the interaction between charges which may be stationary or in motion. Unlike circuit theory, electromagnetism deals with effects at a point in space (and time), and it is therefore able to give a more detailed description and understanding.

It is to be expected that a price has to be paid for this additional information and that electromagnetic theory is necessarily more complicated than circuit theory. Happily the complication is not overwhelming because a convenient notation has been devised known as the vector calculus.

By means of this calculus the sources of the electromagnetic field can be shown to be of two kinds only. The first kind is associated with systems such as electrostatics in which energy is conserved during a cyclic process within the system. The sources of such a conservative or irrotational system are defined by the divergence of the field. The second kind of source is associated with systems in which there is energy transfer in a cyclic process and is defined by the rotation or curl of the field. A field which has curl but no divergence is called solenoidal, a typical case being the magnetic field of a solenoid. In general, electromagnetic fields have both types of sources.

In order to obtain the field from the sources it is necessary to make use of the electromagnetic potentials. A scalar potential is obtained by summing the effects of the divergence sources and a vector potential by summing the effects of the curl sources. Alternatively, the potentials may be obtained by solving Poisson's or Laplace's equations, subject to known boundary conditions. When the potentials are known, the fields can be calculated by adding the gradient of the scalar potential to the curl of the vector potential.

The sequence of thought is illustrated by the following flow chart.

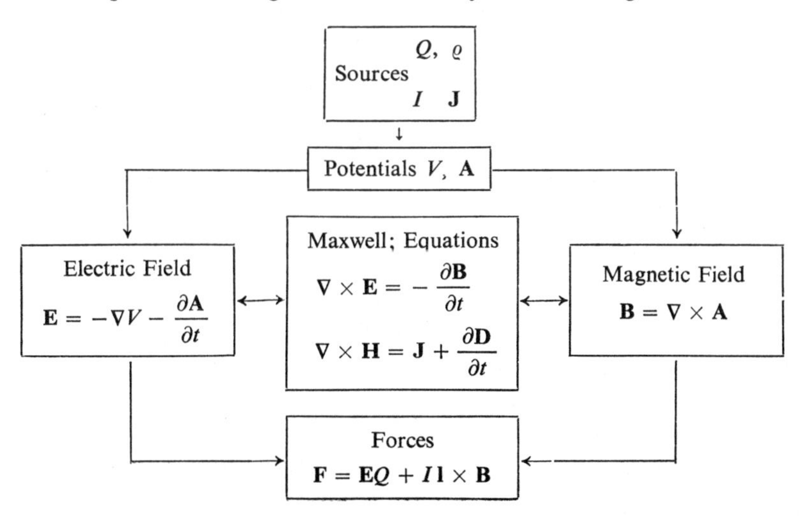

Glossary

Term	*Symbol*	*Explanation*
Scalar product (dot product)	$\mathbf{A} . \mathbf{B}$	$A_x B_x + A_y B_y + A_z B_z$
Vector product (cross product)	$\mathbf{A} \times \mathbf{B}$	$\begin{vmatrix} \mathbf{i} & \mathbf{j} & \mathbf{k} \\ A_x & A_y & A_z \\ B_x & B_y & B_z \end{vmatrix}$
Gradient (the operator ∇ by itself is called del)	∇ (grad)	A vector operator giving the maximum slope of a scalar field. In x, y, z coordinates $\nabla = \left(\mathbf{i}\frac{\partial}{\partial x} + \mathbf{j}\frac{\partial}{\partial y} + \mathbf{k}\frac{\partial}{\partial z}\right)$
Divergence	$\nabla .$ (div)	Outflow per unit volume. In x, y, z coordinates $\nabla . \mathbf{A} = \frac{\partial A_x}{\partial x} + \frac{\partial A_y}{\partial y} + \frac{\partial A_z}{\partial z}$
Curl	$\nabla \times$ (curl)	Circulation per unit area. In x, y, z coordinates $\nabla \times \mathbf{A} = \begin{vmatrix} \mathbf{i} & \mathbf{j} & \mathbf{k} \\ \frac{\partial}{\partial x} & \frac{\partial}{\partial y} & \frac{\partial}{\partial z} \\ A_x & A_y & A_z \end{vmatrix}$

Exercises

3.1. What is the scalar product of any two diagonals through the body of a cube of side a? [Ans.: a^2.]

3.2. In the cube of question 1, what is the vector product between the diagonal from the point (0, 0, 0) to the point (a, a, a) and the diagonal from the point $(a, 0, 0)$ to the point $(0, a, a)$? [Ans.: The vector $(0, -2a^2, 2a^2)$.]

3.3. The force $\mathbf{P} = 3\mathbf{i} + 5\mathbf{j}$ moves its point of application from (1, 2) to (4, 3). What is the work done? [Ans.: 14.]

3.4. Show that the area of a parallelogram of sides $\mathbf{a}$ and $\mathbf{b}$ is given by $\mathbf{A} = \mathbf{a} \times \mathbf{b}$. What is meant by the direction of an area?

3.5. Show that $\mathbf{A} \times (\mathbf{B} \times \mathbf{C}) = \mathbf{B}(\mathbf{A} . \mathbf{C}) - \mathbf{C}(\mathbf{A} . \mathbf{B})$. (This can be remembered by pronouncing it *bac* – *cab*.)

3.6. Explain what is meant by the gradient of a scalar quantity. Express the gradient of the scalar potential V in cylindrical (r, θ, z) and spherical (r, θ, ϕ) coordinates.

$$\left[\text{Ans.: } \frac{\partial V}{\partial r}, \frac{1}{r}\frac{\partial V}{\partial \theta}, \frac{\partial V}{\partial z}; \frac{\partial V}{\partial r}, \frac{1}{r}\frac{\partial V}{\partial \theta}, \frac{1}{r \sin\theta}\frac{\partial V}{\partial \phi}.\right]$$

3.7. The vector $\mathbf{r}$ points from the source located at (x', y', z') to the field point (x, y, z). Show that $\nabla(1/r) = -\nabla'(1/r)$, where ∇ operates on the field coordinates and ∇' on the source coordinates. Determine $\nabla(1/r)$. [Ans.: $-(x-x')/r^3$, $-(y-y')/r^3$, $-(z-z')/r^3$.]

3.8. Explain what is meant by the *divergence* of a vector field. Discuss why an electric field $\mathbf{D}$ can have divergence, but a magnetic field $\mathbf{B}$ cannot have divergence. Express the divergence of $\mathbf{D}$ in spherical coordinates.

$$\left[\text{Ans.: } \frac{1}{r^2}\frac{\partial}{\partial r}(r^2 D_r) + \frac{1}{r\sin\theta}\frac{\partial}{\partial\theta}(D_\theta \sin\theta) + \frac{1}{r\sin\theta}\frac{\partial D_\phi}{\partial\phi}.\right]$$

3.9. Show that Laplace's equation in cylindrical coordinates is

$$\nabla^2 V = \frac{\partial^2 V}{\partial r^2} + \frac{1}{r}\frac{\partial V}{\partial r} + \frac{1}{r^2}\frac{\partial^2 V}{\partial \theta^2} + \frac{\partial^2 V}{\partial z^2} = 0.$$

3.10. Show that $V = \iiint (\varrho/4\pi\varepsilon_0 r)\, dv'$ is the solution of $\nabla^2 V = -\varrho/\varepsilon_0$. [*Hint:* Divide the region between a small sphere around the point $r = 0$ and the rest of space.]

3.11. Show that $\operatorname{div} \mathbf{r} = 3$.

3.12. A vector has components $A_r = k/r$, $A_\theta = 0$, $A_z = 0$. Show that the vector can be derived from a scalar potential and find its value. [Ans.: $V = -k \ln(r)$.]

3.13. A two-dimensional field is given by the relationships

$$E_x = k_1 \cos ax \cosh by,$$
$$E_y = k_2 \sin ax \sinh by.$$

Find the relationship between k_1 and k_2 if the field is irrotational. [Ans.: $bk_1 = ak_2$.]

3.14. Explain what is meant by the *curl* of a vector field. Show that the curl of a vector $\mathbf{A}$ in spherical coordinates is given by

$$\nabla \times \mathbf{A} = \begin{vmatrix} \dfrac{\mathbf{u}_r}{r^2 \sin\theta} & \dfrac{\mathbf{u}_\theta}{r\sin\theta} & \dfrac{\mathbf{u}_\phi}{r} \\ \dfrac{\partial}{\partial r} & \dfrac{\partial}{\partial \theta} & \dfrac{\partial}{\partial \phi} \\ A_r & rA_\theta & r\sin\theta A_\phi \end{vmatrix}$$

3.15. Derive the following relationships:

$$\operatorname{div}(a\mathbf{u}) = a \operatorname{div}\mathbf{u} + \mathbf{u} \cdot \operatorname{grad} a$$
$$\operatorname{curl}(a\mathbf{u}) = a \operatorname{curl}\mathbf{u} - \mathbf{u} \times \operatorname{grad} a$$
$$\operatorname{div}(\mathbf{u} \times \mathbf{v}) - \mathbf{v} \cdot \operatorname{curl}\mathbf{u} - \mathbf{u} \cdot \operatorname{curl}\mathbf{v}.$$

3.16. Show that in rectangular coordinates

$$\text{curl}\,(\text{curl}\,\mathbf{A}) = \text{grad div}\,\mathbf{A} - \nabla^2\mathbf{A},$$

where

$$\nabla^2\mathbf{A} = \mathbf{i}\,\nabla^2 A_x + \mathbf{j}\,\nabla^2 A_y + \mathbf{k}\,\nabla^2 A_z.$$

3.17. Show that

$$\text{curl}\,(\mathbf{u}\times\mathbf{v}) = (\mathbf{v}\,.\,\nabla)\,\mathbf{u} - (\mathbf{u}\,.\,\nabla)\,\mathbf{v} + \mathbf{u}\,\text{div}\,\mathbf{v} - \mathbf{v}\,\text{div}\,\mathbf{u}$$

and

$$\text{grad}\,(\mathbf{u}\,.\,\mathbf{v}) = (\mathbf{v}\,.\,\nabla)\,\mathbf{u} + (\mathbf{u}\,.\,\nabla)\,\mathbf{v} + \mathbf{v}\times\text{curl}\,\mathbf{u} + \mathbf{u}\times\text{curl}\,\mathbf{v}.$$

3.18. Use the divergence theorem to prove that

$$\iiint \text{grad}\, V\,dv = \oiint V\mathbf{n}\,ds,$$

where $\mathbf{n}$ is the unit vector normal to the surface. [*Hint:* In the divergence theorem divide the outflow into three components along the three rectangular coordinate directions. Then multiply each component by the unit vector in that direction.]

3.19. Use the divergence theorem to prove that

$$\iiint \text{curl}\,\mathbf{F}\,dv = \oiint \mathbf{n}\times\mathbf{F}\,ds.$$

[*Hint:* Put $\mathbf{F} = \mathbf{a}\times\mathbf{D}$, where $\mathbf{a}$ is a constant vector. Then use the third expression in Exercise 15.]

3.20. Prove Green's theorem

$$\iiint (U\nabla^2 V - V\nabla^2 U)\,dv = \iint (U\nabla V - V\nabla U)\,.\,d\mathbf{s}$$

[*Hint:* Apply the divergence theorem to the vector $U\nabla V$.]

3.21. Show that in a moving fluid the rate of change of potential is given by

$$\frac{dV}{dt} = \frac{\partial V}{\partial t} + \mathbf{v}\,.\,\nabla V,$$

where $\mathbf{v}$ is the particle velocity of the fluid.

CHAPTER 4

ELECTROSTATICS I

4.1. Source Distributions

Electrostatics deals with the interaction of stationary electric charges. As was pointed out in the last chapter, such problems of interaction can be simplified if they are considered in two stages. First the charges are considered as sources of an electrostatic field and then this field acts on the charge at the field point. Thus in Fig. 4.1, if we want to find the force

FIG. 4.1. Force on a point charge.

on the point charge Q_4 due to the three point charges Q_1, Q_2, Q_3, we first remove Q_4 and calculate the field due to the sources Q_1, Q_2, Q_3 at the field point at which Q_4 was located. We then replace Q_4 and calculate the force on it by multiplying the electric field strength due to Q_1, Q_2, Q_3 by

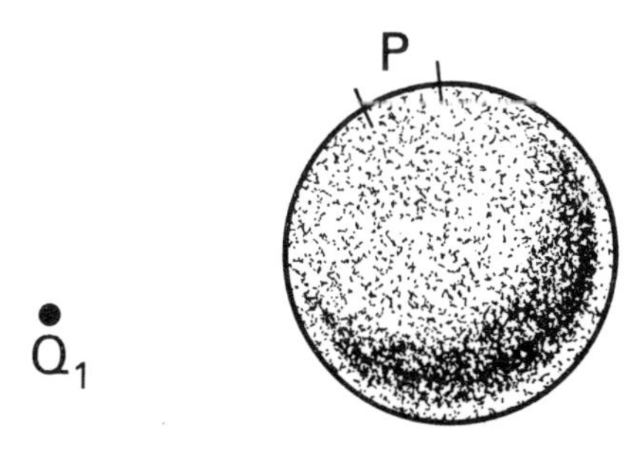

FIG. 4.2. Force on a conducting body.

the charge strength Q_4. Underlying this procedure is the principle of superposition. The electric field is proportional to the charge strength and this enables us to calculate the total field by summation. The application of the process to point charges is straightforward, but care needs to be taken when one is dealing with charged bodies. Consider this with respect to Fig. 4.2. A point charge Q_1 interacts with a conducting body and we require the force at a point P on the surface of the body. In this case we cannot say that the source of the field at P is Q_1 only. We must also take into consideration all the induced charge distribution on the sphere—except the point charge at P. This point charge is the local charge on a very small area at the place at which we want to find the force. We shall revert to this discussion later in this chapter.†

All kinds of charge distributions can be built up from point sources, but it is often convenient to start the synthesis of a problem by using other simple source distributions. We shall, therefore, first of all consider the electric field of certain standard arrangements of sources. Section 4.1 should be treated by the reader as a compendium of useful results.

4.1.1. The Electric Field of a Point Charge

This is the basic "brick" of electrostatics.

$$\mathbf{E} = \frac{Q}{4\pi\varepsilon_0 r^2}\,\hat{\mathbf{r}} = -\frac{Q}{4\pi\varepsilon_0}\,\nabla\left(\frac{1}{r}\right), \tag{4.1}$$

$$V = \frac{Q}{4\pi\varepsilon_0 r}. \tag{4.2}$$

4.1.2. The Electric Field of a Doublet or Dipole of Charge

Such dipoles are used to represent polarized materials. Some materials, such as water, have polar molecules which can be aligned by an electric field. Also in ordinary non-polar molecules, an applied field can displace the electron shells around the atomic nucleus and thus cause polarization. The dipole model which we are discussing is independent of the actual mechanism of polarization.

† See §§ 4.1.5 and 4.2.2.

The potential of a dipole can be obtained from eqn. (4.2) by adding the potentials of a positive and negative charge as shown in Fig. 4.3. The source coordinates are shown primed and we consider a small distance $\delta x'$ between the charges $-Q$ and $+Q$. Instead of x, y, z let us use the subscript notation x_1, x_2, x_3.

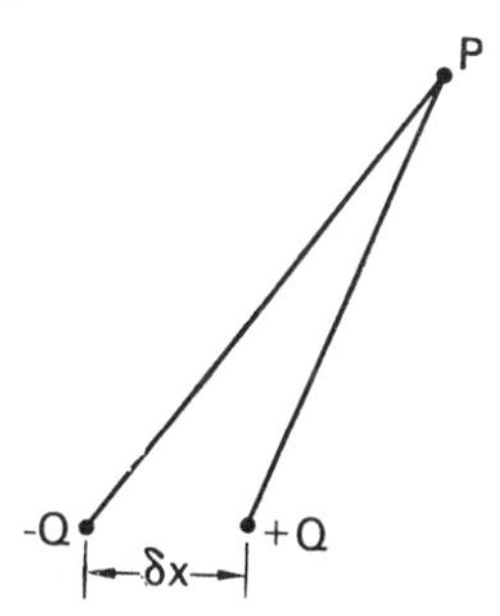

FIG. 4.3. A dipole.

The potential at the typical field point x_1 due to $+Q$ at the source point x_1' is

$$V = \frac{Q}{4\pi\varepsilon_0 r} = \frac{Q}{4\pi\varepsilon_0}\left[1\Big/\sum_{1,2,3}(x_1 - x_1')^2\right]^{1/2}. \tag{4.3}$$

The potential at x_1 due to $+Q$ at $x_1' + \delta x'$ is

$$V + \frac{\partial V}{\partial x_1'}\delta x_1',$$

and the potential due to $-Q$ and $+Q$, as shown in Fig. 4.3, is obtained by subtracting V from $V + (\partial V/\partial x_1')\,\delta x_1'$. This potential is therefore given by

$$\frac{\partial V}{\partial x_1'}\delta x_1' = \frac{Q\,\delta x_1'\,(x_1 - x_1')}{4\pi\varepsilon_0 r^3} = \frac{(Q\,\delta x_1')\cos\theta}{4\pi\varepsilon_0 r^2}, \tag{4.4}$$

where θ is the angle between r and the x axis.

If we write

$$Q\,\delta x_1' = P_1 \tag{4.5}$$

and call $\mathbf{P}$ the dipole moment, we can write the potential of the dipole as

$$V = \frac{1}{4\pi\varepsilon_0}\frac{\mathbf{P}\,.\,\hat{\mathbf{r}}}{r^2}, \tag{4.6}$$

where the vector dot product replaces the cos θ of eqn. (4.4). Then the electric field strength

$$\mathbf{E} = -\nabla V = \frac{1}{4\pi\varepsilon_0}\nabla\left[\mathbf{P}\cdot\frac{\hat{\mathbf{r}}}{r^2}\right]. \tag{4.7}$$

Note that the potential of a dipole decreases as the square of the distance and the electric field strength as the cube of the distance. The action of a dipole is therefore much weaker than the action of a point charge.

4.1.3. The Electric Field of a Line Filament of Charge

A filament is a useful starting point for dealing with such problems as the capacitance of a cable. Consider a filament of constant line density of charge q coulomb/metre as in Fig. 4.4. If the filament is infinitely long the potential cannot be defined as the work done in bringing a unit charge from infinity because the total charge of the filament is infinite and the potential will also be infinite. This does not mean that the electric field

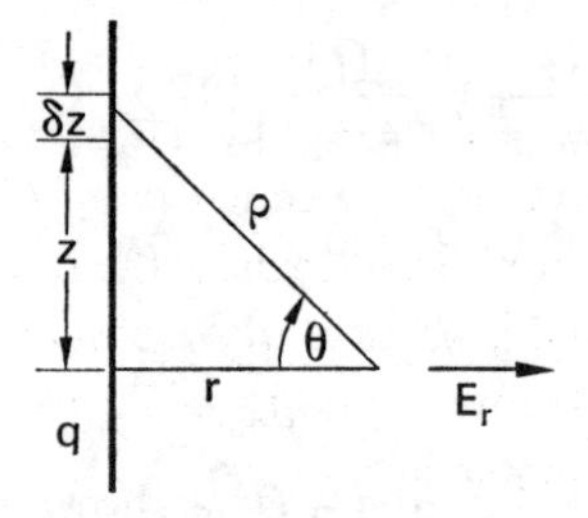

FIG. 4.4. A surface layer of charge.

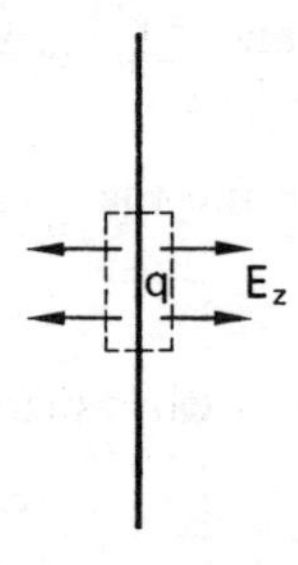

FIG. 4.5. A line filament of charge.

strength will be infinite because the field strength decreases more rapidly with distance than the potential. It is therefore wise to work with the electric field strength. Refer to Fig. 4.5.

$$E_r = \frac{q}{4\pi\varepsilon_0} \int_{-\infty}^{+\infty} \frac{\cos\theta}{\varrho^2}\, dz, \tag{4.8}$$

where

$$z = r \tan\theta,$$

$$\delta z = r \sec^2\theta\, \delta\theta,$$

$$\varrho = r \sec\theta;$$

hence

$$E_r = \frac{q}{4\pi\varepsilon_0} \int_{-\pi/2}^{+\pi/2} \frac{\cos\theta}{r}\, d\theta = \frac{q}{2\pi\varepsilon_0 r}$$

or, vectorially,

$$\mathbf{E} = \frac{q}{2\pi\varepsilon_0 r}\,\hat{\mathbf{r}}. \tag{4.9}$$

Note that the electric field decays as the distance.

This result could have been obtained from Gauss's theorem without the need for an integration. Nevertheless, the integration shows the physical basis very clearly and is particularly useful in deciding how long a real finite filament must be before the solution for an idealized infinite filament can be used with reasonable accuracy.†

The potential can be obtained from eqn. (4.9), since

$$\mathbf{E} = -\nabla V, \quad \text{and} \quad V = -\int_{\infty}^{r} \mathbf{E}\,.\,d\mathbf{r}.$$

Thus

$$V = -\frac{q}{2\pi\varepsilon_0} \ln r. \tag{4.10}$$

This is the part of V which varies with r; there will also be an infinite constant of integration.

† See Exercise 4.2.

4.1.4. The Electric Field of a Filament Doublet

Figure 4.3 can now be used to describe the end-on view of such an arrangement. The potential is obtained from the potential of a single filament by the use of eqn. (4.4):

$$\frac{\partial V}{\partial x'}\,\delta x' = \frac{q\,\delta x'}{2\pi\varepsilon_0}\,\frac{\cos\theta}{r}. \tag{4.11}$$

If $\mathbf{p} = q\,\delta\mathbf{x}'$ is the dipole moment per unit length, this becomes

$$V = \frac{\mathbf{p}\,.\,\hat{\mathbf{r}}}{2\pi\varepsilon_0 r}, \tag{4.12}$$

where V is now the potential of the filament doublet.

The electric field is given by

$$\mathbf{E} = -\frac{1}{2\pi\varepsilon_0}\,\nabla\left(\frac{\mathbf{p}\,.\,\mathbf{r}}{r}\right). \tag{4.13}$$

Here V decays as the distance and $\mathbf{E}$ as the square of the distance.

We shall show later [see eqn. (4.81)] that the external field of a uniformly polarized cylinder is identical with that of a filament doublet.

4.1.5. The Electric Field of a Surface Layer of Charge

This is a case of tremendous importance. In conductors the static charge always appears as a surface layer and it will be shown in § 4.1.7 that dielectrics also can often be represented by surface layers of charge.

If the layer is of infinite extent the potential will be infinite and we must work with the electric field strength. This can be obtained from an integration or from Gauss's theorem. We shall leave the integration as an exercise for the reader.† Gauss's theorem can be applied as in Fig. 4.5. If the charge density is q coulomb/metre^2, the electric field strength is everywhere constant and normal to the surface. It is given by

$$\mathbf{E} = \frac{q}{2\varepsilon_0}\,\hat{\mathbf{z}}. \tag{4.14}$$

† See exercise 4.6.

The potential therefore has a finite part as well as an (infinite) constant part. The finite part is

$$V = -\frac{qz}{2\varepsilon_0}. \tag{4.15}$$

Figure 4.6 sketches the variation of the electric field strength and of the potential in the neighbourhood of the surface on the assumption that the surface is a large disc. It is interesting to note that the potential is con-

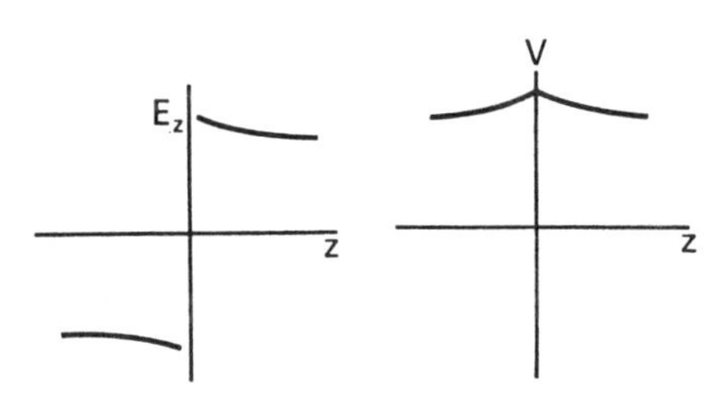

FIG. 4.6. Field of a surface layer.

tinuous at the surface (where $z = 0$) and that the electric field strength of the surface has a discontinuity proportional to the charge density at that point.

It might be thought that a charge layer of infinite extent is a very artificial concept. This is not so. As long as z is small compared with the dimensions of the surface, eqns. (4.14) and (4.15) describe the field very closely. Later in this chapter we shall be considering boundary conditions in the electric field. We shall have to know the field *just* outside a charge surface and we can then make use of the results of this section. Moreover, even a curved surface will appear to be a plane to a small observer close to the surface. At a point *just* outside it any surface can be represented by a plane. All this is, of course, subject to the limits imposed on the study of electromagnetism. Our dimensions must be big enough to enable us to neglect the granular particle structure of electric charge and to treat electricity as a continuous fluid.

4.1.6. The Electric Field of a Double Layer of Charge

This case is included here for completeness. It is actually of greater interest in magnetostatics because a current loop can be represented by a magnetic shell which is a double layer of magnetic pole strength.

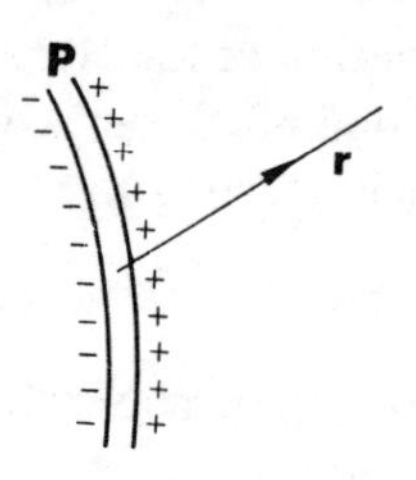

FIG. 4.7. A double surface layer.

Let the double layer have a dipole moment **p** per unit area of surface. The potential can be obtained by the use of eqn. (4.6). For a small element δs of surface we have (see Fig. 4.7)

$$\delta V = \frac{1}{4\pi\varepsilon_0} \frac{\mathbf{p} \cdot \hat{\mathbf{r}}}{r^2} \delta s. \tag{4.16}$$

If **p** is constant on the surface and has a direction normal to the surface, we can take the magnitude of **p** outside the integral needed to obtain V, so that

$$V = \frac{|\mathbf{p}|}{4\pi\varepsilon_0} \iint \frac{\hat{\mathbf{r}} \cdot d\mathbf{s}'}{r^2} = \frac{|\mathbf{p}|}{4\pi\varepsilon_0} \Omega, \tag{4.17}$$

where $|\mathbf{p}|$ denotes the magnitude of **p** and $d\mathbf{s}'$ embodies the direction of **p**. The integral Ω in eqn. (4.17) is called the solid angle. When the field point is close to the surface, the value of Ω is 2π, which is the solid angle of a

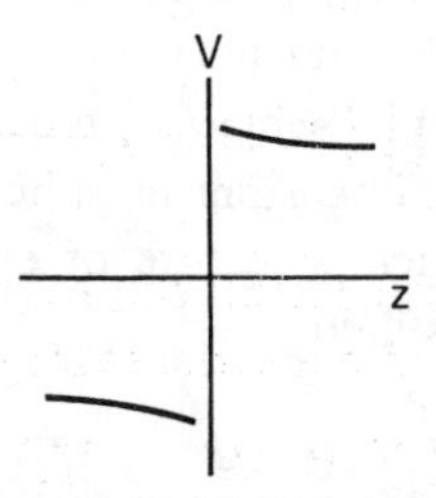

FIG. 4.8. Potential of a double layer.

hemisphere. Hence the solid angle will jump by 4π as the field point crosses the surface and the direction of $\hat{\mathbf{r}}$ is reversed. Thus the potential has a discontinuity of $\mathbf{p}/\varepsilon_0$ at the surface (see Fig. 4.8). On the other hand, the electric field strength is continuous across such a double layer, as can readily be shown by the use of Gauss's theorem because the Gaussian surface will not contain any net charge. We must not draw the surface through the layer which is assumed to be infinitesimally thin.

4.1.7. The Electric Field of a Volume Distribution of Dipoles

Polarizable materials consist of an aggregate or volume distribution of dipoles. A simple expression for the potential of such a distribution is therefore very desirable. Using eqn. (4.6) and writing $\nabla'(1/r)$ for $\hat{\mathbf{r}}/r^2$ since

$$r = \sqrt{[(x - x')^2 + (y - y')^2 + (z - z')^2]},$$

we have

$$V = \frac{1}{4\pi\varepsilon_0}\iiint \mathbf{P} \cdot \nabla'\left(\frac{1}{r}\right) dv', \tag{4.18}$$

where $\mathbf{P}$ is the dipole moment per unit volume and the primed quantities refer to the source coordinates. The vector identity

$$\operatorname{div} u\mathbf{v} = \operatorname{grad} u \cdot \mathbf{v} + u \operatorname{div} \mathbf{v} \tag{4.19}$$

can be used to rewrite the integrand of eqn. (4.18),

$$\mathbf{P} \cdot \nabla'\left(\frac{1}{r}\right) = \nabla' \cdot \left(\frac{\mathbf{P}}{r}\right) - \frac{1}{r}\nabla' \cdot \mathbf{P}, \tag{4.20}$$

whence eqn. (4.18) can be written

$$V = \frac{1}{4\pi\varepsilon_0}\iiint \left(\nabla' \cdot \left(\frac{\mathbf{P}}{r}\right) - \frac{1}{r}\nabla' \cdot \mathbf{P}\right) dv'. \tag{4.21}$$

The divergence theorem [eqn. (3.16)] can be applied to the first term of the right side to convert it into a surface integral. Then

$$V = \frac{1}{4\pi\varepsilon_0}\left[\oint\!\!\oint \frac{\mathbf{P} \cdot d\mathbf{s}'}{r} - \iiint \frac{\nabla' \cdot \mathbf{P}}{r} dv'\right]. \tag{4.22}$$

The meaning of this expression becomes clear when we substitute

$$\left.\begin{aligned} P_n &= q, \\ -\nabla' \,.\, \mathbf{P} &= \varrho. \end{aligned}\right\} \tag{4.23}$$

We now obtain

$$V = \frac{1}{4\pi\varepsilon_0}\left[\oiint \frac{q\,ds'}{r} + \iiint \frac{\varrho\,dv'}{r}\right]. \tag{4.24}$$

The potential consists of two parts: a surface integral of an equivalent charge layer of density $q = P_n$ and a volume integral of an equivalent volume density of charge $\varrho = -\nabla' \,.\, \mathbf{P}$. Thus the effect of the *double* sources can be expressed in terms of equivalent *single* sources. Any polarizable material can be represented by these equivalent sets of sources. Moreover, we shall see later in this chapter (§ 4.2.1) that in many materials there is no internal divergence of dipole moment, so that the volume density ϱ is zero. For such material the polarization can be described completely by replacing the material by a surface layer of charge density equal to the normal component of dipole moment at the surface. Readers who are puzzled by the negative sign in eqn. (4.23) may find it helpful to refer to Fig. (4.9). This shows that a decrease of dipole moment is equivalent to a positive volume density of charge.

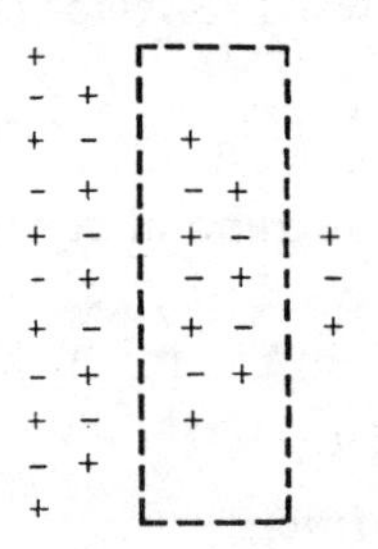

FIG. 4.9. Volume distribution of dipoles.

4.1.8. The Complex Logarithmic Potential

Equipotentials and flux lines are by definition at right angles to each other. If the flux line had a component along an equipotential, there would be a force or potential gradient along the equipotential and that would be absurd.

Such orthogonality is familiar to us from the behaviour of complex numbers, where we are used to associating two orthogonal quantities x and y in the complex quantity $z = x + jy$ in cartesian coordinates or $z = r\,e^{j\theta}$ in cylindrical polar coordinates.

We can, therefore, apply the theory of the complex variable to electrostatics, provided only that we restrict ourselves to two-dimensional problems, because complex numbers can only deal with two coordinates.

Consider the line charge discussed in § 4.1.3. The potential function was given in eqn. (4.10) as

$$V = -\frac{q}{2\pi\varepsilon_0}\ln r.$$

Suppose we replace r by the complex number z so that we obtain the complex potential $W = V + jU$, where

$$\begin{aligned} W &= -\frac{q}{2\pi\varepsilon_0}\ln z \\ &= -\frac{q}{2\pi\varepsilon_0}\ln r\exp(j\theta) \\ &= -\frac{q}{2\pi\varepsilon_0}(\ln r + j\theta). \end{aligned} \tag{4.25}$$

The first part of the complex potential is the ordinary potential function, the second part represents an associated function at right angles and must therefore be the flux. This can easily be checked by reference to Fig. 4.10, which shows the field of a line charge. The flux between two such lines is the flux density D multiplied by the arc $r\theta$. Thus the flux can be derived

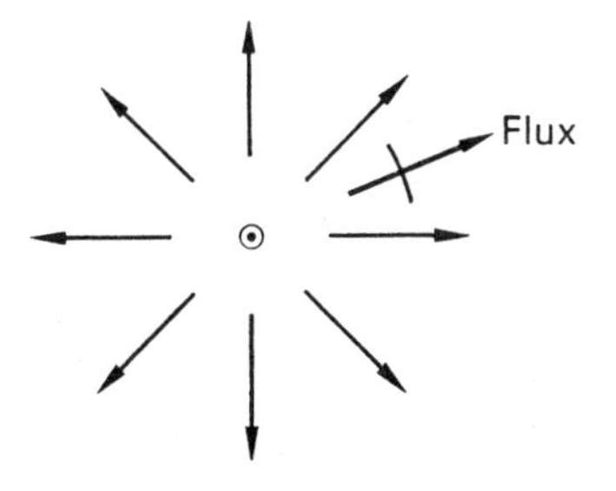

FIG. 4.10. Field of a line charge.

from eqn. (4.9) as

$$\Psi = \frac{q}{2\pi r} r\theta = \frac{q}{2\pi}\theta. \tag{4.26}$$

This is the second part of the complex potential except for the arbitrary negative sign and the scale factor ε_0.

Field plots can be obtained by choosing values for the potential V and the flux Ψ and calculating the corresponding values for r and θ from eqn. (4.25). Figure 4.11 shows such a field plot for a line charge at the

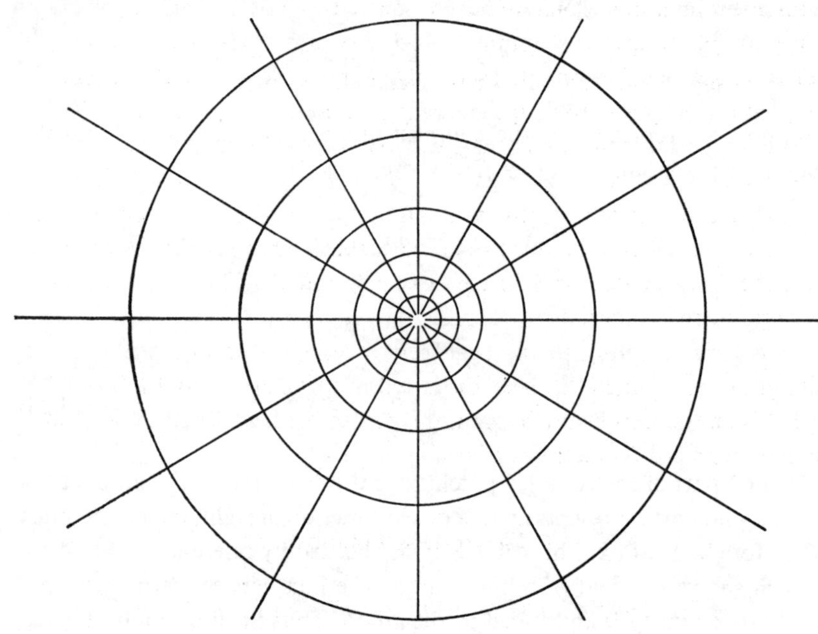

FIG. 4.11. Potential and flux of a line charge.

origin. If the field of several line charges of equal strength is required and if they are located at $z_1, z_2, z_3, \ldots$, we can obtain the total complex potential as

$$W = -\frac{q}{2\pi\varepsilon_0}[\ln(z - z_1) + \ln(z - z_2) + \cdots]$$

$$= -\frac{q}{2\pi\varepsilon_0}\ln[(z - z_1)(z - z_2)\cdots]. \tag{4.27}$$

4.1.9. The Use of Source Distributions in Engineering Design

Some readers may feel that this section on source distributions has been rather theoretical. Since electrical devices are made of copper and other materials which have definite electrical properties, it seems very strange to discuss such abstractions as filaments of charge without specifying the substance on which the charge is located.

This difficulty must be faced squarely. The answer is that as yet the source distributions which we have discussed are not located on anything. They are just charge distributions. This does not mean that such postulated charge distributions are useless figments of the imagination. Indeed, they are a powerful tool in the hands of a design engineer. The design process necessarily has three stages. First comes the specification of the desired electrical field, secondly, the designer equips himself with the knowledge of certain idealized situations. Only when the second stage has been passed can the third stage be tackled. Only at this third stage can the designer deal with the practical problem of how to attach the charges to real materials.

It is very tempting to the beginner to omit the second stage of the design process which has been discussed in this chapter. But the penalty for this omission is heavy both in time and effort because rational design is then replaced by methods of trial and error.

4.2. Electrostatic Polarization

4.2.1. Electric Field Strength and Electric Flux Density

The entire subject of electrostatics can be seen as a direct consequence of the inverse square law of force between point charges. This law is universal in its application and takes no account of the substance in which the charges are located. To put it in another way, the electrical action of any material is completely described by taking account of all the charges to be found there. It has been rightly said that the inverse square law is independent of the material medium and acts "through thick and thin".

Nevertheless, it is often useful to distinguish between two broad classes of substance—the conductors and the insulators or dielectrics. More precisely it is useful to distinguish between "free" charges which take part in conduction processes and "bound" charges which specify electric polarization processes. On a subatomic scale this distinction is almost meaningless, but we remember that classical electromagnetic theory is restricted to macroscopic phenomena.

Since the electrostatic field represents the effect of the charge distributions we can associate the field quantities with the charges in such a manner as to distinguish between the field of the free and the bound charges. We have

$$\varrho_{\text{total}} = \varrho_{\text{free}} + \varrho_{\text{bound}} . \tag{4.28}$$

Let us associate the electric field strength $\mathbf{E}$ with the total charge distribution

$$\nabla . \mathbf{E} = \varrho_{\text{total}}/\varepsilon_0 . \tag{4.29}$$

From eqn. (4.23) we have $-\nabla' . \mathbf{P} = \varrho_{\text{bound}}$. Since there is no ambiguity between source and field coordinates the prime can be dropped and

$$\nabla . \mathbf{P} = -\varrho_{\text{bound}} , \tag{4.30}$$

whence

$$\nabla . (\varepsilon_0\mathbf{E} + \mathbf{P}) = \varrho_{\text{total}} - \varrho_{\text{bound}} = \varrho_{\text{free}} . \tag{4.31}$$

It is now convenient to define the electric flux density $\mathbf{D}$ as

$$\mathbf{D} = \varepsilon_0\mathbf{E} + \mathbf{P} \tag{4.32}$$

so that

$$\nabla . \mathbf{D} = \varrho_{\text{free}} . \tag{4.33}$$

This definition is consistent with the previous definition of $\mathbf{D}$ in free space where we had $\mathbf{D} = \varepsilon_0\mathbf{E}$. It is important to note that although in eqn. (4.32) the flux density $\mathbf{D}$ is given as the *sum* of two terms, $\mathbf{D}$ is really a *partial* field vector since it only deals with free charges. The *total* field is given by $\mathbf{E}$.

In many materials, notably gases, liquids, amorphous solids, and cubic crystals, the polarization is proportional to the electric field $\mathbf{E}$, so that we can write

$$\mathbf{P} = \varepsilon_0\chi_e\mathbf{E} , \tag{4.34}$$

where the constant χ_e is called the *electric susceptibility*. Hence

$$\mathbf{D} = \varepsilon_0 (1 + \chi_e) \mathbf{E} = \varepsilon_0 \varepsilon_r \mathbf{E}, \tag{4.35}$$

where ε_r is the *dielectric constant* or *relative permittivity*. Where there is no free charge $\nabla . \mathbf{D} = 0$. Hence, if ε_r is constant $\nabla . \mathbf{E} = 0$ from eqn. (4.35) and $\nabla . \mathbf{P} = 0$ from eqn. (4.31). Hence in eqn. (4.23) $\varrho = 0$ and the volume integral disappears in eqn. (4.24). This leads to the vitally important result that a substance of constant relative permittivity can be represented by a surface layer of charge. The two constants $\varepsilon_0 \, \varepsilon_r$ are often combined into one and written as ε which is called *permittivity*.

In anisotropic materials eqns. (4.34) and (4.35) have to be replaced by a much more complicated relationship typified by

$$P_\alpha = \varepsilon_0 \chi_{\alpha\beta} E_\beta . \tag{4.36}$$

P is then no longer parallel to **E**, and **D** must be obtained by vector addition.

4.2.2. Boundary Conditions at a Dielectric Surface

Figure 4.12 shows a boundary between two materials of permittivity ε_1 and ε_2. Let us first assume that there are no free charges at the interface. By the application of Gauss's theorem to the vector field **D**, we can see

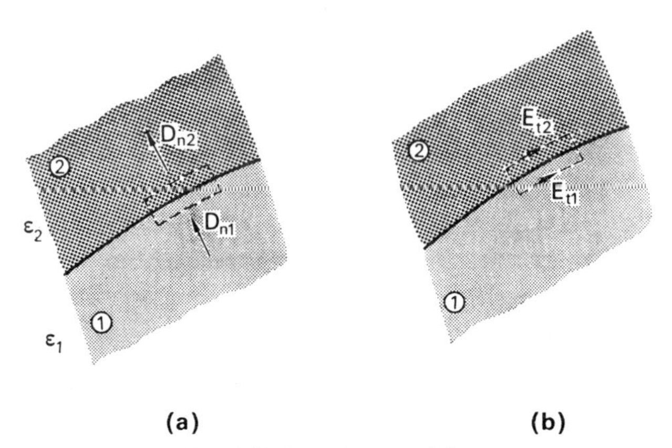

FIG. 4.12. Boundary conditions.

that the normal component of **D** is continuous across the boundary, because **D** has no sources at the boundary. If, however, there are free charges of surface density q at the boundary, the normal component of **D** has a discontinuity of q, so that

$$D_{n_2} - D_{n_1} = q\,. \tag{4.37}$$

To obtain the relationship between the tangential components we remember that the total field is conservative so that

$$\oint \mathbf{E}\,.\,d\mathbf{l} = 0\,. \tag{4.38}$$

If this relationship is applied to the rectangular path in Fig. 4.12 and if the ends of this path shrink to zero, we obtain

$$E_{t_2}\delta l - E_{t_1}\delta l = 0 \tag{4.39}$$

or

$$E_{t_2} = E_{t_1}\,. \tag{4.40}$$

This relationship holds even when the field is non-conservative.† We can note, therefore, that the continuity of tangential electric field strength across a boundary is always true in electromagnetics.

4.2.3. The Electric Field Inside a Material

There is a lot of confusion about this subject, a confusion that is caused by the lack of precision as to the meaning of the word "inside". Let us first consider this on a microscopic scale. Clearly the local atomic field will fluctuate violently with position and with time. We could take the space–time average of this microscopic field as our definition of the macroscopic field. Experimentally such a field could be measured by the deflection of a fast electron on passing through the material. Since the space–time average of the total charge density is the same as the total electrostatic charge density, the field defined in this way will be the electrostatic field **E**. A second possible microscopic definition applied to a polarizable material would be to define the "inside" field as the field observed by a single molecule assuming that the molecules are electric dipoles. The

† A step in tangential E would require a "magnetic current". See § 5.4.

discussion of this case is beyond the scope of this book,† but it can be stated that in terms of the electrostatic field vectors the molecular field is $\mathbf{E} + (\mathbf{P}/3\varepsilon_0)$.

Although these microscopic definitions are useful in showing the relationship of microscopic to macroscopic theory, we are on firmer ground when we restrict the discussion to macroscopic phenomena. To be inside a material then means to be inside a *hole* in the material. Clearly the shape or geometry of the hole will have to be taken into consideration and the field will vary with this geometry.

Consider first a "worm-hole" cavity, the axis of which lies along the field (Fig. 4.13). The ends of this cavity are perpendicular to the polarization vector $\mathbf{P}$ and there will therefore be an equivalent charge distribution $q = P$ on these ends. However, if the hole is long and narrow the

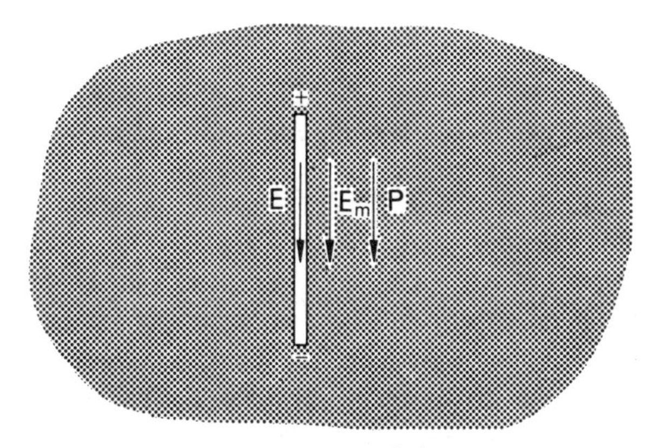

FIG. 4.13. Worm hole.

charge q will be far away and its amount will be slight, so in the limit we can neglect it. If we define the electric field in the material as $\mathbf{E}_m$ and observe the electric field in the cavity as $\mathbf{E}$, then by use of eqn. (4.40)

$$\mathbf{E}_m = \mathbf{E}, \tag{4.41}$$

and we have defined the internal electric field $\mathbf{E}_m$ as the field which is observed in a long narrow cavity which lies along the field.

† See, for instance, Panofsky and Phillips, *Classical Electricity and Magnetism*, Addison Wesley, 1955, chapter 2.

A second important type of cavity is the disc cavity of Fig. 4.14. Here there is a great deal of equivalent charge on the wide surfaces of the

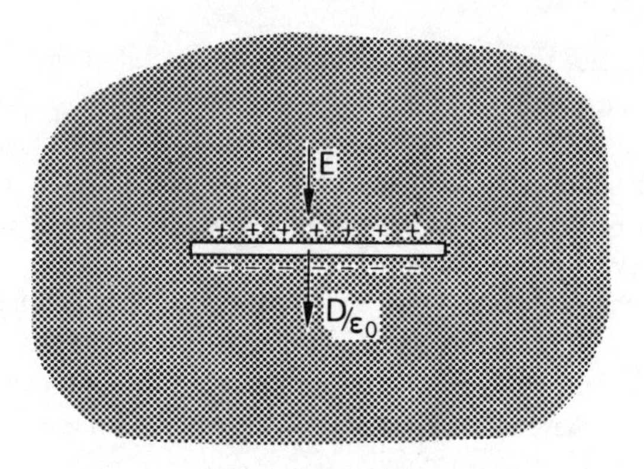

FIG. 4.14. Disc hole.

cavity. The vector **D** is, however, by definition unaffected by this bound charge and so we can make use of the continuity of D_n across the boundary. Hence

$$\varepsilon_0 \varepsilon_r \mathbf{E}_m = \varepsilon_0 \mathbf{E}_2 . \tag{4.42}$$

Thus the field in the disc cavity is $\varepsilon_r \mathbf{E}_m$ or $\mathbf{D}_m/\varepsilon_0$.

The two types of cavity in Figs. 4.15 and 4.16 thus enable both $\mathbf{E}_m$ or $\mathbf{D}_m$ to be defined experimentally. They have also another important advantage over all other shapes because the surface charge which is laid bare in forming the cavities does not affect the material outside the cavities. In the worm-hole this is because the charge itself is negligible and in the disc cavity the effect of the two charges layers cancels everywhere except in the cavity itself. Other cavities such as spheres, ellipsoids, or cavities of irregular shape modify the field, and therefore the polarization, in the material surrounding the cavity. Such problems are known as boundary-value problems, and this type of problem will be discussed in the next section. It should be noted here that we are concerned with the *shape* of the cavity and not with its *size*. In designing experiments it is important to realize that even small cavities will distort the field unless they are of the shape shown in Figs. 4.13 and 4.14.

4.3. Solution of Field Problems by Laplace's and Poisson's Equations

The potential due to any distribution of electric charge can be obtained from an integral of the form

$$V = \frac{1}{4\pi\varepsilon_0} \iiint \frac{\varrho}{r}\, dv' . \tag{4.43}$$

Often, however, the differential form of this statement is more useful as a starting point. We have, from § 3.2.4,

$$\nabla^2 V = -\varrho/\varepsilon_0 \tag{3.21}$$

if there is charge density ϱ at the observation or field point or

$$\nabla^2 V = 0 \tag{3.24}$$

if there is no charge at the field point.

Equation (3.21) is called Poisson's equation and eqn. (3.24) Laplace's equation. The solution of both equations will be very similar, except that Poisson's equation has a particular integral as well as a complementary function. We shall consider Laplace's equation first.

4.3.1. Laplace's Equation in Rectangular Coordinates and Discussion of the Uniqueness of Fields in Closed Volumes

Consider first the two-dimensional case

$$\nabla^2 V = \frac{\partial^2 V}{\partial x^2} + \frac{\partial^2 V}{\partial y^2} = 0 . \tag{4.44}$$

The most powerful way of solving partial differential equations is the method of the separation of variables. We assume a solution in the form $V(x, y) = X(x)\, Y(y)$, where X is a function of x only and Y a function of y only. Then

$$Y \frac{\partial^2 X}{\partial x^2} + X \frac{\partial^2 Y}{\partial y^2} = 0 \tag{4.45}$$

or

$$\frac{1}{X}\frac{\partial^2 X}{\partial x^2} + \frac{1}{Y}\frac{\partial^2 Y}{\partial y^2} = 0. \tag{4.46}$$

Both terms of this equation are functions of one variable only. Since their sum must always be zero, each term must be equal to a constant and we can write

$$\frac{1}{X}\frac{\partial^2 X}{\partial x^2} = -\frac{1}{Y}\frac{\partial^2 Y}{\partial y^2} = \text{constant}. \tag{4.47}$$

Suppose we write the constant as $-n^2$, then eqn. (4.44) will have solutions of the form

$$X = A_n \sin nx + B_n \cos nx, \tag{4.48}$$

$$Y = C_n \sinh nx + D_n \cosh ny, \tag{4.49}$$

where A_n, B_n, C_n, D_n are constants.

The special case $n = 0$ will give solution of the form

$$X = A_0 + B_0 x, \tag{4.50}$$

$$Y = C_0 + D_0 y, \tag{4.51}$$

and V is obtained by forming the products $(XY)_n$ and $(XY)_0$. The constant in eqn. (4.49) is called the separation constant.

The usefulness of this type of solution depends on the manner in which the problem is specified. If the potentials or their gradients are specified on the boundaries, it is often relatively easy to obtain the constant coefficients in eqns. (4.48) to (4.51). An example will illustrate the method. Consider the potential in a rectangular duct as shown in Fig. 4.15. The potentials of the walls of the duct are shown.

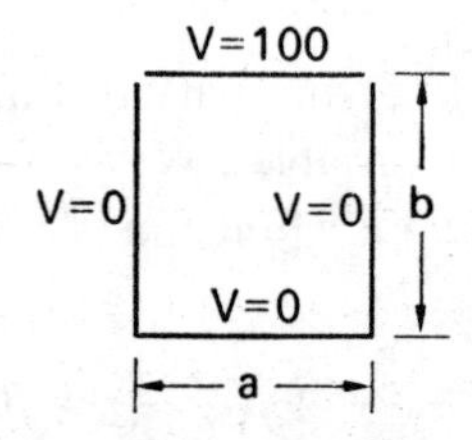

FIG. 4.15. Field in a rectangular duct.

Consideration of eqns. (4.48) to (4.51) suggests that all constants except A_n are zero and the potential can be written

$$V = \sum A_n \sin nx \sinh ny \tag{4.52}$$

since

$$V = 0 \quad \text{at} \quad x = a$$

$$\sin na = 0, \tag{4.53}$$

whence

$$na = k\pi$$

$$n = k\pi/a, \quad \text{where } k = 1, 2, 3 \cdots$$

At $y = b$, $V = 100$, so that

$$100 = \sum_{k=1}^{\infty} A_k \sin k\pi x/a \sinh k\pi b/a, \tag{4.54}$$

where A_n has been replaced by A_k.

A_k can be obtained by multiplying both sides of eqn. (4.54) by $\sin k\pi x/a$ and integrating from 0 to a, i.e. by the usual method of Fourier analysis. Thus

$$100 \int_0^a \sin k\pi x/a \, dx = A_k \sinh k\pi b/a \int_0^a \sin^2 k\pi x/a, \tag{4.55}$$

whence

$$\frac{100a}{k\pi}\left[\cos k\pi x/a\right]_a^0 = A_k \sinh k\pi b/a \, \frac{a}{2}. \tag{4.56}$$

$$\left.\begin{aligned} &\text{If } k \text{ is even,} && A_k = 0. \\ &\text{If } k \text{ is odd,} && A_k = \frac{400}{k\pi}\,\frac{1}{\sinh k\pi b/a}. \end{aligned}\right\} \tag{4.57}$$

Hence, by use of eqn. (4.52),

$$V = \sum_{k=1,3,5\cdots} \frac{400}{k\pi}\,\frac{\sin k\pi x/a \sinh k\pi y/a}{\sinh k\pi b/a}. \tag{4.58}$$

This expression for the potential satisfies the boundary conditions on the walls of the duct. Moreover it describes the field *uniquely* because, by the use of eqn. (3.61), it can be shown that the potential within a closed region is uniquely specified by the specification of the potential, or the normal electric field, at the boundary surface. The proof is left as exercise 4.14 at the end of this chapter, but a formal proof is almost superfluous if we combine the underlying thought of § 3.2.6 with § 4.1. The potential must always depend on the sources and a specification of the sources will uniquely specify the potential. Now the potential on a surface can always be specified by a double layer of charge because, as shown in § 4.1.6, we can introduce any required step of potential by means of a suitable double layer. Similarly, as shown in § 4.1.5, a normal gradient can be specified by a single layer of charge. It is true, of course, that in an actual problem the surface potential may not be due to charges on this surface, but due to other charges outside the closed region. Nevertheless, we can replace all such problems by equivalent ones in which there are only charges on the surface of the closed region considered.†

The reader should ponder deeply the relationship between source distributions and potentials and should avoid regarding the potential as a physical entity apart from source distribution. It is wise to regard the sources as primary and the potentials as secondary concepts.

An important corollary of the uniqueness theorem is the fact that any field whatsoever within a closed volume can be caused by a suitable distribution of surface charges. Consider, for instance, the electrostatic field in a conductor carrying steady current. This field is given by Ohm's law $\mathbf{E} = \varrho \mathbf{J}$, where ϱ is the resistivity. The cause of this field is due to surface charges on the conductor.‡ Volume charges are ruled out because they would immediately disperse to the surface. If we take a wire of constant length and cross-section and apply a constant potential difference between its ends, it will carry a constant field strength $\mathbf{E}$ within the material, according to Ohm's law. If now we coil the wire in any arbitrary manner, but leave the potential difference unchanged, we know by experiment that the current remains unchanged. But clearly the electric field has

† Charges within the closed region have been ignored in this discussion which started with eqn. (4.44) which applies to a region free from charge. If there are charges within the region their potential can be added. Mathematically we need the particular integral of Poisson's equation to add to the complementary function of Laplace's equation.

‡ See the detailed discussion in chapter 4 of *Electromagnetism for Engineers* by the present author.

changed because the direction of the wire has changed and so has the proximity of parts of the wire (Fig. 4.16). Thus the surface charge distribution will have changed. The question now arises whether Ohm's law is

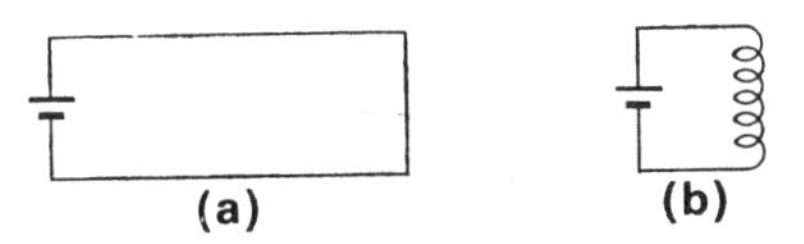

FIG. 4.16. Ohm's law.

always independent of the geometry of the circuit or whether for certain configurations there is no arrangement of surface charge which can give the required field **E**. The uniqueness theorem sets these doubts aside. There is always a suitable charge distribution on a closed surface to give any field whatever in the volume enclosed by the surface.

In three-dimensional rectangular coordinates Laplace's equation becomes

$$\nabla^2 V = \frac{\partial^2 V}{\partial x^2} + \frac{\partial^2 V}{\partial y^2} + \frac{\partial^2 V}{\partial z^2} = 0. \tag{4.59}$$

We again seek a solution of the form

$$V(x, y, z) = X(x)\, Y(y)\, Z(z). \tag{4.60}$$

Hence

$$\frac{1}{X}\frac{\partial^2 X}{\partial x^2} + \frac{1}{Y}\frac{\partial^2 Y}{\partial y^2} + \frac{1}{Z}\frac{\partial^2 Z}{\partial z^2} = 0. \tag{4.61}$$

Each term of this equation must be equal to a constant, so that

$$\left.\begin{aligned} \frac{1}{X}\frac{\partial^2 X}{\partial x^2} &= a \\ \frac{1}{Y}\frac{\partial^2 Y}{\partial y^2} &= b \\ \frac{1}{Z}\frac{\partial^2 Z}{\partial z^2} &= c \end{aligned}\right\} \tag{4.62}$$

and

$$a + b + c = 0. \tag{4.63}$$

At least one of these constants must be positive and at least one must be negative. If we choose $a = -m^2$ as negative and $b = n^2$ as positive,

$$X = A_m \sin mx + B_m \cos mx,$$

$$Y = C_n \sinh ny + D_n \cosh ny.$$

If $|a| > |b|$, c can be written as $c = +p^2$ and

$$Z = E_p \sinh pz + F_p \cosh pz.$$

If $|a| < |b|$, c can be written as $c = -p^2$ and

$$Z = E_p' \sin pz + F_p' \cos pz.$$

The special cases when $a = 0$ or $b = 0$ or $c = 0$ or $a = b = c = 0$ are easily found. For the general case it is important to notice explicitly that there must always be at least one set of circular functions and one set of hyperbolic functions in the product solution.

4.3.2. Laplace's Equation in Cylindrical Coordinates

Cylindrical coordinates are often useful because the shape of many engineering devices is cylindrical. It is not immediately obvious how $\nabla^2 V$ should be written in any curvilinear coordinates. One might be tempted to use δr, $r\,\delta\theta$, and δz as equivalent to δx, δy, and δz and to write

$$\nabla^2 V = \frac{\partial^2 V}{\partial r^2} + \frac{1}{r^2}\frac{\partial^2 V}{\partial \theta^2} + \frac{\partial^2 V}{\partial z^2},$$

but this is wrong. It is necessary to return to the definition of ∇^2 as div grad [eqn. (3.20)],

$$\text{grad } V = \frac{\partial V}{\partial r}\hat{\mathbf{r}}, \quad \frac{1}{r}\frac{\partial V}{\partial \theta}\hat{\boldsymbol{\theta}}, \quad \frac{\partial V}{\partial z}\hat{\mathbf{z}}$$

by definition of a gradient. Divergence is defined in eqn. (3.15) as outflow per unit volume for an element of volume. Consider this with reference to Fig. 4.17. The inflow of a vector field $\mathbf{E}$ (E_r, E_θ, E_z) is

$$E_r r\,\delta\theta\,\delta z + E_\theta\,\delta r\,\delta z + E_z r\,\delta\theta\,\delta r\,,$$

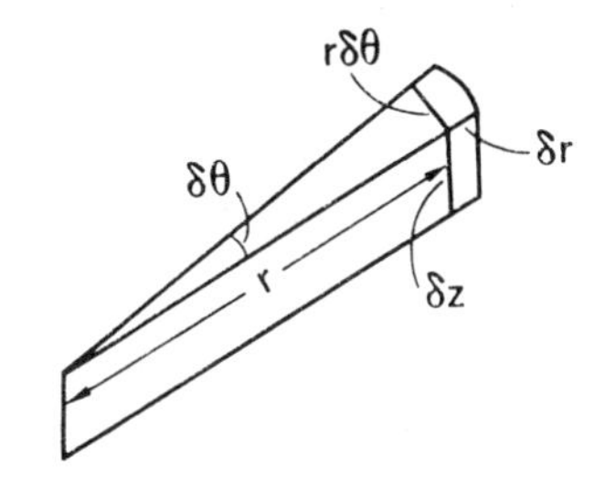

FIG. 4.17. Divergence in cylindrical coordinates.

and the elemental volume is $r\,\delta\theta\,\delta r\,\delta z$. Hence the net outflow per unit volume is

$$\frac{\partial/(\partial r)\,(E_r r\,\delta\theta\,\delta z)\delta r + \partial/(\partial\theta)\,(E_\theta\,\delta r\,\delta z)\delta\theta + \partial/(\partial z)\,(E_z r\,\delta\theta\,\delta r)\delta z}{r\,\delta\theta\,\delta r\,\delta z}$$

$$= \frac{\partial E_r}{\partial r} + \frac{E_r}{r} + \frac{1}{r}\frac{\partial E_\theta}{\partial\theta} + \frac{\partial E_z}{\partial z} = \text{div}\,\mathbf{E}\,. \qquad (4.64)$$

Substituting $\mathbf{E} = -\text{grad}\,V$ we have

$$-\text{div}\,\mathbf{E} = +\text{div grad}\,V = \nabla^2 V$$

$$= \frac{\partial^2 V}{\partial r^2} + \frac{1}{r}\frac{\partial V}{\partial r} + \frac{1}{r^2}\frac{\partial^2 V}{\partial\theta^2} + \frac{\partial^2 V}{\partial z^2} = 0\,. \qquad (4.65)$$

We try a solution which separates the coordinates in the form

$$V(r, \theta, z) = R(r)\,\Theta(\theta)\,Z(z)\,, \qquad (4.66)$$

$$\frac{1}{R}\frac{\partial^2 R}{\partial r^2} + \frac{1}{rR}\frac{\partial R}{\partial r} + \frac{1}{r^2\Theta}\frac{\partial^2\Theta}{\partial\theta^2} + \frac{1}{Z}\frac{\partial^2 Z}{\partial z^2} = 0\,. \qquad (4.67)$$

For simplicity let us restrict attention to two-dimensional problems, so that there is no variation with z. Then

$$\frac{1}{R}\frac{\partial^2 R}{\partial r^2} + \frac{1}{rR}\frac{\partial R}{\partial r} + \frac{1}{r^2\Theta}\frac{\partial^2\Theta}{\partial\theta^2} = 0\,,$$

or

$$\frac{r^2}{R}\frac{\partial^2 R}{\partial r^2} + \frac{r}{R}\frac{\partial R}{\partial r} + \frac{1}{\Theta}\frac{\partial^2\Theta}{\partial\theta^2} = 0\,. \tag{4.68}$$

To obtain Θ in circular functions it is convenient to put

$$\frac{1}{\Theta}\frac{\partial^2\Theta}{\partial\theta^2} = -n^2 \quad \text{and} \quad \frac{r^2}{R}\frac{\partial^2 R}{\partial r^2} + \frac{r}{R}\frac{\partial R}{\partial r} = n^2\,,$$

where n^2 is the separation constant. Then

$$\left.\begin{aligned}\Theta &= A_n \sin n\theta + B_n \cos n\theta\,,\\ R &= C_n r^n + D_n/r^n\,.\end{aligned}\right\} \tag{4.69}$$

For the special case $n = 0$,

$$\left.\begin{aligned}\Theta &= A_0 + B_0\theta\,,\\ R &= C_0 + D_0 \ln r\,.\end{aligned}\right\} \tag{4.70}$$

The potential V can be obtained from the products of the solutions in eqns. (4.69) and (4.70). As in the previous section, we notice that there are an infinite number of possible solutions. At first this seems strange and unhelpful when we are trying to solve the field pattern of a device which can be uniquely specified. The difficulty is resolved when we consider the physical content of Laplace's equation. This content is minimal. All that we have said by stating this equation is that there shall be no sources in the region under consideration. We cannot except a single answer with so vague and general a specification. It is sometimes said that the central problem of electrostatics is the solution of Laplace's equation. That is a misleading statement. Laplace's equation by itself cannot be solved in any meaningful way. What is lacking is physical information and such "solutions" as eqns. (4.69) and (4.70) show the beginning of a problem not the end. It is foolish to grumble at this, because the mathematical structure of science is kept deliberately general, so that one formulation can cover an infinity of applications.

Although Laplace's equation does not say much, we can derive some help from the formal statement of eqns. (4.69) and (4.70). The separation constant n appears in the expressions for both θ and r. Thus r^2 and $1/r^2$ are always linked with $\sin 2\theta$ and $\cos 2\theta$. Terms such as $r^3 \cos 4\theta$ or $(1/r)^5 \sin 4\theta$ are not possible. This knowledge can be put to good use as is shown in the following example of the field of a line charge (Fig. 4.18).

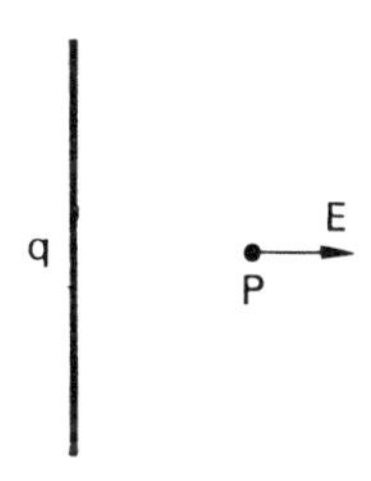

FIG. 4.18. Field of a line charge.

The potential of such a filament located at the origin was found in § 4.1 as

$$V = -\frac{q}{2\pi\varepsilon_0} \ln r. \tag{4.10}$$

We can use this expression to find the potential at P in Fig. 4.19. This is given by

$$V = -\frac{q}{2\pi\varepsilon_0} \ln (r - c). \tag{4.71}$$

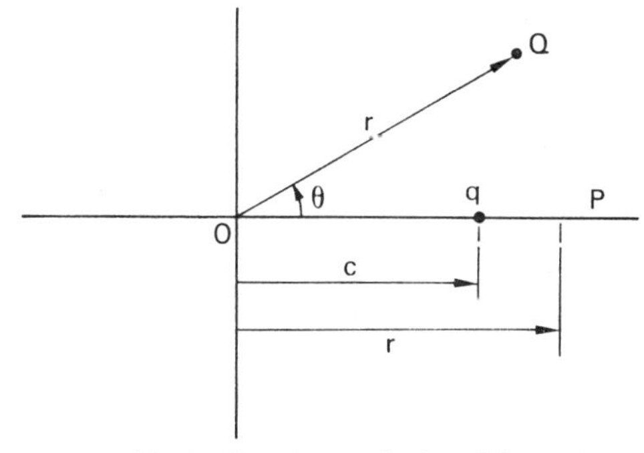

FIG. 4.19. Field of a line charge displaced from the origin ($r > c$).

For $r > c$ we can derive an expansion in powers of c/r,

$$V = -\frac{q}{2\pi\varepsilon_0}\left[\ln r + \ln\left(1 - \frac{c}{r}\right)\right]$$

$$= -\frac{q}{2\pi\varepsilon_0}\left[\ln r - \frac{c}{r} - \frac{1}{2}\left(\frac{c}{r}\right)^2 - \cdots - \frac{1}{n}\left(\frac{c}{r}\right)^n - \cdots\right]. \quad (4.72)$$

Consider now the potential at some point $Q\ (r, \theta)$. Every power of r must be associated with a circular function. A possible solution is

$$V = -\frac{q}{2\pi\varepsilon_0} \times$$

$$\times\left[\ln r - \frac{c}{r}\cos\theta - \frac{1}{2}\left(\frac{c}{r}\right)^2\cos 2\theta - \cdots - \frac{1}{n}\left(\frac{c}{r}\right)^n\cos n\theta \cdots\right]. \quad (4.73)$$

To find whether this is a complete solution we consider the other possible types of eqns. (4.69) and (4.70). Symmetry about the line $\theta = 0$ forbids the use of sine functions or of θ itself. The distant field of the filament must be the same as if the filament were at the origin, hence no terms in r^{+n} are allowable. Hence the solution of eqn. (4.70) is complete except for an arbitrary constant. Knowledge of the form of permissible solutions has enabled us to deduce the complicated expression of eqn. (4.73) from the simple expression (4.71).

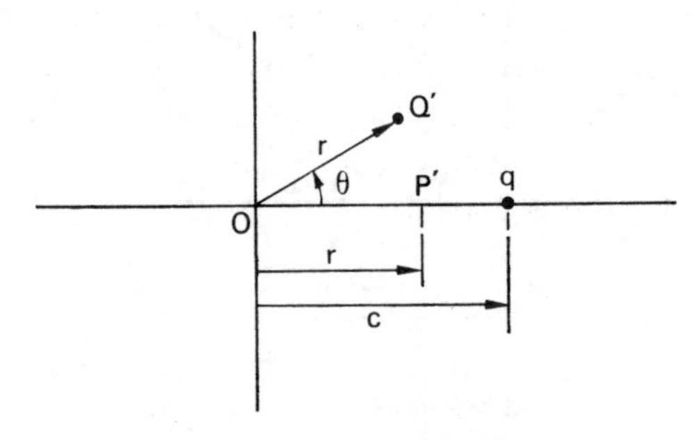

FIG. 4.20. Field of a line charge displaced from the origin ($r < c$).

For $r < c$ the potential at P' (Fig. 4.20),

$$V = -\frac{q}{2\pi\varepsilon_0} \ln (c - r), \tag{4.74}$$

$$V = -\frac{q}{2\pi\varepsilon_0}\left[\ln c + \ln\left(1 - \frac{r}{c}\right)\right]$$

$$= -\frac{q}{2\pi\varepsilon_0}\left[\ln c - \frac{r}{c} - \frac{1}{2}\left(\frac{r}{c}\right)^2 - \cdots - \frac{1}{n}\left(\frac{r}{c}\right)^n - \cdots\right]. \tag{4.75}$$

The first term is a constant and can be left out of the discussion. Hence for the point Q' (r, θ)

$$V = \frac{q}{2\pi\varepsilon_0} \times$$

$$\times \left[\frac{r}{c}\cos\theta + \frac{1}{2}\left(\frac{r}{c}\right)^2 \cos 2\theta + \cdots + \frac{1}{n}\left(\frac{r}{c}\right)^n \cos n\theta + \cdots\right]. \tag{4.76}$$

This is the complete solution apart from a constant, because $\sin n\theta$ and θ are excluded by symmetry, $\ln r$ is excluded because it becomes infinite at the origin and terms like r^{-n} are excluded for the same reason.

The power of this approach is shown in a slightly different context in the example of Fig. 4.21. A cylinder of radius a and relative permittivity ε_r is placed transversely to an electric field E_0 which was uniform before the cylinder was introduced. The field E_0 is due to uniformly charged plates of large area as shown in the figure. It can be assumed that the plates are so far away from the cylinder that the charges on them are unaffected by the cylinder. We require to find the potential inside and outside the cylinder. Taking the origin of coordinates at the axis of the cylinder we can derive the field E_0 from a potential such that

$$E_0 = -\frac{\partial V_0}{\partial x},$$

$$V_0 = -E_0 x$$

$$= -E_0 r \cos\theta. \tag{4.77}$$

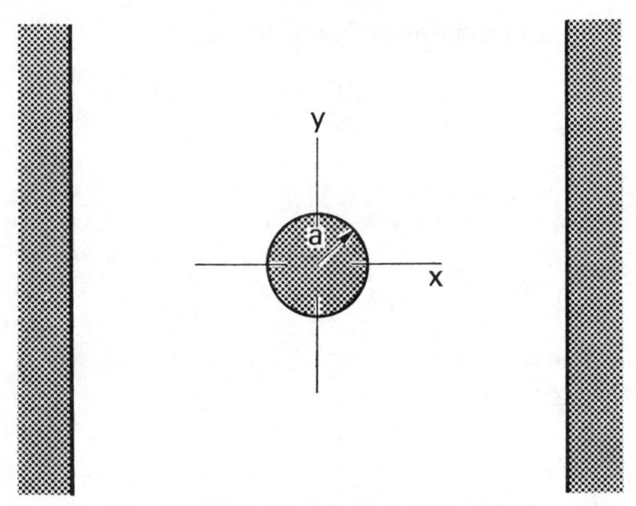

FIG. 4.21. Dielectric cylinder in uniform field.

V_0 is the exciting potential and is due to the charges on the plates. In order to obtain the total potential we must add the potential due to the (bound) charges in the cylinder. Moreover, the system is linear because the relative permittivity of the cylinder is constant, which means that the induced charges are proportional to the inducing charges. Hence the θ variation must be $\cos\theta$ for all the potentials. Outside the cylinder the bound charges will, therefore, produce a potential $(A/r)\cos\theta$, and inside $Br\cos\theta$. At $r = a$ the potential will be continuous and the normal component of flux density will be continuous. Thus

$$Ba\cos\theta - E_0 a\cos\theta = \frac{A}{a}\cos\theta - E_0 a\cos\theta \qquad (4.78)$$

and

$$\varepsilon_r(B\cos\theta - E_0\cos\theta) = -\frac{A}{a^2}\cos\theta - E_0\cos\theta. \qquad (4.79)$$

Hence

$$\left.\begin{aligned} A &= E_0 a^2 \frac{\varepsilon_r - 1}{\varepsilon_r + 1}, \\ B &= E_0 \frac{\varepsilon_r - 1}{\varepsilon_r + 1}. \end{aligned}\right\} \qquad (4.80)$$

The potential V_1 outside the cylinder is

$$V_1 = \frac{E_0 a^2 \cos\theta\,(\varepsilon_r - 1)}{r\,(\varepsilon_r + 1)} - E_0 r \cos\theta\,. \qquad (4.81)$$

The first term of this equation should be compared with eqn. (4.12). It shows that the outside effect of the cylinder is the same as that of a filament doublet of dipole moment per unit length

$$\mathbf{p} = 2\pi\varepsilon_0 E_0 a^2 \left(\frac{\varepsilon_r - 1}{\varepsilon_r + 1}\right).$$

The inside potential V_2 is

$$V_2 = -\frac{2E_0 r \cos\theta}{(\varepsilon_r + 1)}\,. \qquad (4.82)$$

The surface density of charge equivalent to the volume distribution of dipoles in the cylinder (see § 4.1) can be obtained from Gauss's theorem

$$\begin{aligned} \frac{q}{\varepsilon_0} &= (E_{r1} - E_{r2})_{r=a} \\ &= \left(-\frac{\partial V_1}{\partial r} + \frac{\partial V_2}{\partial r}\right)_{r=a} \\ &= E_0 \cos\theta \,\frac{2\varepsilon_r}{\varepsilon_r + 1} - E_0 \cos\theta \,\frac{2}{(\varepsilon_r + 1)} \\ &= 2E_0 \cos\theta \,\frac{\varepsilon_r - 1}{\varepsilon_r + 1}\,. \end{aligned} \qquad (4.83)$$

It is useful to notice that a uniform field induces a $\cos\theta$ distribution of surface charge on a cylinder. Also, since V_2 has the same form as the exciting field V_0, except for a constant multiplier, we note that a $\cos\theta$ distribution of surface charge gives a uniform field inside a cylinder. The effect of a conducting cylinder can be obtained by letting $\varepsilon_r \to \infty$. Then $V_2 \to 0$, so that there is no field inside the cylinder.

4.3.3. Poisson's Equation

A simple example of the use of Poisson's equation is provided by the study of the potential distribution in a space-charge limited diode.

Consider a plane diode of large surface area and neglect edge effects so that there is variation of potential only in the direction perpendicular to the electrodes. Poisson's equation becomes

$$\frac{d^2V}{dx^2} = -\frac{\varrho}{\varepsilon_0}. \tag{4.84}$$

There is steady current flow, so that

$$J = \varrho u \tag{4.85}$$

where J is current per unit area and u is the velocity of the charge.

From considerations of energy, assuming that electrons leave the cathode with negligible velocity,

$$\tfrac{1}{2}\, mu^2 = eV, \tag{4.86}$$

where m and e are electronic mass and charge.

Combining these three equations we have

$$\frac{d^2V}{dx^2} = -\frac{J}{\varepsilon_0}\left(\frac{m}{2e}\right)^{1/2} V^{-1/2}. \tag{4.87}$$

Multiplying both sides by dV/dx and integrating, we obtain

$$\frac{1}{2}\left(\frac{dV}{dx}\right)^2 = -\frac{J}{\varepsilon_0}\left(\frac{m}{2e}\right)^{1/2} 2V^{1/2}. \tag{4.88}$$

The negative sign can be omitted by noting that the current flow is from anode to cathode

$$\frac{dV}{dx} = 2\left(\frac{J}{\varepsilon_0}\right)^{1/2}\left(\frac{mV}{2e}\right)^{1/4}. \tag{4.89}$$

Equation (4.89) becomes, on integration,

$$V^{3/4} = \frac{3}{2}\left(\frac{J}{\varepsilon_0}\right)^{1/2}\left(\frac{m}{2e}\right)^{1/4} x,$$

$$V = \left(\frac{3}{2}\right)^{4/3}\left(\frac{J}{\varepsilon_0}\right)^{2/3}\left(\frac{m}{2e}\right)^{1/3} x^{4/3}. \tag{4.90}$$

This is the particular integral of Poisson's equation. To complete the solution we must add the complementary function which is the solution of Laplace's equation,

$$\frac{d^2V}{dx^2} = 0, \tag{4.91}$$

whence

$$V = Ax + B. \tag{4.92}$$

Since $V = 0$ when $x = 0$, we have $B = 0$, and since $dV/dx = 0$ when $x = 0$, $A = 0$.

Both these conditions are inherent in eqn. (4.86). First V is there taken as the potential difference between the cathode and the point x. Secondly, electrons emerge with zero velocity from the cathode, and this implies that the field $dV/dx = 0$ at the cathode. Equation (4.90) is therefore the complete solution. It is generally written in terms of the current density

$$J = \frac{4}{9}\varepsilon_0 \left(\frac{2e}{m}\right)^{1/2} V^{3/2} x^{-2} \tag{4.93}$$

and is known as the Child–Langmuir equation.

4.3.4. The Method of Images

In § 4.3.1 we discussed the uniqueness of an electrostatic field inside a closed volume and we noticed that such a field can be uniquely specified by specifying the potential (or normal potential gradient) at the surface surrounding the volume. We felt that this statement was reasonable because we could always obtain the required potential (or gradient) by removing the actual sources outside the volume and replacing them by an equivalent surface layer of sources.

In this section we reverse the reasoning. Suppose we have an actual surface distribution of charges, the field of which is rather cumbersome to compute. Let us replace the actual charge distribution by a simple equivalent set of sources outside the volume which is easier to deal with. As long as the potential (or gradient) at the boundary surface is the same for the actual and equivalent sources, the process is perfectly in order.

This reasoning is the basis of the method of images which we shall now discuss briefly. Consider Fig. 4.22. A line charge of density q lies parallel to a conducting plate and the perpendicular distance between the plate and line charge is a. Since the plate is conducting, its potential must be constant at all points on it. We can, within an arbitrary constant, say

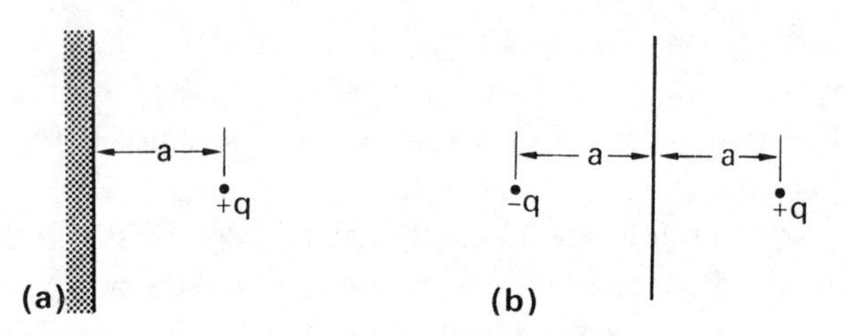

FIG. 4.22. Image of line charge in conducting plate.

that this potential is zero. Thus the induced charges produce at the plate a potential equal and opposite to that of the inducing charge. But just such an effect would be caused by a line charge of equal strength and opposite sign at a distance a on the far side of the plate. We can use this "image" charge to calculate the field anywhere on the near side of the plate. This is clearly a great simplification.

To achieve success with this method one has to be able to guess a simple equivalent charge distribution. If the equivalent charges are just as complicated as the actual induced charges, the method is useless. Under what circumstances shall we be able to expect a simple image distribution? This question can be answered by looking at the form of the potential. This always contains the distance of the source from the field point. A simple image is obtainable if a point can be found (the image point) whose distance from the surface is in a constant ratio to the distance of the source from the surface. In the case of the plane this ratio is unity. With a curved surface we require the source and image to be at inverse points with respect to the surface. In two-dimensional problems this is only possible where the surface is cylindrical and in three dimensions the surface has to be spherical.

The method can also be used when the boundary consists of several planes. See Fig. 4.23, where the boundary has been replaced by three image charges. If the boundary consists of intersecting planes the angle between planes must be π/n, where n is an integer.

The method can be extended to boundaries made of dielectric material, but only where there are single images. The reason for this limitation is that each image is reduced in magnitude because on a dielectric the in-

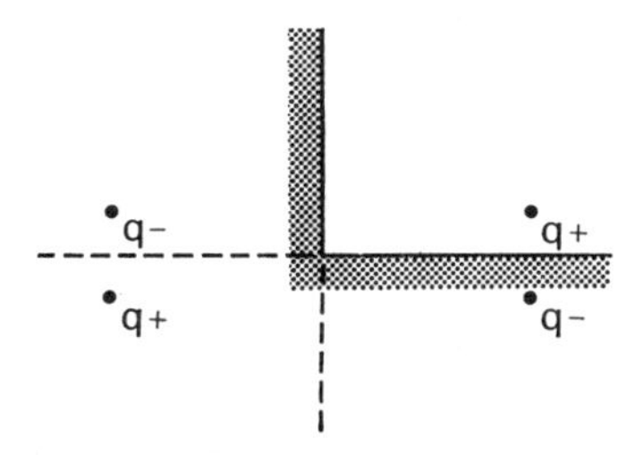

FIG. 4.23. Image of line charge in two conducting plates at right angles.

duced charge is always less than the inducing charge. This does not matter where there is only one image to consider, but the method fails when there are several images. Consider, for instance, Fig. 4.24 where there are multiple reflections. Here the images form an infinite series. With dielectric boundaries the magnitude of the terms in the infinite series

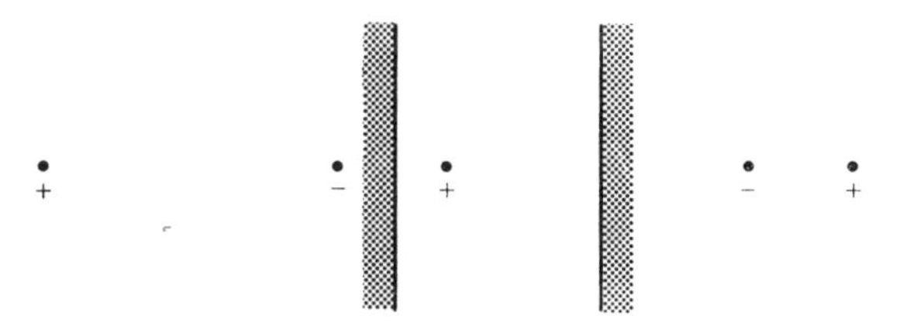

FIG. 4.24. Multiple images in two parallel conducting plates.

would all be different, whereas all the images are of constant strength if the boundaries are conductors. Fourier series, which have a constant amplitude, can be used to calculate this latter field, but not the field of images in dielectric boundaries. Hence the method of images is not very useful when there is more than one dielectric boundary.

Summary

In this chapter we have dealt with the interaction of stationary electric charges. Any general distribution of charges can be built up by superposition of point charges and we have derived the potentials of certain useful standard charge distributions. We have paid particular attention to electric dipoles because polarizable materials consist of aggregates of such dipoles. We have proved that such an aggregate can be replaced in its electrical effects by a surface layer of charge and a volume distribution of charge. This volume distribution is zero in materials of constant permittivity. Dielectrics of constant ε_r can therefore be represented by charge layers on their surfaces.

In order to distinguish between the bound charges of a dielectric and the free charges of a conductor we have defined a total electric field **E** and a partial field **D**, where **D** arises from free charges only.

We have examined the notion of fields inside materials and have concluded that a clear experimental definition of internal fields is impossible. It is, however, possible and helpful to define the field inside certain cavities inside a material and we have shown that the field in a "worm-hole" corresponds to **E** and in a disc-shaped hole corresponds to **D**.

Field problems in regions with defined boundary conditions can often be solved most easily by the method of differential equations. We have examined Laplace's and Poisson's equations and have shown how a knowledge of the general structure of these equations can shorten the labour of finding solutions in particular cases. We have also briefly mentioned how certain boundary conditions can be replaced by "image" sources.

Throughout the chapter we have noticed the close relationship between source charges and potentials in determining the electrostatic field. We have seen that these two concepts provide alternative means of specifying a problem. This is particularly clear in boundary value problems, where the boundary conditions can either be introduced as constants of integration in the solution of Laplace's or Poisson's equations, or where alternatively the boundary surface can be replaced by a distribution of surface charges. Charges and potentials are aspects of the same phenomenon and do not exist independently of each other.

Exercises

4.1. Four equal charges are placed at the corners of a square of area 1 m^2. The force on each charge is 1 N. Determine the value of each charge. [Ans.: 7·6 μC.]

4.2. A filament of charge q C/m is l m long. The electric field strength around the central portion of the filament is to be examined and compared with the field of a filament of infinite length. At what distances from the filament will the actual field be within (a) 10%, and (b) 1% of the field of an infinite filament? [Ans.: (a) 0·24l; (b) 0·071l.]

4.3. What is the greatest charge which can be carried by a metal sphere of 0·5 m diameter surrounded by air of break-down strength 3×10^6 V/m? [Ans.: 20·8 μC.]

4.4. A linear "quadrupole" of electric charges consists of three charges $+q$, $-2q$, and $+q$ placed along the x-axis. There is a space Δx between charges 1 and 2 and between charges 2 and 3. Show that the potential of the quadrupole is given by

$$V = \frac{q\,(\Delta x)^2}{4\pi\varepsilon_0}\,\frac{(3\cos^2\theta - 1)}{r^3},$$

where θ is the angle between r and the x-axis. [*Hint:* Use the method of § 4.1.2 to add the potentials of two dipoles.]

4.5. Two long conducting wires of diameter d are placed parallel to each other with a distance D between their axes. One wire carries a positive charge q C/m and the other a negative charge of the same amount. By considering the charge distribution on the wires show that the force between the wires can be expressed as

$$F = a/D + b/D^2 + c/D^3 + \cdots \text{ N/m}.$$

Calculate a and estimate the largest value for b. [Ans.: $q^2/2\pi\varepsilon_0$; $q^2D/2\pi\varepsilon_0$.] [*Hint:* Use the result of § 4.1.4.]

4.6. Calculate the potential at a point distant z above the centre of a horizontal disc of radius a which carries a uniform charge q C/m^2. What is the largest value of z if the electric field strength of the disc is to be within 1% of the field above a charge distribution of infinite extent?

$$\left[\text{Ans.: } \frac{qa}{2\varepsilon_0}\left(1 - \frac{z}{a} + \frac{1}{2}\frac{z^2}{a^2}\cdots\right), \quad z = 0{\cdot}01a.\right]$$

4.7. Estimate the total electric field **E** inside a thin plate of relative permittivity ε_r inserted across a uniform field E_0. [Ans.: E_0/ε_r.]

4.8. Estimate the partial electric field **D** inside a long cylinder of relative permittivity ε_r inserted along a uniform field E_0. [Ans.: A sketch will show how the cylinder gathers flux to itself. **D** may become very large.]

4.9. Derive an expression for the capacitance of two concentric spheres, if the larger sphere has an inner radius a and the smaller sphere an outer radius b and if there is a material of permittivity ε_r between the spheres. Show that for a given radius a and p.d. V the greatest electric field strength will be a minimum when $b = a/2$.

$$\left[\text{Ans.: } 4\pi\varepsilon_0\varepsilon_r\Big/\left(\frac{1}{b} - \frac{1}{a}\right).\right]$$

4.10. Examine the possible forms of solution of Laplace's equation in two-dimensional polar coordinates. A dielectric cylinder of radius a in a two-dimensional field has on its surface an induced charge distribution $q \sin n\theta$ C/m^2, where θ is the angular coordinate. Determine the potential both inside and outside the cylinder. [Ans.: $qr^n \sin n\theta/2\varepsilon_0 na^{n-1}$, $qa^{n+1} \sin n\theta/2\varepsilon_0 nr^n$.]

4.11. The density of charge on a long cylinder of radius R is given by the expression $q = f(\theta)$, where θ is the angular coordinate. Show that the normal electric field E_n at the surface of the cylinder is independent of θ and depends only on the total charge, and that for a dielectric cylinder $E_n = 0$.

4.12. Explain what is meant by an electrostatic *image*. A line charge of strength q C/m is placed at a distance a from the plane surface of a large slab of dielectric of relative permittivity ε_r. Determine the image charge in the dielectric. Determine also the equivalent charge outside the dielectric which gives the correct total electric field in the dielectric. [Ans.: line charge $-(\varepsilon_r - 1)/(\varepsilon_r + 1)\, q$ at point a below surface of dielectric, line charge $2/(\varepsilon_r + 1)\, q$ at point of original charge.]

4.13. A charge Q is situated at the point $x = 3, y = 4, z = 0$. The planes Oxz and Oyz are conductors and O is the origin of coordinates. Use the method of images to determine the density of surface charge at the point $x = 3, y = 0, z = 0$. Show that the method of images can be applied to planes intersecting at an angle θ if π/θ is an integer. [Ans.: $-0{\cdot}0083\, Q$ C/m^2.]

4.14. Show that the potential inside a closed volume is uniquely determined by the sources in the volume and the potential (or the normal gradient of the potential) on the enclosing surface. [*Hint:* Assume that two different potentials are possible. Then apply eqn. (3.61) to the difference of the two potentials as in § 3.2.7.]

CHAPTER 5

ELECTROSTATICS II

IN THE previous chapter we have been concerned with the electric field due to various charge distributions. Books on electromagnetism often give the impression that a problem is solved once the field has been calculated. This is true if the object of the exercise is to obtain mathematical solutions. But engineers have to make things that will be useful. Their interest in electrostatics is practical in wanting to harness the electric field. Such a consideration leads to the desire to know the forces which can be exerted by electrostatic devices and the energy which can be stored in them. The criterion of successful design will often be to have the largest possible forces and energy in a given volume. In this chapter we shall therefore investigate the energy and the forces associated with electrostatic fields.

5.1. Assembly Work and Field Energy

In Chapter 2 we introduced the notion of potential in terms of the work which had to be done to bring a set of charges within a finite distance of each other. The term potential is an abbreviation for "potential energy per unit charge" and defines the work done in bringing a unit charge to the point at which the potential is to be measured. We shall now use the potential to determine the *total* potential energy of a system.

Consider first the work W† done in bringing the charges Q_1, Q_2 within a distance r_{12} of each other. This work is

$$W_2 = \frac{Q_1 Q_2}{4\pi\varepsilon_0 r_{12}} = Q_1 V_{12}, \tag{5.1}$$

where V_{12} is the potential due to Q_2 at the point occupied by Q_1. Now

† The letter W was used in § 4.1.8 to designate the complex potential, but the context should obviate any confusion. Here W is work done.

let another charge Q_3 be brought to a position r_{13} from Q_1 and r_{23} from Q_2. The extra work will be

$$W = \frac{Q_1 Q_3}{4\pi\varepsilon_0 r_{13}} + \frac{Q_2 Q_3}{4\pi\varepsilon_0 r_{23}} = Q_1 V_{13} + Q_2 V_{23}. \tag{5.2}$$

Since $V_{mn} = V_{nm}$, the total assembly work of n charges is

$$W = \frac{Q_1}{2}(V_{12} + V_{13} + V_{14} + \cdots V_{1n})$$

$$+ \frac{Q_2}{2}(V_{21} + V_{23} + V_{24} + \cdots V_{2n}) + \cdots = \frac{1}{2}\sum_{i=1}^{n} Q_i V_i, \tag{5.3}$$

where V_i is the potential of all the charges except the ith charge, at the point occupied by the ith charge. Let us now try to express the assembly work in terms of the electric field itself. To do so we first express each point charge as a small volume distribution of charge.

$$W = \frac{1}{2}\sum_{i=1}^{n} QV = \frac{1}{2}\sum_{i=1}^{n}\iiint_{v'} V\varrho\, dv'. \tag{5.4}$$

Also $\operatorname{div} \mathbf{E} = \varrho/\varepsilon_0$, so that

$$W = \frac{\varepsilon_0}{2}\sum_{i=1}^{n}\iiint_{v'} V \operatorname{div} \mathbf{E}\, dv'. \tag{5.5}$$

This expression can be transformed by using the vector identity

$$\operatorname{div}(V\mathbf{E}) = V \operatorname{div} \mathbf{E} + \mathbf{E} \,.\, \operatorname{grad} V. \tag{5.6}$$

Thus

$$\iiint_{v'} V \operatorname{div} \mathbf{E}\, dv' = \iint_{s'} V\mathbf{E} \,.\, ds' - \iiint_{v'} \mathbf{E} \,.\, \operatorname{grad} V\, dv'. \tag{5.7}$$

We can drop the surface term by noting that we can make v' and s' arbitrarily large, since for each point charge Q_i there is charge density only at the point. For a very large surface s' the surface integral disappears because V varies as $1/r$, $\mathbf{E}$ varies as $1/r^2$ and s' as r^2.

Hence the assembly work is given by

$$W = -\frac{\varepsilon_0}{2}\sum_{i=1}^{n}\iiint [\mathbf{E} \,.\, \operatorname{grad} V]\, dv'. \tag{5.8}$$

The summation sign adds up all regions of charge distribution and the integration sums the charges within such regions. Both these processes can be included in the integration sign, so that we can write

$$W = -\frac{\varepsilon_0}{2}\iiint [\mathbf{E}\,.\,\mathrm{grad}\,V]\,dv'. \tag{5.9}$$

Consider now the expression $\mathbf{E}\,.\,\mathrm{grad}\,V$. As in eqn. (5.3), V is the potential due to all charges except the one at the place where V is observed. Since $\mathbf{E} = -\mathrm{grad}\,V$ we have

$$\mathbf{E}\,.\,\mathrm{grad}\,V = -\sum_{i=1}^{n}\sum_{j=1}^{n}\mathbf{E}_i\,.\,\mathbf{E}_j, \tag{5.10}$$

where $i = j$ is to be omitted from the summation.

But

$$\mathbf{E} = \sum_{i=1}^{n}\mathbf{E}_i$$

and therefore

$$E^2 = \sum_{i=1}^{n}E_i^2 + \sum_{i=1}^{n}\sum_{i=1}^{n}\mathbf{E}_i\,.\,\mathbf{E}_j. \tag{5.11}$$

Whence, finally,

$$W = \frac{\varepsilon_0}{2}\iiint E^2\,dv' - \frac{\varepsilon_0}{2}\iiint E_i^2\,dv' \tag{5.12}$$

or more simply

$$W = U - U_s, \tag{5.13}$$

where

$$U = \frac{\varepsilon_0}{2}\iiint E^2\,dv \tag{5.14}$$

and

$$U_s = \frac{\varepsilon_0}{2}\iiint E_i^2\,dv. \tag{5.15}$$

The prime on the volume integration has been omitted because the integration is to be carried out over all space and there need be no distinction between field coordinates and source coordinates.

U is called the *field energy* and U_s the *self-energy* of the system of charges. The self-energy has to be subtracted from the field energy to obtain the assembly work W because we have assumed that the charges $Q_1, Q_2, \ldots, Q_n$ already existed before they were assembled. If they had to be condensed from a cloud of charge density ϱ, extra assembly work would have to be done. Once the charges have been brought into existence, U_s does not enter into any calculations because experimentally only changes of energy can be observed.

The expressions so far derived describe the electric energy of a set of charges in free space. If there are polarizable materials present we can deal with them in two alternative ways. Either we can treat these materials as additional sources, so that **E** and ϱ are taken as the total electric field and the total charge density. In such an approach all space is free space and no further discussion is necessary. Alternatively, we can retain **E** as the total electric field but confine ϱ to the density of free charges. We then define a partial field **D** by means of the equation $\nabla . \mathbf{D} = \varrho$ [eqn. (4.33)] and then derive an expression for the field energy in terms of **E** and **D**.

As before,

$$W = \tfrac{1}{2} \sum_{i=1}^{n} Q_i V_i, \tag{5.3}$$

but Q_i now describes the free charges only. Instead of eqn. (5.5) we have

$$\iiint_{v'} V\varrho \, dv' = \iiint_{v'} V \operatorname{div} \mathbf{D} \, dv', \tag{5.16}$$

hence

$$W = -\tfrac{1}{2} \Sigma \iiint \mathbf{D} \, . \operatorname{grad} V \, dv'. \tag{5.17}$$

Reference to the argument of eqns. (5.9)–(5.12) shows that we now face a difficulty. Before we can proceed we need to know the relationship between **D** and **E**. In this section we shall assume that

$$\mathbf{D} = \varepsilon_0 \varepsilon_r \mathbf{E}, \tag{4.35}$$

where ε_r is a constant of the material. This is a reasonable assumption for dielectrics, but when in the next chapter we come to investigate the parallel case of magnetic energy we shall have to face the fact that μ_r is very far from constant.

Using eqn. (4.35) we can now write

$$W = \frac{\varepsilon_0}{2} \iiint \varepsilon_r E^2 \, dv' - \frac{\varepsilon_0}{2} \iiint \varepsilon_r E_i^2 \, dv'$$

$$= \frac{1}{2} \iiint \mathbf{E} \cdot \mathbf{D} \, dv' - \frac{1}{2} \iiint \mathbf{E}_i \cdot \mathbf{D}_i \, dv'. \tag{5.18}$$

Once again we can write

$$W = U - U_s \tag{5.13}$$

where now

$$U = \tfrac{1}{2} \iiint \mathbf{E} \cdot \mathbf{D} \, dv \tag{5.19}$$

and

$$U_s = \tfrac{1}{2} \iiint \mathbf{E}_i \cdot \mathbf{D}_i \, dv. \tag{5.20}$$

Eqns. (5.14) and (5.15) can be taken as a special case of eqns. (5.19) and (5.20) for $\varepsilon_r = 1$. It is important to notice that ε_r does not have to have the same value throughout space. The requirement is that ε_r shall be constant and independent of the field at every point of space.

If ε_r is not independent of the field there is no linear relationship between the electric field (or potential) and the free charge distribution. The assembly work has now to be obtained by integration and instead of eqn. (5.4) we have

$$\delta W = \Sigma \iiint V \, \delta\varrho \, dv', \tag{5.21}$$

where ϱ is the volume density of free charge.

Also

$$\iiint V \, \delta\varrho \, dv' = \iiint V \, \delta \, (\text{div } \mathbf{D}) \, dv'. \tag{5.22}$$

The vector identity eqn. (5.6) can now be written

$$\text{div } (V \, \delta\mathbf{D}) = V \, \delta \text{ div } \mathbf{D} + \delta\mathbf{D} \cdot \text{grad } V, \tag{5.23}$$

whence by the use of the argument of eqn. (5.7) we have

$$\delta W = \Sigma \iiint \mathbf{E} \cdot \delta\mathbf{D} \, dv' \tag{5.24}$$

and

$$\delta U = \iiint \mathbf{E} \cdot \delta\mathbf{D} \, dv'. \tag{5.25}$$

The field energy is therefore given by

$$U = \iiint \left[\int_0^D \mathbf{E} \, . \, d\mathbf{D} \right] dv'. \tag{5.26}$$

In eqn. (5.13) we have equated work and internal energy. In thermodynamic terms this implies an *adiabatic* process. Such processes in general produce a change of temperature, which we have so far ignored. Experiment shows that the permittivity is often very dependent on temperature and we must therefore specify a constant temperature or *isothermal* process so that the permittivity remains constant. In an isothermal process the assembly work will cause heat transfer as well as change in internal energy. The field energy cannot therefore be the internal energy of the system. But we shall see that it corresponds to the so-called "free" energy of the system. Let this free energy be defined as

$$F = U_{\text{th}} - TS, \tag{5.27}$$

where U_{th} is the thermodynamic internal energy, T the temperature, and S the entropy.

$$dF = dU_{\text{th}} - T\,ds - S\,dT. \tag{5.28}$$

For an isothermal process,

$$dF = dU_{\text{th}} - T\,ds = dU_{\text{th}} - dQ, \tag{5.29}$$

where dQ is the change of heat in a reversible process. By the first law of thermodynamics

$$dQ = dU_{\text{th}} - dW, \tag{5.30}$$

where dW is the increment of work done on the system.

Thus

$$dW = dF, \tag{5.31}$$

and since $dW = dU$ we can identify the field energy with the free energy. Since the free energy gives the maximum work that can be extracted from the system, we can determine this maximum work by considering changes in the field energy.

The concept of field energy is very useful. Instead of thinking about the complicated interaction of the charges we can focus attention on the local field. But a word of caution is needed. It is true that the energy is given in terms of the field and from there it is only a small step to saying that the energy is *stored* in the field. This is, however, only a form of speaking. A man on top of a mountain possesses more potential energy than when he is at the bottom, but it would be meaningless to say that potential energy is stored at the top of the mountain. Gravitational potential energy is always a *mutual interaction* between masses and similarly electrostatic potential energy is the *mutual energy* of electric charges. The expression for U is a consequence of the assembly work W and is not based on any separate physical observation.

The integrand $\frac{1}{2}\mathbf{E}\,.\,\mathbf{D}$ in eqn. (5.19) is often called the electrostatic *energy density*. This statement also needs to be treated with caution. The integral of the expression does of course represent the free energy. But just as it is impossible to localize this energy, so also is it impossible to associate energy with each part of a volume. There are no experiments which can determine the distribution of energy. We are again forced to the conclusion that the field concept is secondary to the concept of sources. The field energy is the potential energy of the sources.

5.2. Field Energy and Tubes of Flux

The expression for field energy [eqn. (5.19)] focuses attention on the electric field strength and electric flux density throughout all space. It is true that we derived the expression by a consideration of the work which has to be done to assemble a set of charges and at that time we thought of interaction rather than of the field as such. But now that it has been shown that the energy can be expressed as a function of the field without explicit reference to the position of the charges, it is very instructive to change our point of view and to see whether eqn. (5.19) could have been derived without discussing the assembly work.

We remind ourselves that every point in space can be associated with an electric field strength E defined as a force on a unit charge. This statement is true for any arbitrary law of interaction between charges. Since, however, the law is that of the inverse square of the distance, we can also associate with every point in space an electric flux density D. In other words we can describe the field in terms of tubes of flux, each of which contains a constant amount of this flux. Such a three-dimensional flux map contains information about force E (volts/metre or newtons/coulomb) and quantity of electricity D (coulombs/metre^2), and we shall show in § 5.4 that this information is sufficient (and necessary) to enable the mechanical force on charged bodies to be determined. It is similarly reasonable that the potential energy should be calculable from the field E and D and this is what we proved in the last section. Dimensionally the product ED is joules/metre^3 and its volume integral is joules.

Consider then an element of a tube of flux (Fig. 5.1). This element can be replaced by two charges as shown in Fig. 5.2, forming a small capacitor

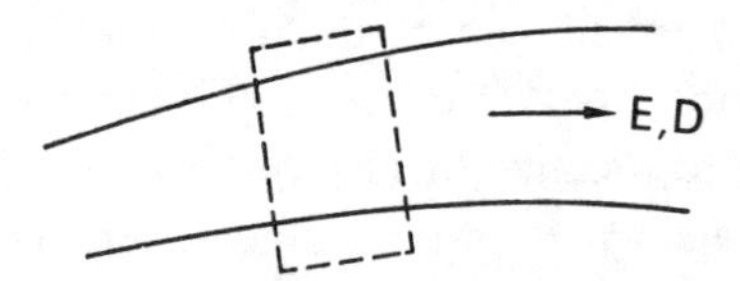

FIG. 5.1. An element of a tube of flux.

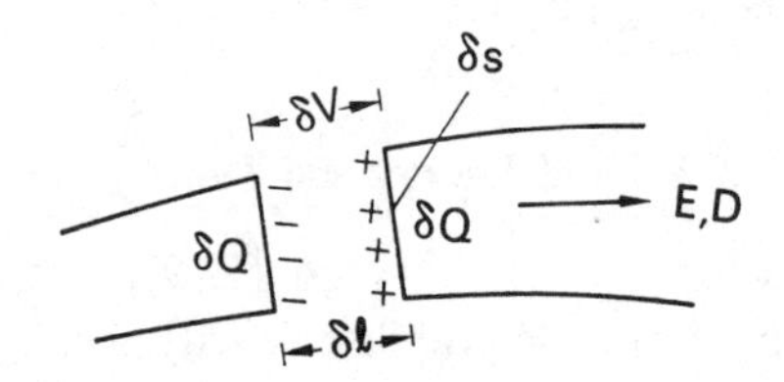

FIG. 5.2. Element of tube replaced by capacitor.

of charge δQ and potential difference δV. Now $\delta Q = D\,\delta s$ and $\delta V = E\,\delta l$. The potential energy is $\frac{1}{2}\delta Q\,\delta V = \frac{1}{2}ED\,\delta v$, where δv is an element of volume. Hence the potential energy is $\frac{1}{2}ED$ per unit volume and the energy of the system is as before

$$U = \tfrac{1}{2}\iiint \mathbf{E}\,.\,\mathbf{D}\,dv, \tag{5.19}$$

where the integration has to be carried out throughout all space occupied by tubes of flux. The tubes of flux are associated with the energy. The energy expended in establishing the flux is the same as that expended in assembling the charges.

We could of course have expressed the energy in terms of capacitance. Thus for the small capacitor of Fig. 5.2

$$\delta C = \delta Q/\delta V = D\,\delta s/E\,\delta l.$$

The energy is $\frac{1}{2}\delta C\,(\delta V)^2 = \frac{1}{2}ED\,\delta v$.

Each little volume of flux can be replaced by a small capacitor δC and this represents a circuit analogue of the electric field.

This short derivation of eqn. (5.19) is a direct consequence of the inverse square law and is as rigorous as that of the previous section. Experimentally no distinction can be made between the two methods.

5.3. Energy and Force

Energy transfer lies at the heart of electrical engineering. The discussion in §§ 5.1 and 5.2 gives insight into the criteria of energy storage in electrostatic systems. Equation (5.19), for instance, indicates the importance of a large relative permittivity, if the stored energy per unit volume is to be as large as possible. We shall now examine the forces and torques associated with energy transfer. Since W is the assembly work we can write at once

$$F_x = -\frac{\partial W}{\partial x} \tag{5.32}$$

or, as a vector equation,

$$\mathbf{F} = -\text{grad}\,W. \tag{5.33}$$

The torque is given by

$$T_\theta = -\frac{\partial W}{\partial \theta}. \tag{5.34}$$

Often it is more convenient to use U rather than W. Since

$$W = U - U_s, \tag{5.13}$$

where U_s is constant we can write

$$F_x = -\frac{\partial U}{\partial x} \tag{5.35}$$

and

$$T_\theta = -\frac{\partial U}{\partial \theta}. \tag{5.36}$$

The negative sign in these equations states that the force or torque acts to *reduce* the energy. This is a characteristic of all systems of potential energy.

The forces and torques act on the charges making up the electrostatic system. If these charges are attached to each other or to some otherwise uncharged body, then the force will be transmitted by "non-electric" mechanical pressures and tensions to the body as a whole. For instance, the force on the plate of a capacitor acts on the surface charges and these transfer the force to the plate. If we want to determine the force on a dielectric we can think in terms of either the bound charges or dipoles, or we can use the relative permittivity in the expression for the field energy as an alternative description.

5.3.1. *Volume Distribution of Force*

In all branches of electromagnetism we seek to find the action at a *point* and we shall now try to obtain expressions for force distribution. The force per unit volume is defined by the equation

$$\mathbf{F} = \iiint \mathbf{f}\, dv. \tag{5.37}$$

In free space we have by definition of **E** the straightforward result

$$\mathbf{f} = \mathbf{E}\varrho. \tag{5.38}$$

In a charged dielectric the bound charges have to be included in the calculation so that

$$\mathbf{f} = \mathbf{E}\varrho_{\text{total}} = \mathbf{E}\,(\varrho_{\text{free}} + \varrho_{\text{bound}}) \tag{5.39}$$

or

$$\mathbf{f} = \varepsilon_0 \mathbf{E}\,(\nabla \cdot \mathbf{E}). \tag{5.40}$$

If we use ϱ for the free charge density and derive ϱ_{bound} from the dipole moment per unit volume by eqn. (4.30), this expression becomes

$$\mathbf{f} = \mathbf{E}\varrho - \mathbf{E}\,(\nabla \cdot \mathbf{P}). \tag{5.41}$$

Unfortunately this is not the only possible force distribution. We can for instance work out the force on the dipole moment **P**. Consider this with reference to Fig. 5.3. A typical component of the force is

$$q\frac{\partial E_y}{\partial x}\,\delta x = P_x\frac{\partial E_y}{\partial x},$$

where q is the charge at the ends of the dipole.
Considering only the component P_x of the vector **P** the force is

$$\mathbf{i}P_x\frac{\partial E_x}{\partial x} + \mathbf{j}P_x\frac{\partial E_y}{\partial x} + \mathbf{k}P_x\frac{\partial E_z}{\partial x}.$$

Hence the force on **P** can be written $(\mathbf{P}\,.\,\nabla)\,\mathbf{E}$, so that

$$\mathbf{f} = \mathbf{E}\varrho + (\mathbf{P}\,.\,\nabla)\,\mathbf{E}. \qquad (5.42)$$

Comparison with eqn. (5.41) shows that the second term in each equation is altogether different.

It is at first sight surprising that there should be various different expressions for **f** and one may ask which of these expressions is the correct one. This question has no answer. All the expressions give the correct total force when integrated throughout a body. None of them give a uniquely true force distribution. There are two interlinked reasons for this unsatisfactory state of affairs.

The first was mentioned in § 4.2.3. It is not possible to define what is meant by being "inside" a material and so there can be no unique field strength at a point in a material. The electric field strength **E** is either the field in a "worm-hole" or the space–time average of the local fields.

The second reason arises from the lack of unique definition of the sources of the polarization in a material. Consider this with reference to Figs. 5.3 and 5.4 in which an electric dipole is immersed in an electric field **E**. If we regard the dipole as a single entity, the force on it is $(\mathbf{P}\,.\,\nabla)\,\mathbf{E}$

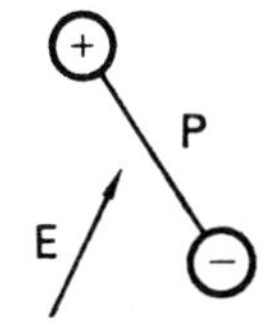

FIG. 5.3. Force on a dipole.

as in eqn. (5.42). If, on the other hand, we find the force on one of the charges making up the dipole, the local force is $\mathbf{E}q$ in the direction of the field plus $q^2/(4\pi\varepsilon_0\,|d\mathbf{l}|^2)$ along the dipole. This second force component is due to the interaction of the two charges in the dipole. Of course this

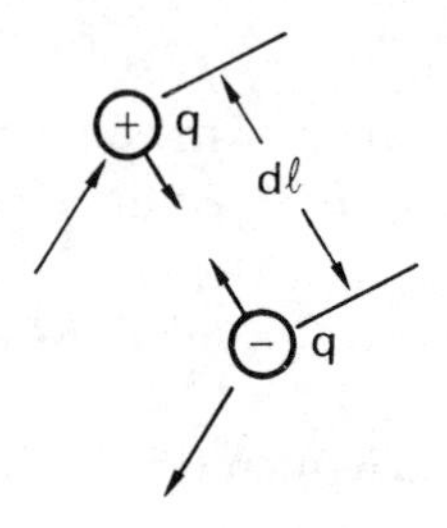

FIG. 5.4. Force on the ends of a dipole.

interaction will cancel if we consider both the charges; nevertheless, the local force will depend on our choice of fundamental particle. If we take the dipole as a single particle we disregard its internal electrical force. More accurately we postulate an internal non-electric pressure which keeps the dipole from collapsing and which equilibrates the internal electrical force.

We are now at the heart of the problem. In seeking a unique electrical force distribution inside a material we have ignored the other forces which may be due to many effects such as elasticity, viscosity, gravitation, and heat flow. It is true that on a microscopic scale some of these effects are electric in origin, but our macroscopic theory has nothing to say about molecular forces. In discussing *electric* forces inside a material we are attempting to isolate a *partial* set of forces. The difference between the expressions for the force **f** which we have derived lies in the different distribution of electrical and non-electrical forces, as in the case of the dipole which we have just discussed. Only the *total* force on a complete body can be observed and only this force has a unique value. The force distribution **f** can be used as a tool in calculation, but has no other significance.

So far we have made no assumptions about the mechanism of polarization. Such a mechanism is implied if we use the permittivity instead of the polarization or the bound charges, as was shown in § 4.2.1. If the permittivity is independent of field strength but varies from place to place we can write

$$\mathbf{D} = \varepsilon_0 \varepsilon_r \mathbf{E}. \tag{4.35}$$

Substituting in eqn. (5.40) gives

$$\mathbf{f} = \mathbf{E}\left(\nabla \cdot \frac{\mathbf{D}}{\varepsilon_r}\right) = \mathbf{E}\frac{\nabla \cdot \mathbf{D}}{\varepsilon_r} - \frac{\mathbf{E}(\mathbf{D} \cdot \nabla\varepsilon_r)}{\varepsilon_r^2} = \mathbf{E}\frac{\varrho}{\varepsilon_r} - \frac{\varepsilon_0}{\varepsilon_r}\mathbf{E}(\mathbf{E} \cdot \nabla\varepsilon_r). \tag{5.43}$$

In a region of the material where $\nabla\varepsilon_r$ is zero, the expression reduces to

$$\mathbf{f} = \mathbf{E}\frac{\varrho}{\varepsilon_r}. \tag{5.44}$$

Some writers deduce from this expression that the force in a dielectric is reduced in the ratio of the relative permittivity. Such a statement is unhelpful because it throws no light on the meaning of ε_r, and ignores the fact that any expression for **f** is arbitrary to some extent. Moreover, eqn. (5.44) is useless in calculation because a complete body must have a surface at which $\nabla\varepsilon_r$ is not zero. Equation (5.44) is then incorrect and eqn. (5.43) must be used.

5.3.2. Electrostriction†

Before we leave this difficult question of a volume force density we shall derive one other useful expression for **f**, in which the relative permittivity is related to the density of the material.

We assume that the material is displaced by some arbitrary amount $\mathbf{l}(x, y, z)$, where $\mathbf{l}$ varies within the material. We can relate the displacement to the change of the energy U and obtain by the use of eqn. (5.19)

$$\iiint \mathbf{f} \cdot \mathbf{l}\, dv = -\delta U = -\tfrac{1}{2}\delta \iiint \mathbf{E} \cdot \mathbf{D}\, dv. \tag{5.45}$$

Using eqn. (4.35) we can write $\mathbf{E} = \mathbf{D}/\varepsilon_0\varepsilon_r$. Hence

$$\delta U = \iiint \mathbf{E} \cdot \delta\mathbf{D}\, dv + \frac{1}{2\varepsilon_0}\iiint D^2 \delta\left(\frac{1}{\varepsilon_r}\right) dv. \tag{5.46}$$

† It is recommended that §§ 5.3.2 and 5.3.3 should be omitted on a first reading of this book.

Now the first term can be written

$$\iiint \mathbf{E} \,.\, \delta\mathbf{D}\, dv = -\iiint \nabla V \,.\, \delta\mathbf{D}\, dv, \tag{5.47}$$

which, by the use of the vector identity eqn. (5.6) and by the omission of the surface term as in eqn. (5.7), can be written

$$\iiint \mathbf{E} \,.\, \delta\mathbf{D}\, dv = \iiint V\nabla \,.\, \delta\mathbf{D}\, dv = \iiint V\delta\nabla \,.\, \mathbf{D}\, dv = \iiint V\delta\varrho\, dv. \tag{5.48}$$

So that

$$\delta U = \iiint V\delta\varrho\, dv - \frac{\varepsilon_0}{2}\iiint E^2\delta\varepsilon_r\, dv. \tag{5.49}$$

The equation of continuity of electric charge under the displacement $\mathbf{l}$ is given by

$$\iiint \delta\varrho\, dv = -\oiint \varrho\mathbf{l} \,.\, d\mathbf{s} = -\iiint \nabla \,.\, \varrho\mathbf{l}\, dv, \tag{5.50}$$

where the charge is bound together with the volume. Hence

$$\delta\varrho = -\nabla \,.\, (\varrho\mathbf{l}). \tag{5.51}$$

Similarly, the continuity of mass gives

$$\delta g = -\nabla \,.\, (g\mathbf{l}), \tag{5.52}$$

where g is the mass volume density.

If we assume that ε_r is a unique function of g, an assumption which is certainly true for gases and liquids,

$$\delta\varepsilon_r = \frac{d\varepsilon_r}{dg}\delta g = -\frac{d\varepsilon_r}{dg}\nabla \,.\, (g\mathbf{l}). \tag{5.53}$$

Using eqn. (5.6) again,

$$V\delta\varrho = -V\nabla \,.\, (\varrho\mathbf{l}) = -\nabla \,.\, (V\varrho\mathbf{l}) + \varrho\mathbf{l} \,.\, \nabla V$$

and

$$E^2\,\delta\varepsilon_r = -E^2\frac{d\varepsilon_r}{dg}\nabla \,.\, g\mathbf{l} = -\nabla \,.\, \left(E^2\frac{d\varepsilon_r}{dg}g\mathbf{l}\right) + g\mathbf{l} \,.\, \nabla\left(E^2\frac{d\varepsilon_r}{dg}\right). \tag{5.54}$$

Omitting the surface integrals as in eqn. (5.7), we arrive at

$$\delta U = -\iiint \left[\varrho E + \frac{\varepsilon_0}{2} g \nabla \left(E^2 \frac{d\varepsilon_r}{dg} \right) \right] . \mathbf{l}\, dv, \tag{5.55}$$

whence from eqn. (5.45) and omitting the arbitrary $\mathbf{l}$ from the force expression

$$\mathbf{f} = \varrho \mathbf{E} + \frac{\varepsilon_0}{2} g \nabla \left(E^2 \frac{d\varepsilon_r}{dg} \right). \tag{5.56}$$

Also, since

$$\nabla \varepsilon_r = \frac{d\varepsilon_r}{dg} \nabla g, \tag{5.57}$$

$$\mathbf{f} = \mathbf{E}\varrho - \frac{\varepsilon_0}{2} E^2 \nabla \varepsilon_r + \frac{\varepsilon_0}{2} \nabla \left(E^2 \frac{d\varepsilon_r}{dg} g \right). \tag{5.58}$$

Let us now examine this expression. The first term gives the force on the free charges. The second term is a convenient expression for the force that arises at a place where ε_r varies in space, particularly of course at a surface of a dielectric. The third term is known as the electrostriction effect. Since for any arbitrary function V

$$\iiint \nabla V \, dv = \iint V \mathbf{n} ds \tag{5.59}$$

(see exercise 3.18), the electrostriction effect makes no contribution to the force if the integration is carried out over a large enough volume so that the surface is in a field-free region or in a region where g is zero.

5.3.3. *Pressure in Polarized Liquids and Gases*†

In liquids and gases in equilibrium it is possible to define a single "hydrostatic" pressure. Every volume force $\mathbf{f}$ will be associated with a pressure p by the relationship

$$\mathbf{f} = \nabla p. \tag{5.60}$$

This equation can be combined with eqn. (5.56) or eqn. (5.58) to calculate the pressure difference between two places inside the fluid. We

† It is recommended that this section should be omitted on a first reading of the book.

have, by eqn. (5.56), in the absence of free charge

$$\int_1^2 \frac{dp}{g} = \frac{\varepsilon_0}{2}\int_1^2 \nabla\left(E^2\frac{d\varepsilon_r}{dg}\right) = \frac{\varepsilon_0}{2}\left[E_2^2\left(\frac{d\varepsilon_r}{dg}\right)_2 - E_1^2\left(\frac{d\varepsilon_r}{dg}\right)_1\right]. \quad (5.61)$$

If the fluid is incompressible, g is constant and

$$p_2 - p_1 = \frac{g\varepsilon_0}{2}\left[E_2^2\left(\frac{d\varepsilon_r}{dg}\right)_2 - E_1^2\left(\frac{d\varepsilon_r}{dg}\right)_1\right]. \quad (5.62)$$

5.4. Electric Field Stresses

We come now to the most powerful method for the calculation of forces on charged and polarized bodies in an electrostatic field. This is the method of field stresses. The concept of stress is taken from ordinary mechanics, and we have here another example of the close and fruitful relationship between mechanics and the theory of electricity.

Let us first remind ourselves what is meant by mechanical stress. The concept arises in the discussion of forces within a body. In order to find such local forces we imagine a cut to be made in the body. The material on one side of the cut is left undisturbed, but on the other side of the cut the material is removed. In order to maintain equilibrium a system of forces has then to be applied on the free surface of the cut, and these are the same forces which were exerted by the material before it was removed. The surface forces therefore represent the internal forces in the material before the cut was made. In order to obtain more detailed information the surface forces are expressed as force/area or stress. In general a surface will experience direct stress and shear stress.

As a simple example consider the beam in Fig. 5.5. To find the stress at the section AA we imagine the beam cut and the material on one side

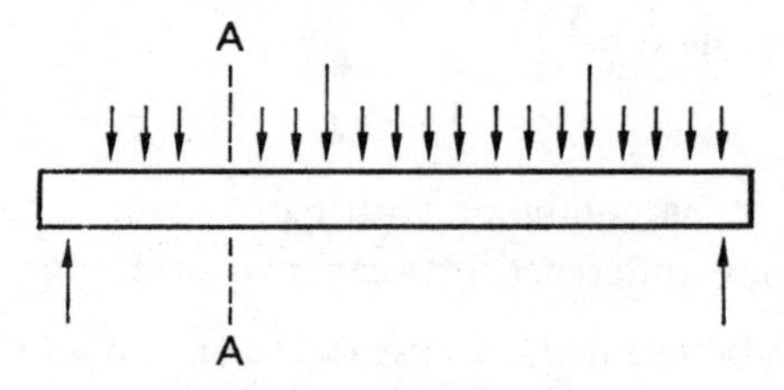

FIG. 5.5. A loaded beam.

removed. A force F and couple M, the shear force and bending moment, must then be applied at the section to keep the remaining part of the beam in equilibrium (Fig. 5.6). Such forces and couples represent the local effect in the material due to all the forces and couples applied to the beam anywhere along its length. When the forces and couples are known, the direct and shear stresses can be calculated. Of course the whole process is conceptual; no actual cuts need be made. But equally the notion of stress is conceptual. Stress has been invented to account for the manner in which the forces and torques are transmitted through the material.

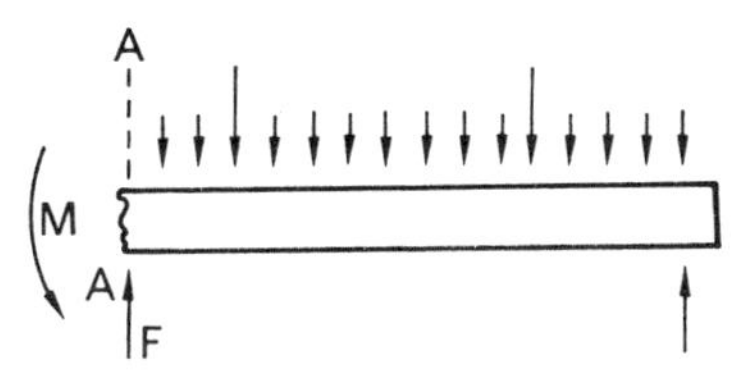

FIG. 5.6. Beam cut to find stresses.

We now want to apply the idea of stress to a system of electric forces. If we consider stresses in a material body, there should be no difficulty. But we have become used to the notion that electric forces act through empty space and this involves us in talking about a stress in the absence of any material. Many readers will have strong misgivings about such talk and rightly so. The idea of a stress in empty space is either meaningless or seems to commit us to the view that space is filled with an elastic substance which transmits stress. This is the old *aether theory* which was in vogue until experiments showed that velocity relative to the aether could not be observed. The resulting relativity theory led most physicists to abandon the idea of aether altogether.

It will have been observed that in this book we are committed to the view that only those properties of nature which can be measured have a place in the theory. We have no intention of reintroducing the notion of an all-pervading aether. We do not discuss how charges act on each other across empty space; we merely observe that they do, and seek to construct a conceptual framework which enables us to calculate such interaction in a simple and straightforward manner. With this assurance let the reader overcome his repugnance to field stresses at least until he has seen the use to which they may be put. To allay fears we shall proceed very cautiously until we arrive at the result which is one of the most important results in electromagnetism and which will speak for itself.

Let us first consider a system of point charges as illustrated in Fig. 5.7. To find the force on the charge at P we calculate the electric field there due to all other charges. The force is then obtained by multiplying this field strength by the magnitude of the charge at P. This is the method used in § 5.1 to calculate the assembly work of such a set of charges. Consider next the more complicated problem illustrated in Fig. 5.8. We now have to deal with charged (or polarized) bodies rather than with point charges. One method of dealing with this problem is to split each charged body up into minute regions containing point charges or dipoles and then to proceed as before.

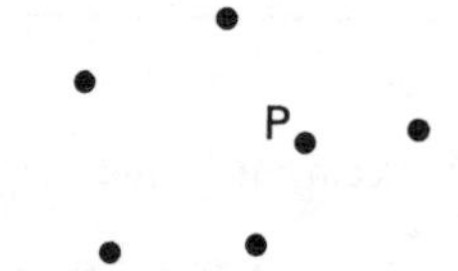

FIG. 5.7. System of point charges.

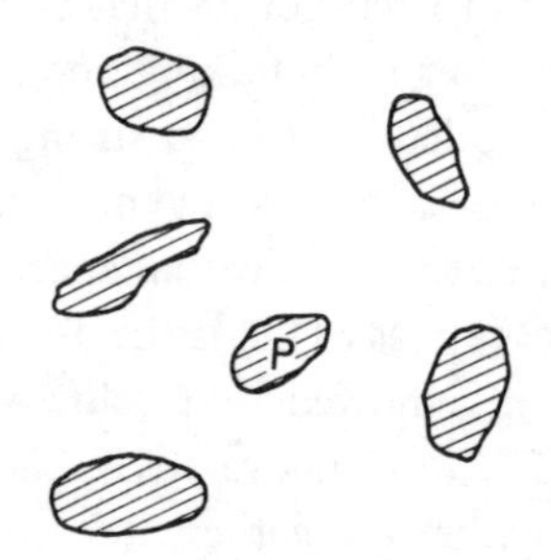

FIG. 5.8. System of charged bodies.

Thus the force on the body marked P in Fig. 5.8 can be calculated by the resultant of all the forces on the particles making up the body. This we attempted to do by finding the local force distribution $\mathbf{f}$ in § 5.3. The attempt was beset with difficulties and we had to content ourselves with various expressions for $\mathbf{f}$, none of which had any claim to uniqueness.

Let us now be a little less ambitious and content ourselves with a knowledge of the total electric force on the body P without seeking to know the distribution of this force inside the body. This can be done by calculating the field energy U and obtaining the force by the use of eqn. (5.35) which typifies the statement $\mathbf{F} = -\text{grad}\, U$. The calculation would involve the integration of the field energy by the use of the relationship

$$U = \tfrac{1}{2} \iiint \mathbf{E} \cdot \mathbf{D}\, dv \tag{5.19}$$

and the subsequent differentiation of U in a virtual displacement to determine the gradient. This is a very roundabout sort of process. In principle a knowledge of the field should be sufficient to determine the force without having to resort to arbitrary displacements, since the force between charges depends on their position and not in changes on position. The same thought was expressed in § 5.2. Since all interaction can be described in terms of the electric field and since there is no interaction without a field, we must be able to find the interaction force by mapping the field on any arbitrary surface enclosing a body. This is a direct consequence of the inverse square law which associates every element of charge with an indestructible tube of flux. Gauss's theorem [eqns. (2.13) and (2.14)] has already taught us that the total charge inside a closed volume can be obtained by adding the tubes of flux emerging from the volume. Thus the field gives information about the quantity of charge as well as the force per unit charge. We are now feeling our way to a very important result which is the consequence of the information given in the field map: the force on a body inside a closed volume must be calculable by observing the field on the enclosing surface. Referring to Fig. 5.9, we should be able to calculate the force on the body P by knowing the field on any such surface as that indicated by the dotted line. We propose to

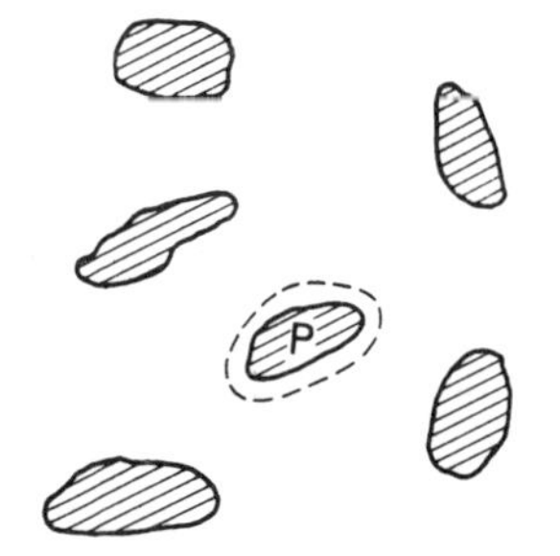

FIG. 5.9. Force on charged body.

call the force per unit area of such a surface the *electric stress*. We do not say that the force acts on the surface, but on the body enclosed by the surface. Moreover our argument shows that we can choose any arbitrary surface surrounding the body. Hence the field stress will not be unique, only its summation over a complete surface will be unique. Since this is the only thing we can measure, nothing more can be expected.

Let us now define an electric field stress in the same manner as the mechanical stress is defined.

Consider a tube of flux as in Fig. 5.10. To find the stress in the direction of the flux let us make a cut as shown in the figure and remove the flux

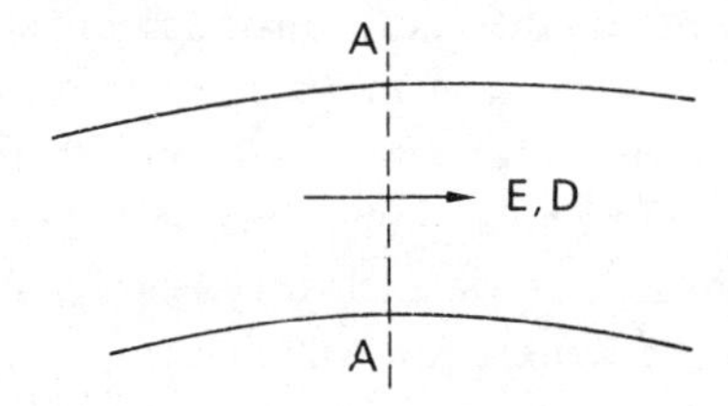

FIG. 5.10. Transverse cut in tube of flux.

on one side of the cut. In a mechanical system we can cut the material with a saw and carry it away, but what sort of saw will cut an electric field? Such a field can only be cut (or terminated) by a layer of surface charge as shown in Fig. 5.11. The charge density required is given by Gauss's theorem as $q = D$. The stress can now be obtained from the force acting on the charge-layer to keep it in equilibrium. Let us calculate this stress.

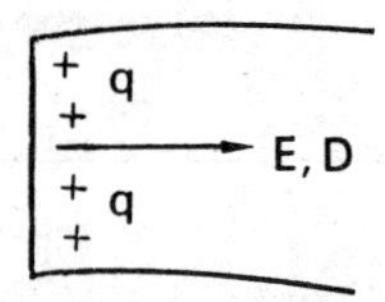

FIG. 5.11. Removal of flux by surface charge.

Since the field is terminated at the surface we can think of a conducting body with q as a surface charge. The flux density D outside the conductor is the total flux density due to the local q and all other charges. The flux density due to q only will by symmetry be $D/2$ (Fig. 5.12), hence the

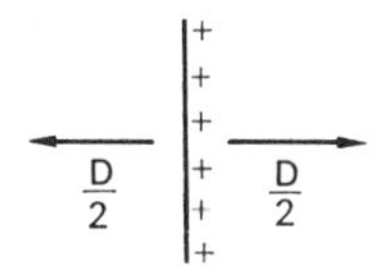

FIG. 5.12. Field of surface charge.

remaining $D/2$ must be due to the other charges. The force on q will be the interaction force between the local and distant charges. It can be obtained by multiplying the electric field strength of the distant charges and multiplying it by the local charge. Thus the force per unit area, or the stress, is given by

$$\sigma = \frac{E}{2}q = \frac{ED}{2}, \tag{5.63}$$

where σ, E, and D are all normal to the surface (Fig. 5.13). Reverting to our consideration of field stresses we note that a tensile stress of $ED/2$ acts along every tube of electric flux. Since we have been considering flux in free space, we can write the stress as $\varepsilon_0 E^2/2$. Figure 5.14 illustrates

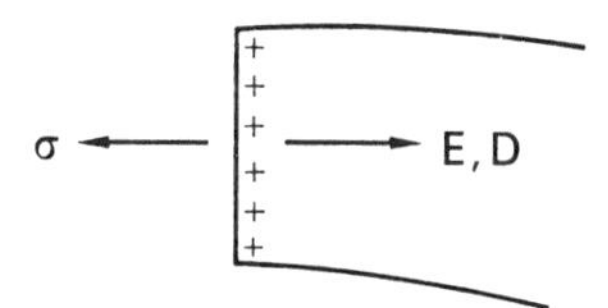

FIG. 5.13. Field stress.

how the force between two unlike charges can be calculated by considering the field stress on the mid-plane AA between the charges. This fits in with the discussion of the method of images in § 4.3.4, only in this case the surface charge replaces the real point charge. Once again it is clear that to find the force on a charge it is sufficient to know the field on a surface enclosing it. We do not need to know the field at the charge itself.

It is now necessary to examine whether there is any stress perpendicular to a tube of flux. Again we remove part of the field by cutting (Fig. 5.15). The normal component of flux density can be removed by a surface charge, but how can we terminate a tangential electric field

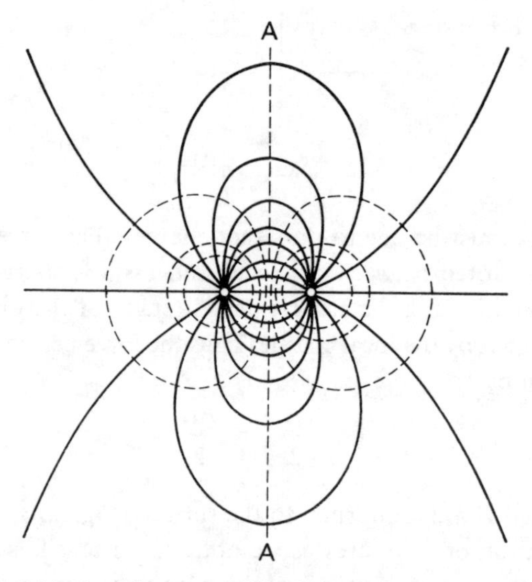

FIG. 5.14. Force between unlike charges of equal magnitude.

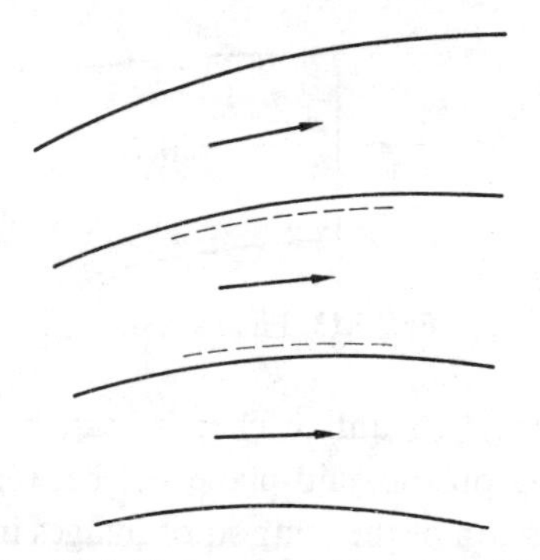

FIG. 5.15. Tangential cut in tube of flux.

on a surface? In magnetostatics we know that a tangential magnetic field can be removed by a surface electric current. By symmetry in electrostatics we should need a magnetic current, but such a thing does not exist in nature. However, this need not deter us. Stress after all is a concept and not a thing that can be measured. We can invent the magnetic current as a conceptual saw with which to cut the electric field.

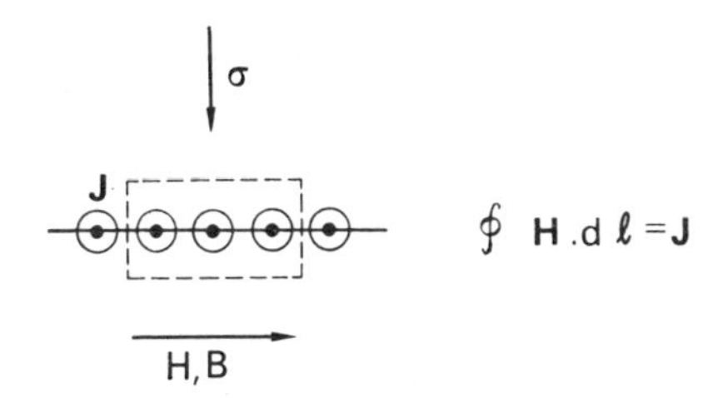

FIG. 5.16. Termination of tangential magnetic field.

In the magnetic field **H, B** an electric current of line density **J** A/m will terminate a tangential field **H** A/m if **J** = **H** (see Fig. 5.16). The force per unit area on such a current sheet will be perpendicular to **J** and **B** and will be given by

$$\sigma = \tfrac{1}{2}\,\mathbf{J} \times \mathbf{B} = \tfrac{1}{2}\,\mathbf{H} \cdot \mathbf{B}. \tag{5.64}$$

Figure 5.17 shows the parallel electric case where **J***† is the line density of magnetic current and **E, D** the tangential electric field. By symmetry,

$$\sigma = -\tfrac{1}{2}\,\mathbf{J}^* \times \mathbf{D} = \tfrac{1}{2}\,\mathbf{E} \cdot \mathbf{D}. \tag{5.65}$$

The factor $\frac{1}{2}$ in these expressions arises for the same reason as that in eqn. (5.63).

We have now established the fact that tubes of flux push against each other with a pressure of $ED/2$. Since this pressure acts in all directions perpendicular to a tube it is sometimes called a *hydrostatic* pressure.

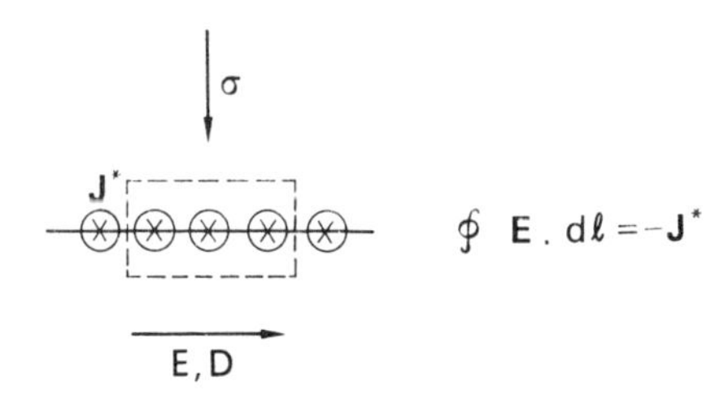

FIG. 5.17. Termination of tangential electric field.

† The sign of **J*** is chosen opposite to that of **J** in order to match the change of sign between the m.m.f. equation $\oint H \,.\, dl = I + d\Psi/dt$ and the e.m.f. equation $\oint E \,.\, dl = -\,d\phi/dt$.

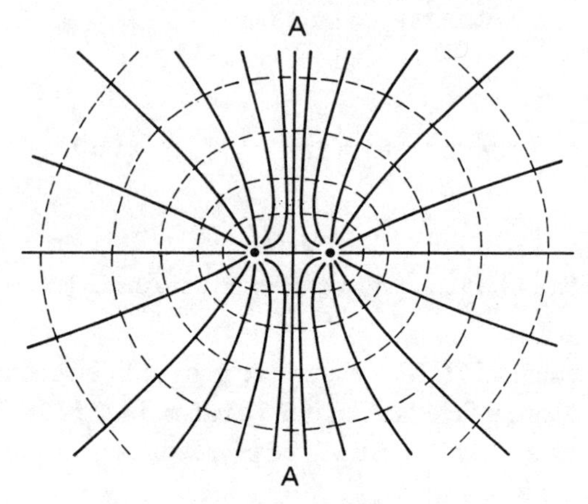

FIG. 5.18. Force between like charges of equal magnitude.

Figure 5.18 illustrates the effect by considering the force between two like charges, which can be calculated by the hydrostatic pressure on the mid-plane *AA*.

Figures 5.14 and 5.18 were chosen to illustrate the separate effects of the tensile stress and the hydrostatic pressure. In general both effects will occur together. The complete stress system comprises a tensile stress $ED/2$ along the direction of flux and a compressive stress $ED/2$ at right angles to the flux. Readers with experience in the theory of elasticity will have noticed that these are the "principal" stresses of the system. At a surface inclined at some arbitrary angle to the direction of the flux there

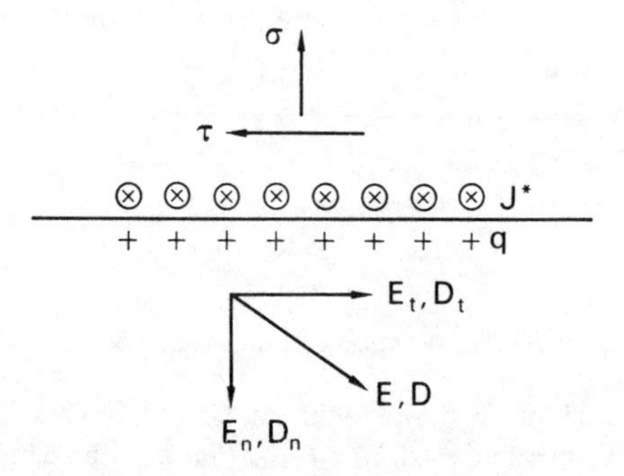

FIG. 5.19. Tensile and shear stress.

will also be a shear stress. Its magnitude is easily obtained by resolving the flux density along the plane and at right angles to it. Consider the two-dimensional system of Fig. 5.19. To terminate the field we need layers of q and J^* such that

$$q = D_n, \quad J^* = E_t, \tag{5.66}$$

where n and t signify the normal and tangential direction respectively. The tensile stress will then be

$$\sigma = \frac{E_n}{2} q - \frac{D_t}{2} J^* = \frac{E_n D_n}{2} - \frac{E_t D_t}{2}, \tag{5.67}$$

and the shear stress will be

$$\tau = \frac{E_t}{2} q + \frac{D_n}{2} J^* = \frac{E_t D_n}{2} + \frac{E_t D_n}{2} = E_t D_n. \tag{5.68}$$

The stress system which we have now derived provides a powerful tool for the calculation of forces as an alternative to the ordinary method of finding the force on charges. Consider the problem of the simple electrostatic voltmeter in Fig. 5.20. We wish to determine the force on the mov-

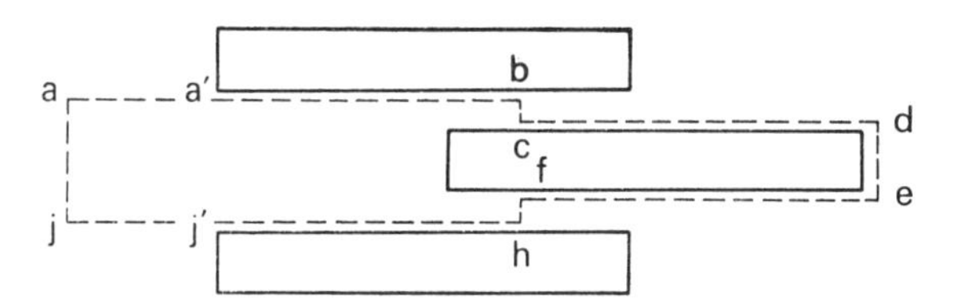

FIG. 5.20. Simple electrostatic voltmeter.

ing vane AB. This force can be calculated by finding the field stresses over any arbitrary surface enclosing AB. A possible surface is shown by the dotted line. Since both the stationary and the moving vanes are conductors, the field on their surface will be perpendicular to the surface. There will therefore be no shear stress on such surfaces and no lateral force. The only lateral force will be on the surface bc and fh in a region where the field can be assumed to cross the gap in a straight line. Let the field be E in this region. The normal stress will be [by use of eqn. (5.67)] $\sigma = -\varepsilon_0 E^2/2$. If the total gap length is $2g$ and we assume unit depth into the paper, the inward force will be $F = \varepsilon_0 E^2 g$. If the voltage $V = Eg$, we have $F = \varepsilon_0 V^2/g$. Suppose $V = 1000$ volts and $g = 1$ mm we have $F = 8{\cdot}8 \times 10^{-3}$ Nm. This calculation assumes that the surfaces *a′ajj′* and *de* are outside the field. If greater accuracy is required, the field all around the surface can be determined and the force can be obtained by a numerical integration of the surface stress.

We have now achieved our aim of finding a method for the calculation of force by a system of stresses on a surface completely surrounding a body. There has been the tacit assumption that the enclosing surface is in free space. Can the method be extended to include material bodies?

In principle the method depends only on a knowledge of the electric field strength **E** and flux density **D** at every point on a surface. The presence of matter by itself is irrelevant. However, there must be no charges on the surface, otherwise **E** and **D** will be discontinuous there. What about the field in a polarized material? Here we are back amongst the difficulties of § 4.2. **E** and **D** are not true point-functions in such a material. Nevertheless **D**, which now is $\varepsilon_r \varepsilon_0 \mathbf{E}$, does correspond to the field in a cut across a tube of flux and **E** to the tangential field in a cut along a tube of flux. These are the cuts which have to be made to define the field stress. Hence **D** and **E** can be used as before. Of course the field stress must not be taken as a unique local stress in the material. That would lead us to postulate a unique volume force, which we have seen in § 5.3 to be impossible. Other "mechanical" stresses will also be acting in a material and it will not be possible to isolate an "electrical" component of stress. Lastly we note that eqns. (5.63) and (5.65) have been derived on the assumption that **E** and **D** are in the same direction. If this is not so, these equations must be derived anew from first principles making use of the surface source distributions q and $\mathbf{J}^*$. Used with care the method of field stresses can be applied to material bodies just as well as to free space.

Summary

The interaction forces between electric charges can be used to store potential energy or to exert mechanical pressure. The potential energy arises from the work done in assembling the charges. Since the interaction of charges can be described in terms of tubes of flux it is also possible to express the potential energy in terms of this field. Forces and torques can be obtained from the energy expressions by differentiation.

It is difficult to obtain the force distribution inside bodies, because "mechanical" forces cannot be uniquely separated from "electrical" forces. Only the total force can be measured.

Probably the best way of calculating forces is by means of a system of field stresses, in which the forces are associated with the tubes of flux.

Exercises

5.1. A parallel-plate capacitor has plates of area 0·04 m^2 at a distance 0·5 cm apart from each other. Estimate (a) the capacitance, and when the capacitor is charged to a p.d. of 1000 V estimate, (b) the charge, (c) the stored energy, and (d) the force of attraction between the plates. [Ans.: (a) 70·8 pF; (b) $7{\cdot}08\times10^{-8}$ C; (c) $3{\cdot}54\times10^{-5}$ J; (d) $7{\cdot}08\times10^{-3}$ N.]

5.2. Compare the amount of energy stored in a lead–acid cell giving 2 V and 100 A hr with that in a 1 μF capacitor charged to 10 kV. Explain why capacitors are the most generally used energy storage devices used in electronic circuits. [Ans.: 14,400 : 1.]

5.3. Derive expressions for the energy in an electrostatic system (a) in terms of charges and potentials, and (b) in terms of electric field strength and electric flux density. Discuss the validity and limitations of these expressions. A capacitor of two parallel plates of area A distant x apart is connected to a source of constant p.d. V. The plates are allowed to approach each other slowly until the distance is $x/3$. Estimate the energy supplied by the source. [Ans.: $2V^2\varepsilon A/x$, half of which is stored in the electric field.]

5.4. Explain carefully what is meant by the stress in an electrostatic field. Show that forces on bodies can be correctly predicted by assuming a stress system consisting of a direct tensile stress $\frac{1}{2}\varepsilon_0\varepsilon_r(E_n^2 - E_t^2)$ and a shear stress $\varepsilon_0\varepsilon_r E_n E_t$, where n and t indicate the normal and tangential directions. Show also that the principal stresses are a tensile stress of $\frac{1}{2}\varepsilon_0\varepsilon_r E^2$ along tubes of flux and a "hydrostatic" pressure of the same amount perpendicular to tubes of flux.

5.5. Figure 5.21 shows a long cylindrical conductor of radius a partially inside a cylindrical conducting tube of inner radius b. A potential difference V is applied between the conductors. Use the field stress to estimate the forces which must be applied to the conductors to keep them in equilibrium. Does the force depend on the sign of the p.d.? [Ans.: Axial forces $\pi\varepsilon_0 V^2/\ln(b/a)$; no.]

5.6. Examine the change of electric stress which occurs in an electric field across a boundary between a dielectric material and free space. Show that equilibrium requires the action of other (non-electric) stresses.

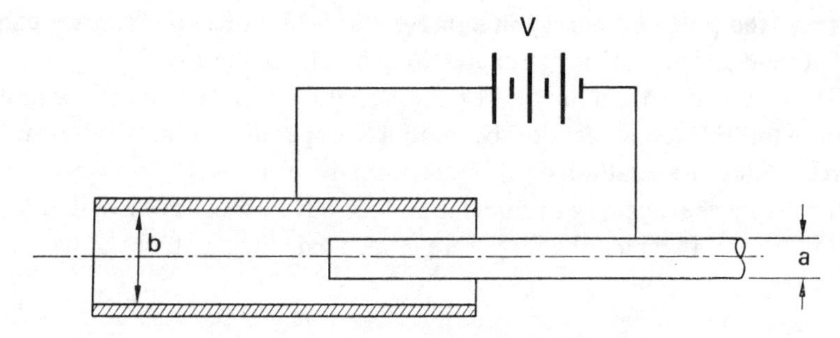

FIG. 5.21. Force between conducting rod and tube.

5.7. Show that the force due to electrostriction is zero for an isolated body and also that it is zero for a part of a body enclosed by a surface on which there is no electric field. Show that in the absence of electrostrictive effects pressure differences in a dielectric liquid can be calculated from the expression

$$p_2 - p_1 = \frac{\varepsilon_0}{2} \int_2^1 E^2 \nabla \varepsilon_r \, . \, d\mathbf{l}.$$

5.8. Figure 5.22 shows a capacitor, the plates of which are dipping into a dielectric liquid of relative permittivity ε_r and density g. Show that the height of the liquid be-

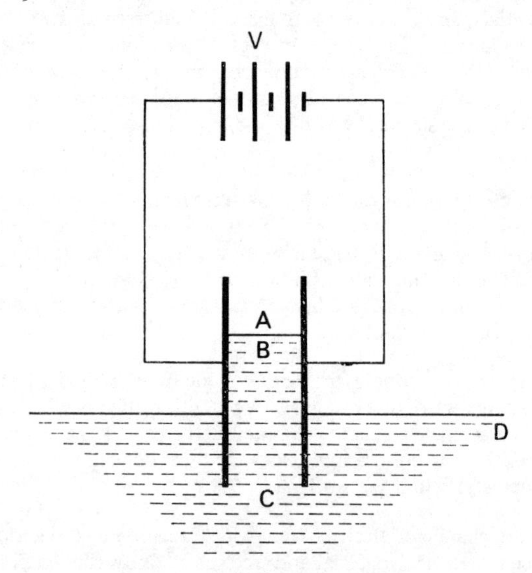

FIG. 5.22. Capacitor in liquid dielectric.

tween the plates relative to the free surface is given by

$$h = \frac{\varepsilon_0 (\varepsilon_r - 1)}{2gG} (E_x^2 + \varepsilon_r E_y^2),$$

where G is the acceleration due to gravity and E_x, E_y is the total electric field in the liquid between the plates just below the surface. [*Hint:* Use the result of Exercise 5.7.]

5.9. Assuming that the electric field is given by $E_x = E$, $E_y = 0$ at the surface of the liquid between the plates (Fig. 5.22) show that there is a pressure rise at the surface given by

$$p_B - p_A = \frac{\varepsilon_0}{2} E^2 \left[g \frac{d\varepsilon_r}{dg} - (\varepsilon_r - 1) \right].$$

5.10. Show that if $\mathbf{f}$ is the electrical force per unit volume and $\mathbf{G}$ is the acceleration due to gravity, the pressure p in the liquid dielectric in Fig. 5.22 is given by the equation

$$\nabla p = \mathbf{f} + g\mathbf{G}.$$

Sketch the pressure distribution between A and D.

5.11. A conducting sphere of mass M floats on oil of relative permittivity ε_r. When the sphere is uncharged, one-quarter of it is immersed in the oil. The sphere is then given a charge q and this causes it to sink until half of it is immersed. Estimate the total charge on the sphere. [Ans.: $\sqrt{[32\pi a^2 \varepsilon_0 MG (\varepsilon_r - 1)]}$, where G is the acceleration due to gravity.]

CHAPTER 6

MAGNETOSTATICS I

6.1. The Sources of a Steady Magnetic Field

We have already touched on this topic in § 4 of the first chapter and readers may like to refer to that section. We remind ourselves that Ampère discovered the equivalence of current loops and magnetic dipoles, from which he inferred that magnetic dipoles consist of molecular current loops. It is a sound scientific principle to reduce the number of concepts to the minimum number required. Philosophers attribute this principle to William of Ockham† and call it "Ockham's razor". If dipoles and current loops are identical we can dispense with one of these notions. Since the notion of current is linked to that of charge, it seems best to retain current loops and dismiss magnetic dipoles.

There are, however, two objections to this seeming simplification. The first of these will seem invalid to philosophers but may readily be accepted by pragmatic engineers. Dipoles, or rather poles, are a convenient tool in calculation. Even if they had no physical significance, they would be "invented" by somebody clever enough to realize that the inverse square law could be applied to magnetic poles and that all the results of electrostatics could then at once be translated into magnetostatics.

But there is an even stronger objection to the indiscriminate use of Ockham's razor in this instance. This second objection is based on the physical existence of magnetic materials. There is no room in this book for a detailed discussion,‡ but a few comments may not be out of place. Ampère's guess of molecular currents is very close to the truth but does not give the complete story. According to the modern view, electrons do

† An English Franciscan friar who lived in the first half of the fourteenth century and laid the foundations of the study of dynamics.

‡ The reader may like to refer to F. Brailsford, *Physical Principles of Magnetism*, Van Nostrand, 1966.

not only move in orbits around the atomic nucleus, they also have the property of spin about their own axes. In other words, electrons possess an intrinsic dipole moment as well as electric charge. Just when we had hoped that modern atomic theory would banish magnetic dipoles, we find that a magnetic moment is built into the very nature of matter.

In most materials the magnetic moments inside atoms and molecules cancel each other. The only external magnetic effect of such materials is a weak diamagnetic effect. If a diamagnetic material is subjected to an external magnetic field, the orbital and spin motions alter in such a manner as to reduce the field. This is the behaviour one would expect from small current loops in accordance with Lenz's law. The polarization, or magnetic susceptibility, of diamagnetic materials is negative and is also very small.

If all materials were diamagnetic there would be no strong reason for using dipoles in the description of their behaviour. However, the diamagnetic effect is overwhelmed in paramagnetic substances by the opposite sort of behaviour. In such materials the atoms or molecules have a magnetic moment. If they are subjected to a magnetic field, the elementary magnets turn in a direction to strengthen the field. The magnetic susceptibility is positive.

Outstanding among paramagnetic materials are the *ferromagnetics* such as iron, nickel, and cobalt, and the *ferrimagnetics* such as the ferrites, which consist of various iron oxides. In these materials the alignment forces between the elementary magnets are sufficient at normal temperatures to overcome the disordering forces caused by thermal motion. These materials can therefore produce large magnetic fields without any associated ohmic losses. In terms of current loops they could be regarded as super-conductors at room temperatures. The behaviour is, however, a complicated one, and the internal structure of these substances is best explained in terms of domains or small regions of magnetic alignment. These domains are to all intents and purposes magnetic dipoles. One word of caution is needed because mathematical dipoles have infinitesimal length. We met this difficulty before in talking about point charges. In that context we had to limit the application of the theory to distances larger than 10^{-7} m. Domains are often 10^{-6} m wide and in single crystals may be as large as 10^{-3} m. Therefore domains must in general be regarded as assemblies of dipoles rather than individual dipoles. This is not a serious limitation, and the modern theory of magnetism relies heavily on the idea of magnetic dipoles. The physical theory of magnetism reinforces the mathematical argument that dipoles are more easily usable than current loops.

This means that it is convenient to postulate two types of sources of the magnetic field: currents of electric charge and magnetic dipoles.

6.2. Some Useful Source Distributions

6.2.1. The Magnetic Field of a Dipole

This can be derived by inspection from the field of an electric dipole (§ 4.1.2). If V^* is the magnetic potential and $\mathbf{P}^*$ the dipole moment,

$$V^* = \frac{1}{4\pi\mu_0}\frac{\mathbf{P}^* . \hat{\mathbf{r}}}{r^2} \tag{6.1}$$

and

$$\mathbf{H} = -\nabla V^* = \frac{1}{4\pi\mu_0}\nabla\left[\mathbf{P}^* . \nabla\left(\frac{1}{r}\right)\right]. \tag{6.2}$$

6.2.2. The Magnetic Field of a Small Current Loop

Ampère's experimental law of equivalence between magnetic dipoles and current loops was discussed in § 2.2, eqns. (2.28) and (2.29). If the small area of the loop is s, the current I, and if the direction of the vector $\mathbf{s}$ is defined as the third axis of a right-handed set determined by the circulation of the current,

$$\mu_0 I\mathbf{s} = \mathbf{P}^*, \tag{6.3}$$

where the dimensional constant μ_0 serves to define dipole moment in terms of electric current. (In SI units $\mu_0 = 4\pi \times 10^{-7}$ H/m.)

The potential of a current loop is, therefore,

$$V^* = \frac{I\mathbf{s} . \hat{\mathbf{r}}}{4\pi r^2}. \tag{6.4}$$

It should be noted that this is the potential at a distance r from a small current loop and that it cannot be used to find the field *inside* a current loop. Ampère's equivalence applies only at a distance. The internal structure of dipoles and current loops is not the same.

6.2.3. *The Field of a Magnetic Shell*

A double layer of magnetic pole strength on a surface is called a magnetic shell. If $\mathbf{p}^*$ the dipole moment per unit area is constant and normal to the surface, we can use eqn. (4.17),

$$V^* = \frac{|\mathbf{p}^*|}{4\pi\mu_0}\iint\frac{\hat{\mathbf{r}}\cdot d\mathbf{s}'}{r^2} = \frac{|\mathbf{p}^*|}{4\pi\mu_0}\Omega, \tag{6.5}$$

where Ω is the solid angle subtended by the magnetic shell at the field point at which the potential is measured.

The work done in passing a unit magnetic pole through the shell is p^*/μ_0 and this work is recovered by carrying the unit pole around the outside of the shell (Fig. 6.1). This result can be deduced either by remember-

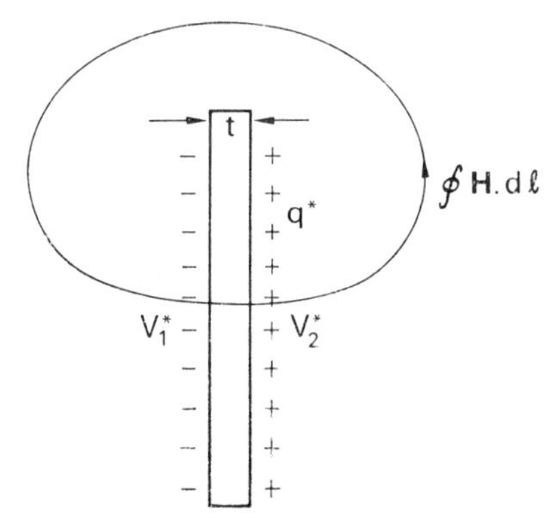

FIG. 6.1. Work done in carrying unit pole around a magnetic shell.

ing that the solid angle changes by 4π in going through a complete loop, or more directly by finding the potential difference across the double layer of pole strength q^*, which can be treated as the magnetic equivalent of a parallel plate capacitor. We have $p^* = q^*t$, where t is the thickness,

$H = q^*/\mu_0$, and $V_1^* - V_2^* = Ht = p^*/\mu_0$. Notice that the potential V^* is unaffected by the shape of the magnetic shell. V^* depends on the solid angle and this depends only on the perimeter of the shell.

6.2.4. The Magnetic Field of a Large Current Loop

Each element of a magnetic shell can be replaced by a small current loop. If all the currents are combined they will cancel everywhere except

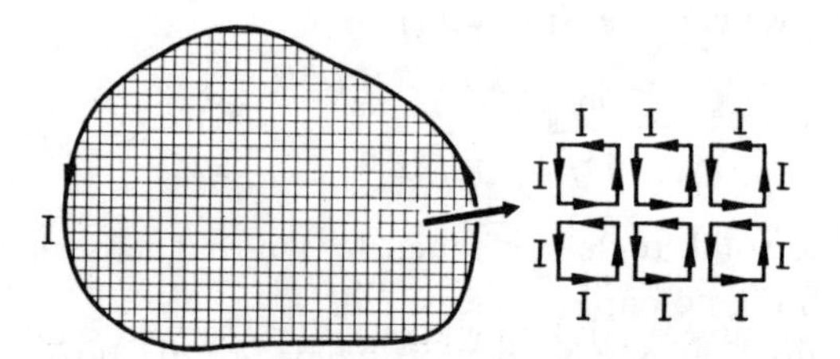

FIG. 6.2. Magnetic shell.

at the perimeter (Fig. 6.2). Thus the field of a *large* current loop is the same as that of a magnetic shell of the same perimeter. Using eqn. (6.3) and dividing both sides by the small area s we obtain

$$\mu_0 I = |\mathbf{p}^*|, \tag{6.6}$$

it being remembered that $\mathbf{p}^*$ is the dipole moment per unit area. We therefore have, by eqn. (6.5),

$$V^* = \frac{I\Omega}{4\pi}. \tag{6.7}$$

However, there is a difference between the field of a magnetic shell and a current loop, because the work done in moving a unit pole around a closed loop linking the current is not zero. An electromotive force would be induced by such a movement in the coil carrying the current, and the field is no longer conservative. Since Ω changes by 4π on each revolution, the potential will change by I every time a closed path is traced out. The potential is therefore multi-valued at every point, and instead of eqn. (6.7) we can write

$$V^* = \frac{I\Omega}{4\pi} \pm nI, \tag{6.8}$$

where n is an integer. Since only potential gradient or potential difference can be observed experimentally this does not affect the measurement of the magnetic field. In terms of magnetic field strength the difference between §§ 6.2.3 and 6.2.4 can be expressed as follows:

In the case of a magnetic shell,

$$\oint \mathbf{H} \cdot d\mathbf{l} = 0, \tag{2.31}$$

and in the case of a current loop,

$$\oint \mathbf{H} \cdot d\mathbf{l} = I \tag{2.33}$$

if the path links the current. Where there is no linkage between path and current there is no means of knowing whether the field is due to a shell or a current. For all such paths eqn. (2.31) holds and the potential is single-valued even if it is due to a current.

6.2.5. *The Magnetic Field of a Volume Distribution of Dipoles*

We can make use of the results of § 4.1.7. Equation (4.23) becomes

$$P_n^* = q^*, \tag{6.9}$$

$$-\nabla \cdot \mathbf{P}^* = \varrho^*, \tag{6.10}$$

where P_n^* is the component of dipole moment normal to the surface enclosing the volume.

Equation (4.24) becomes

$$V^* = \frac{1}{4\pi\mu_0}\left[\oiint \frac{q^*}{r}\, ds' + \iiint \frac{\varrho^*}{r}\, dv'\right]. \tag{6.11}$$

This means that the *dipole* distribution can be replaced by a surface distribution, and a volume distribution, of *poles*. We saw in § 4.2.1 that in dielectric materials having constant permittivity ε_r there is no volume distribution of charge. The electrostatic field of such materials can therefore be described solely in terms of surface charge. In magnetic materials the permeability μ_r is not constant and we cannot therefore put ϱ^* equal to zero. In terms of domains we note that the presence of domain boundaries in the material implies the presence of pole strength in the material.

But although a volume distribution of polarity exists, it is overshadowed by the dominant effect of the surface poles. Equations (6.9) and (6.10) show that the pole strength arises from the divergence of dipole strength. Clearly this divergence is strongest where iron meets air at the surface of the material. Inside the iron there will always be adjacent iron, unless the material is full of slag and blow-holes. It is to be expected that the polarity associated with ϱ^* is an order of magnitude less than that associated with q^* and for the purpose of calculation ϱ^* can be neglected. This knowledge is tremendously useful in determining the magnetic field of transformers and machines, because the iron can be replaced by a layer of poles on its surface.

6.2.6. *The Magnetic Field of a Surface Layer of Poles*

The discussion of the last section has led us to replace volume distributions of dipoles by surface layers of poles. We can make use of § 4.1.5. If the pole density is q^* Wb/m², the magnetic field strength in A/m is normal to the surface and is given by

$$\mathbf{H} = \frac{q^*}{2\mu_0}\,\mathbf{z}. \tag{6.12}$$

The potential has a finite part

$$V^* = -\frac{q^* z}{2\mu_0}. \tag{6.13}$$

The idea of a pole layer of infinite extent is a very useful one if one is calculating the field near an iron surface. (See § 6.4.2 and compare the parallel electric case in § 4.1.5.) Figure 6.3 shows the variation of magnetic field strength and potential near such a layer.

6.2.7. *The Magnetic Field of a Current Element*

The source distributions which we have so far considered have had a firm physical foundation. A current element is a more shadowy concept because steady current can flow only in closed circuits. Heaviside faced

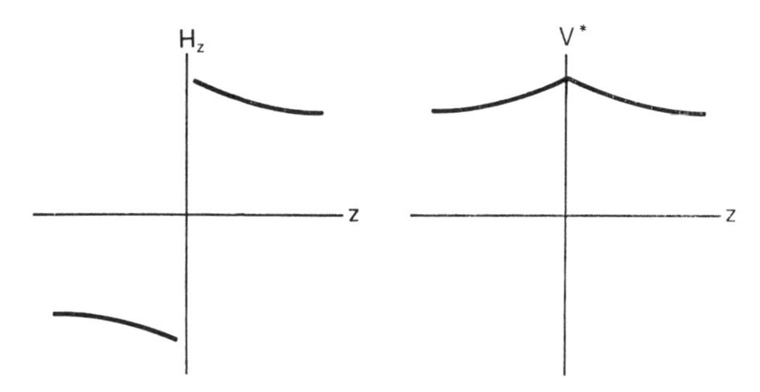

FIG. 6.3. Magnetic field and potential near surface layer of pole strength.

this problem by devising his "rational" current element immersed in a conducting liquid. This idea was briefly discussed in § 2.2. The importance of the current element is as a tool in calculation. Care must be taken to apply it only to closed circuits, otherwise contradictory results may appear. For instance the forces between current elements do not obey Newton's third law of action and reaction, although the forces between closed current loops do.

The magnetic field of a Heaviside current element was derived in § 2.2 and is given by

$$\mathbf{H} = I\frac{\delta\mathbf{l} \times \hat{\mathbf{r}}}{4\pi r^2}. \tag{2.37}$$

It is not possible to associate a scalar potential with this field, which is non-conservative. Current flows everywhere because of the presence of the conducting fluid. Hence curl $\mathbf{H} = \mathbf{J}$ at the field point and we cannot write $\mathbf{H}$ as the gradient of a potential.

6.2.8. *The Magnetic Field of a Long Current Filament*

It is instructive to obtain this by integration using eqn. (2.37). Consider Fig. 6.4. By symmetry the only component at the field point P is

$$H_\theta = \frac{I}{4\pi}\int_{-\infty}^{+\infty}\frac{\cos\psi}{\varrho^2}\,dz, \tag{6.14}$$

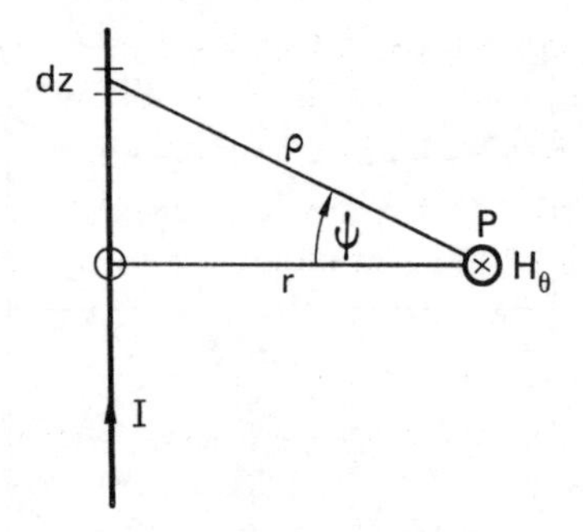

FIG. 6.4. Magnetic field of a long current filament.

where

$$\left.\begin{aligned} z &= r\tan\psi, \\ \delta z &= r\sec^2\psi\,\delta\psi, \\ \varrho &= r\sec\psi; \end{aligned}\right\} \tag{6.15}$$

hence
$$H_\theta = \frac{I}{2\pi r}. \tag{6.16}$$

Since the current stretches from $-\infty$ to $+\infty$, the filament forms a closed circuit and the field of the *complete* current filament can be derived from a scalar potential

$$\mathbf{H} = -\nabla V^*. \tag{6.17}$$

In cylindrical polar coordinates,

$$H_\theta = -\frac{1}{r}\frac{\partial V^*}{\partial\theta}, \tag{6.18}$$

$$V^* = -\frac{I}{2\pi}\theta + \text{constant}. \tag{6.19}$$

V^* is multi-valued. Every time a unit pole encircles the filament, θ changes by 2π and V^* by I. This is, of course, as it should be, since for a closed path $\oint \mathbf{H}\,.\,d\mathbf{l} = I$. This is true for every path encircling the filament. For any other path $\oint \mathbf{H}\,.\,d\mathbf{l} = 0$ and the potential is single-valued. The angle θ changes by 2π only if the path encircles the current. If it is important to have a single-valued potential, this can be ensured by drawing a boundary as in Fig. 6.5, which shows the filament in plan view. No path can then encircle the current. The boundary is equivalent to a magnetic shell which cannot be crossed. The field of the current is now equivalent to the field of a shell with its other edge a long way away.

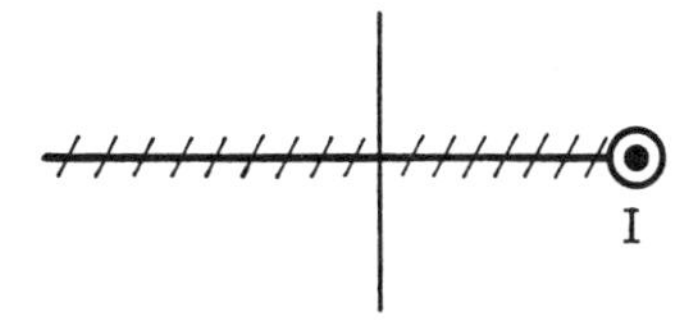

FIG. 6.5. Single valued potential for current filament.

It will have been noticed by the reader that eqn. (6.16) could have been derived more quickly by choosing a path along which H is known to be constant.

$$\oint \mathbf{H} \,.\, d\mathbf{l} = H \times 2\pi r = I. \tag{6.20}$$

Whenever H is constant around a path the integral $\oint \mathbf{H} \,.\, d\mathbf{l}$ becomes $H \times l$ and we can find H directly. But often H is not constant and then the formula for the current element, eqn. (2.37), is a useful starting point.

6.2.9. The Magnetic Field of a Surface Layer of Current

Figure 6.6 illustrates the flow of current in a large surface. Such an arrangement is called a current sheet. The sheet is infinitely wide in the x-direction and infinitely long in the z-direction. It is very thin in the y-direction. The current line density is specified as J_z A/m. By considering

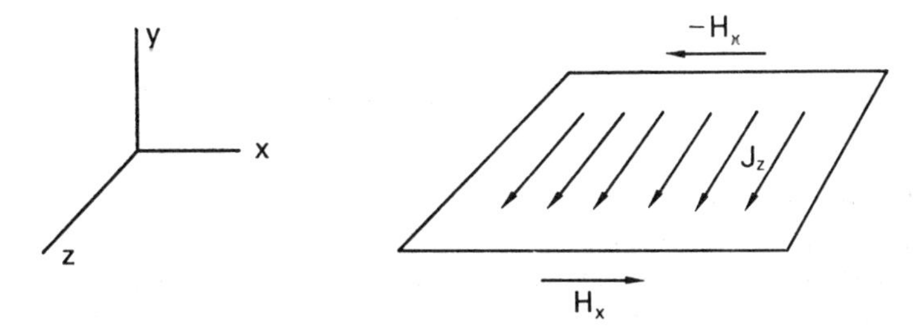

FIG. 6.6. Magnetic field of a current sheet.

a typical symmetrical path as shown in Fig. 6.6 we have below the sheet $H_x = J/2$ and above it $H_x = -J/2$. $H_z = 0$ because there can be no magnetic field parallel to a current, and $H_y = 0$ by symmetry. Hence the magnetic field of a sheet of constant line density of current is parallel to the sheet and has a discontinuity of J at the sheet. We shall need this result in the discussion of boundary conditions later in this chapter.

6.3. The Use of the Vector Potential in Calculating Magnetic Fields

As long as we deal with dipoles as the sources of magnetic fields, we are dealing with conservative fields. In mathematical terms $\oint \mathbf{H} . d\mathbf{l} = 0$ for any path whatsoever and therefore **H** can be derived from a single-valued potential V^* by the relationship $\mathbf{H} = -\nabla V^*$. Such an approach enables us to use the results of electrostatics by substituting **H** for **E** and V^* for V. Section 6.2 provides ample evidence of the power of this method.

However this method breaks down when we deal with the magnetic field of currents. In such cases we know that for steady currents $\oint \mathbf{H} . d\mathbf{l} = I$. The field is not conservative. From an engineering point of view this is excellent news. Conservative fields are no use if energy is to be interchanged in cyclic processes. The useful application of magnetic fields in motors and generators is only possible because magnetic fields are non-conservative. So as engineers we can rejoice in this fact, although as people who wish to calculate fields in the simplest possible manner we deplore it.

The scalar potential V^* is too useful a tool to be easily relinquished. We can regard Ampère's invention of the magnetic shell as a sort of rear-guard action to save the scalar potential. In replacing a current by a magnetic shell we convert the non-conservative field of the current into the conservative field of the shell. The price we pay is seen in § 6.2.8 and Fig. 6.5 where we had to insert a boundary in the field, which must not be crossed. Since the boundary does not affect the calculation, it is a price worth paying.

The magnetic shell is a great help as long as we are not interested in a region where current is actually flowing. If we are *immersed* in a current we cannot any longer pretend that there is no current, but that there is a dipole layer in another place. The scalar potential is then no use, because we are in a "curl" field. We must use the vector potential, which was discussed in § 3.2.5.

We have

$$\oint \mathbf{H} \,.\, d\mathbf{l} = I \tag{3.26}$$

and at a point,

$$\text{curl } \mathbf{H} = \mathbf{J}\,, \tag{3.32}$$

where $\mathbf{J}$ is the current density in A/m².

The magnetic field $\mathbf{H}$ can be derived from a vector potential $\mathbf{A}$ by the relationship

$$\mu_0 \mathbf{H} = \text{curl } \mathbf{A}\,. \tag{3.39}$$

This, with the additional specification div $\mathbf{A} = 0$, leads to the expression

$$\mathbf{A} = \frac{\mu_0}{4\pi} \iiint \frac{\mathbf{J}}{r}\, dv' \tag{3.44}$$

as was shown in § 3.2.5, to which the reader may wish to refer. The sequence for calculation therefore proceeds from the sources $\mathbf{J}$ to the vector potential $\mathbf{A}$ and from $\mathbf{A}$ to $\mathbf{H}$, whereas we previously proceeded from dipole moment $\mathbf{P}^*$ to V^* and then to $\mathbf{H}$. Instead of a single scalar V^* we now have the three components of the vector $\mathbf{A}$. As an example consider the magnetic field of a current element which was discussed in §§ 2.2 and 6.2.7. Refer to Fig. 6.7. Equation (3.44) states that $\mathbf{A}$ is par-

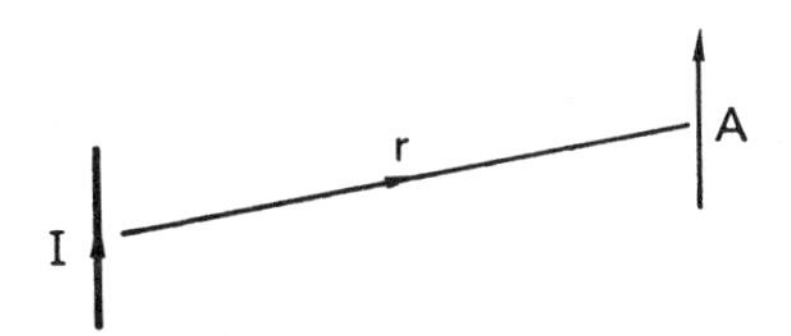

FIG. 6.7. Field of current element.

allel to $\mathbf{J}$. If we use that equation and remove two of the integral signs by replacing current density by current, we have

$$\mathbf{A} = \frac{\mu_0}{4\pi} \int \frac{I\,\delta\mathbf{l}'}{r}\,. \tag{6.21}$$

To find **H** we need the curl of **A**. We can make use of the vector identity

$$\text{curl}\ \phi\mathbf{v} = \phi\ \text{curl}\ \mathbf{v} + \text{grad}\ \phi \times \mathbf{v}. \tag{6.22}$$

Hence

$$\text{curl}\ \mathbf{A} = \frac{\mu_0 I}{4\pi}\ \text{curl}\ \frac{\delta\mathbf{l}'}{r} = \frac{\mu_0 I}{4\pi}\left[\frac{1}{r}\ \text{curl}\ \delta\mathbf{l}' + \text{grad}\left(\frac{1}{r}\right) \times \delta\mathbf{l}'\right]. \tag{6.23}$$

The first term in the bracket is zero because curl operates on the field coordinates and $\delta\mathbf{l}'$ is in source coordinates. Hence

$$\mu_0\mathbf{H} = \text{curl}\ \mathbf{A} = \frac{\mu_0 I}{4\pi}\ \frac{\delta\mathbf{l}' \times \hat{\mathbf{r}}}{r^2}, \tag{6.24}$$

and dropping the prime on $\delta\mathbf{l}'$ we have. as before,

$$\mathbf{H} = I\ \frac{\delta\mathbf{l} \times \hat{\mathbf{r}}}{4\pi r^2}. \tag{2.37}$$

The reader may wonder why the μ_0 was included in eqn. (3.44) only to be cancelled in eqn. (6.24). The reason for this is that when we come to deal with magnetic materials we shall want to define

$$\text{curl}\ \mathbf{A} = \mathbf{B} = \mu_0\mu_r\mathbf{H}. \tag{6.25}$$

Our discussion here is for the field in free space where $\mu_r = 1$. This can be treated as a special case of eqn. (6.25).

The vector potential provides a useful alternative description for magnetic flux and inductance. The flux can be written as

$$\Phi = \iint \mathbf{B} \,.\, d\mathbf{s}. \tag{2.25}$$

Using eqn. (6.25) we have

$$\Phi = \iint \text{curl}\ \mathbf{A} \,.\, d\mathbf{s}. \tag{6.26}$$

By Stokes's theorem [eqn. (3.34)] this can be transformed to

$$\Phi = \oint \mathbf{A} \,.\, d\mathbf{l}. \tag{6.27}$$

In eqn. (2.25) the flux is obtained by counting tubes of flux passing through a surface. In eqn. (6.27) the view is quite different. Here the flux is counted by walking around the perimeter and measuring the line-integral of the vector **A**.

Let us apply this view to the calculation of the mutual inductance between two coils as illustrated in Fig. 6.8. The mutual inductance is de-

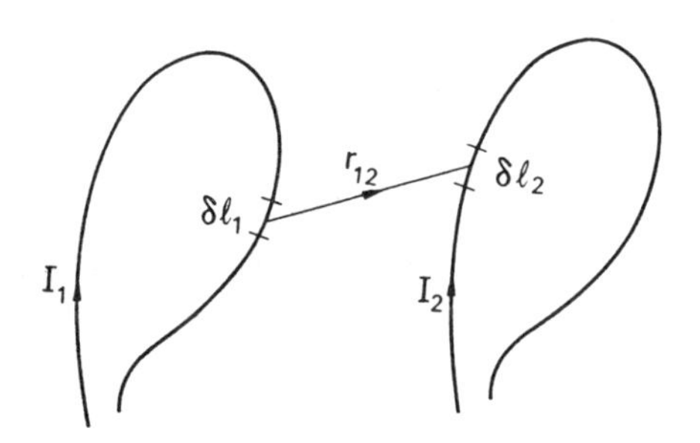

FIG. 6.8. Mutual inductance of two coils.

fined as the flux linked with coil 2 per unit of current in coil 1. Using eqn. (6.21) we can find the vector **A** at the perimeter of coil 2

$$\mathbf{A} = \frac{\mu_0}{4\pi} \oint I_1 \frac{d\mathbf{l}_1}{r_{12}}. \tag{6.28}$$

Using eqn. (6.27) the flux linked is

$$\Phi = \frac{\mu_0}{4\pi} \oint \oint I_1 \frac{d\mathbf{l}_1 \cdot d\mathbf{l}_2}{r_{12}} \tag{6.29}$$

and the mutual inductance

$$M = \frac{\Phi}{I} = \frac{\mu_0}{4\pi} \oint \oint \frac{d\mathbf{l}_1 \cdot d\mathbf{l}_2}{r_{12}}. \tag{6.30}$$

The symmetry of the integral provides a proof for the well-known statement $M_{12} = M_{21}$, i.e. the mutual inductance is the same whichever coil is thought of as the source of the flux. This statement is true for currents in air. The presence of iron provides additional sources for **A** [see eqn. (6.25)] and the integral of eqn. (6.30) may then be no longer symmetrical. The statement $M_{12} = M_{21}$ is not in general true when there is iron in the magnetic field.

We have calculated the mutual inductance eqn. (6.30) on the assumption that the coils can be treated as filaments of negligible thickness. Unfortunately it is not possible to obtain an expression for the self-inductance in this way, because the integral then has to contain the point $r_{12} = 0$ and this makes the inductance infinite. Here, as often in electromagnetism, we find that the mathematical difficulty has a physical cause. It is just not possible for a finite current to flow through an infinitely thin wire. The thinner the wire the more energy will have to be expended. In the formula for mutual inductance this difficulty does not arise, because at a distance the cross-section of the wire does not matter. The effect is a local one and concerns the self-inductance only.

We can overcome the difficulty by treating the conductor like a stranded cable. Each strand has a constant current density, so that a very thin strand carries negligible current. The procedure is, however, rather cumbersome and it is simpler to work in terms of energy.

It will be shown in the next chapter that we can write the magnetic field energy as

$$U^* = \tfrac{1}{2} \iiint \mathbf{H} \,.\, \mathbf{B} \, dv \,. \tag{7.49}$$

The self-inductance can then be obtained from the expression

$$\tfrac{1}{2} LI^2 = \tfrac{1}{2} \iiint \mathbf{H} \,.\, \mathbf{B} \, dv \,. \tag{6.31}$$

In the next chapter we shall also derive a useful alternative expression for the energy in terms of the vector $\mathbf{A}$

$$U^* = \tfrac{1}{2} \iiint \mathbf{A} \,.\, \mathbf{J} \, dv \tag{7.53}$$

whence

$$L = \frac{1}{I^2} \iiint \mathbf{A} \,.\, \mathbf{J} \, dv \,. \tag{6.32}$$

In two-dimensional problems it is useful to define a self-inductance per unit length as

$$l = \frac{1}{I^2} \iint \mathbf{A} \,.\, \mathbf{J} \, ds \,, \tag{6.33}$$

and if the current density $\mathbf{J}$ is constant $I = Js$, where s is the cross-section of the conductor

$$l = \frac{1}{Is} \iint A \, ds \,. \tag{6.34}$$

In this discussion the vector potential **A** has been used as a tool in calculation. The physical nature of this quantity will become much clearer when in Chapter 9 we discuss time-varying electromagnetic processes.

6.4. Magnetic Materials

6.4.1. Magnetic Field Strength, Magnetic Flux Density, and Relative Permeability

A magnetic material, and by this we generally mean iron, affects the magnetic field around it by the action of the dipoles in the material. We have seen in § 6.2.5 that these dipoles can be represented by an equivalent pole distribution of volume density ϱ^* where

$$\varrho^* = -\nabla \cdot \mathbf{P}^* \tag{6.10}$$

and $\mathbf{P}^*$ is the polarization vector, or dipole moment per unit volume. In the language of § 4.2.1 this pole density ϱ^* is a distribution of "bound" polarity. Following the argument of that section we can define a partial field vector **B**, which takes no notice of these bound sources. Since in magnetism there are no "free" poles, the statement equivalent to eqn. (4.33) becomes

$$\nabla \cdot \mathbf{B} = 0, \tag{6.35}$$

where **B** is the magnetic flux density. The statement equivalent to eqn. (4.32) becomes

$$\mathbf{B} = \mu_0\mathbf{H} + \mathbf{P}^* \tag{6.36}$$

so that

$$\nabla \cdot \mathbf{H} = \varrho^*/\mu_0\,. \tag{6.37}$$

The total magnetic field strength **H** is defined by the additional relationship

$$\nabla \times \mathbf{H} = \mathbf{J}, \tag{3.32}$$

and since $\nabla\times\mathbf{P}^* = 0$ because the field of dipoles is by definition a conservative field, we can also write

$$\nabla \times \mathbf{B} = \mu_0\mathbf{J}\,. \tag{6.38}$$

Thus **H** is the *total* field arising from currents and poles, whereas **B** is the *partial* field of the currents only.

The division of the sources into conduction currents and magnetic poles is particularly useful in the study of electrical machines and transformers, or indeed of any device where the currents and the magnetic material each occupy well-defined regions. Most parts of the magnetic field can then be described in terms of a scalar magnetic potential, which in general is easier to handle than the vector potential.

If, however, there are distributed currents or if for reasons of geometry or because of high-frequency effects (see § 10.1) it becomes necessary to use the vector potential, then it is more convenient to describe the action of magnetic materials in terms of additional currents rather than poles. The reason for this stems from the fact that the vector potential can be obtained most readily by summing currents, whereas the scalar potential is calculated from the pole strength.

We retain the statement

$$\nabla \times \mathbf{H} = \mathbf{J}, \tag{3.32}$$

where **J** is, as before, the conduction current. Since there is now no pole strength,

$$\nabla \cdot \mathbf{H} = 0 \tag{6.39}$$

and also, as before,

$$\nabla \cdot \mathbf{B} = 0. \tag{6.35}$$

But now **B** has additional sources from the magnetization currents $\mathbf{J}_m$, so that

$$\nabla \times \mathbf{B} = \mu_0 (\mathbf{J} + \mathbf{J}_m). \tag{6.40}$$

Thus

$$\nabla \times \mathbf{B} = \mu_0 \nabla \times \mathbf{H} + \mu_0 \mathbf{J}_m. \tag{6.41}$$

If now we put

$$\nabla \times \mathbf{P}^* = \mu_0 \mathbf{J}_m \tag{6.42}$$

we obtain

$$\nabla \times \mathbf{B} = \mu_0 \nabla \times \mathbf{H} + \nabla \times \mathbf{P}^*, \tag{6.43}$$

and since there are no divergence sources

$$\mathbf{B} = \mu_0 \mathbf{H} + \mathbf{P}^* \tag{6.36}$$

as before. Thus **H** is now the *partial* field due to conduction current only and **B** is the *total* field due to conduction current and magnetization current.

The polarization $\mathbf{P}^*$ will depend on the total field **H**, but the relationship between the two quantities is a complicated one depending on the domain structure. The hysteresis effect causes $\mathbf{P}^*$ to depend on the previous magnetic state, or magnetic history, as well as on the applied field. In spite of this we can, of course, write an equation like eqn. (4.34),

$$\mathbf{P}^* = \mu_0 \chi_m \mathbf{H}, \tag{6.44}$$

where χ_m is the magnetic susceptibility, but χ_m will be neither constant nor single-valued. Equation (6.36) leads to a relationship between **B** and **H**,

$$\mathbf{B} = \mu_0 (1 + \chi_m) \mathbf{H} = \mu_0 \mu_r \mathbf{H} = \mu \mathbf{H}, \tag{6.45}$$

where μ_r is the relative permeability. Experiment shows that μ_r is not a constant and eqn. (6.45) must always be treated with caution. The reader should note explicitly that the fundamental relationship between **B** and **H** is one involving the addition of sources as shown in eqn. (6.36). Equation (6.45) is merely a concise notation. Relative permeability is not to be treated as a mysterious property of "the medium" but as a means of summing the effect of sources in the iron. On the other hand, we cannot afford to neglect eqn. (6.45). The simple statement $\mathbf{B} = \mu_0 \mu_r \mathbf{H} = \mu \mathbf{H}$, where μ is a constant, is very convenient in the mathematical analysis of magnetic fields. It is virtually impossible to allow for the hysteresis effect in analytical work and it is even difficult to use a **B–H** curve which is single-valued but exhibits saturation (Fig. 6.9). If it is essential to allow for the variation of μ_r, and very often it is essential, recourse must be had to numerical methods as shown in Chapter 13. The use of a constant value for the permeability lies in obtaining quick approximate solutions.

If μ_r is constant, we can write,

$$\nabla \times \mathbf{H} = \mathbf{J}, \tag{3.32}$$

$$\mathbf{B} = \mu \mathbf{H}, \tag{6.45}$$

$$\nabla \cdot \mathbf{B} = 0. \tag{6.35}$$

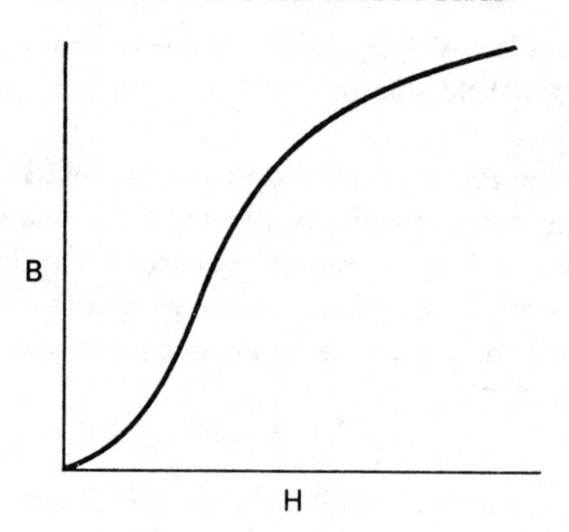

FIG. 6.9. Saturation curve.

Because **B** has zero divergence a vector potential **A** can be defined by

$$\nabla \times \mathbf{A} = \mathbf{B}, \tag{6.25}$$

$$\nabla \cdot \mathbf{A} = 0. \tag{6.46}$$

Hence

$$\nabla \times \mathbf{B} = \nabla \times \nabla \times \mathbf{A} = \mu_0 (\mathbf{J} + \mathbf{J}_m) = \mu_0 \mu_r \mathbf{J}. \tag{6.47}$$

But

$$\nabla \times \nabla \times \mathbf{A} = \nabla (\nabla \cdot \mathbf{A}) - \nabla^2 \mathbf{A} = \nabla^2 \mathbf{A} \tag{3.41}$$

by use of eqn. (6.46).

Hence

$$\nabla^2 \mathbf{A} = -\mu_0 \mu_r \mathbf{J}. \tag{6.48}$$

This vector differential equation was discussed in § 3.2.5. On integration we obtain

$$\mathbf{A} = \frac{\mu_0}{4\pi} \iiint \mu_r \frac{\mathbf{J}}{r} \, dv', \tag{6.49}$$

and if μ_r is constant,

$$\mathbf{A} = \frac{\mu_0 \mu_r}{4\pi} \iiint \frac{\mathbf{J}}{r} \, dv'. \tag{6.50}$$

The description of magnetic sources in terms of magnetization current and the use of the permeability $\mu = \mu_0 \mu_r$ enables us to apply to magnetic materials the same form of solution for the vector potential as for currents in free space.

6.4.2. *Boundary Conditions at a Magnetic Surface*

Figure 6.10 shows such a boundary surface. Since the magnetic flux density has zero divergence, we can use Gauss's theorem to show that the normal component of **B** is continuous across the boundary. Similarly, if

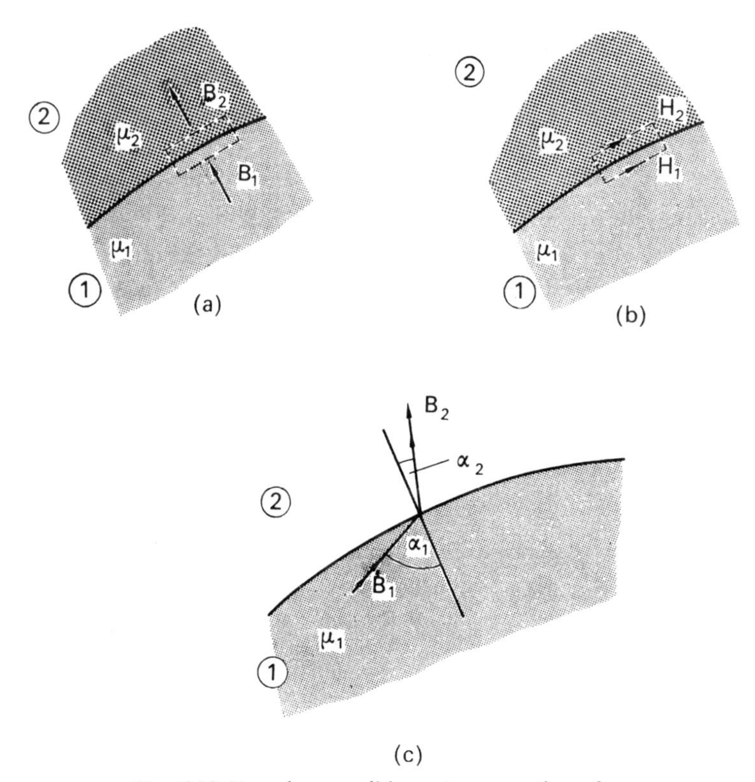

FIG. 6.10. Boundary conditions at a magnetic surface.

Gauss's theorem is applied to the normal component of the total magnetic field strength **H**, there is a discontinuity due to the pole strength q^* at the surface. We have

$$\hat{\mathbf{n}} \, . \, (\mathbf{H}_2 - \mathbf{H}_1) = \frac{q^*}{\mu_0}, \tag{6.51}$$

where $\hat{\mathbf{n}}$ is the unit vector normal to the surface. It is very helpful to make use of § 6.2.6 and to regard $\mathbf{H}_2$ and $\mathbf{H}_1$ as each being made up of two components. Close to the surface we can divide the magnetic sources into (i) a large local surface layer of poles of density q^*, and (ii) all other distant poles and currents. Let the magnetic field strength due to the second set of sources be $\mathbf{H}$. We can now write

$$\left.\begin{aligned}\hat{\mathbf{n}}\,.\,\mathbf{H}_2 &= \hat{\mathbf{n}}\,.\,\mathbf{H} + \frac{q^*}{2\mu_0},\\ \hat{\mathbf{n}}\,.\,\mathbf{H}_1 &= \hat{\mathbf{n}}\,.\,\mathbf{H} - \frac{q^*}{2\mu_0}.\end{aligned}\right\} \tag{6.52}$$

These equations are consistent with eqn. (6.51). So far we have not mentioned the relative permeability μ_r. The description in terms of a pole layer q^* is an *alternative* to the description in terms of μ_r. Indeed, the concept of permeability was derived in § 6.4.1 by the use of the polarization vector which involved q^*. Let us now derive the relationship between q^* and μ_r.

We have seen that the normal component of $\mathbf{B}$ does not have a discontinuity at the surface. Thus

$$\hat{\mathbf{n}}\,.\,\mathbf{B}_2 = \hat{\mathbf{n}}\,.\,\mathbf{B}_1\,. \tag{6.53}$$

If the region **1** is filled by a material of relative permeability μ_r and the region **2** is free space, eqn. (6.53) can be written

$$\hat{\mathbf{n}}\,.\,\mu_0\mathbf{H}_2 = \hat{\mathbf{n}}\,.\,\mu_0\mu_r\mathbf{H}_1 \tag{6.54}$$

and

$$\hat{\mathbf{n}}\,.\,\mathbf{H}_2 = \mu_r\hat{\mathbf{n}}\,.\,\mathbf{H}_1\,. \tag{6.55}$$

Using eqn. (6.51) we obtain

$$\left[\hat{\mathbf{n}}\,.\,\mathbf{H} + \frac{q^*}{2\mu_0}\right] = \mu_r\left[\hat{\mathbf{n}}\,.\,\mathbf{H} - \frac{q^*}{2\mu_0}\right], \tag{6.56}$$

whence

$$\frac{q^*}{2\mu_0} = \frac{\mu_r - 1}{\mu_r + 1}\,\hat{\mathbf{n}}\,.\,\mathbf{H}\,. \tag{6.57}$$

Now μ_r in iron is generally of the order of 1000, so that the ratio $(\mu_r - 1)/(\mu_r + 1)$ is never far from unity. This means that the pole strength

q^* is almost independent of the relative permeability. Even where μ_r varies from place to place in the iron, the actual value of μ_r does not affect q^* to any extent. Thus the assumption of constant μ_r is not as restrictive as might be thought. It is very helpful to bear in mind that if we use q^* in calculations, a wide range of μ_r is quite acceptable as long as $\mu_r \gg 1$.

We have examined the boundary conditions of the normal component of the magnetic field vectors. Now we must examine the tangential components. If there is no electric current on the surface, the application of $\oint \mathbf{H} \,.\, d\mathbf{l} = 0$ (Fig. 6.10b) gives

$$\hat{\mathbf{n}} \times \mathbf{H}_2 = \hat{\mathbf{n}} \times \mathbf{H}_1 \,. \tag{6.58}$$

If there is a surface current of line density $\mathbf{J}$ A/m,

$$\hat{\mathbf{n}} \times (\mathbf{H}_2 - \mathbf{H}_1) = \mathbf{J} \,. \tag{6.59}$$

Figure 6.10c shows that in the absence of a surface current the external flux density vector is almost perpendicular to the iron. Let α_1 and α_2 be the angles to the normal in regions **1** and **2**

$$\left.\begin{aligned} \tan \alpha_1 &= \frac{\text{tangential } B_1}{\text{normal } B_1} \\ \tan \alpha_2 &= \frac{\text{tangential } B_2}{\text{normal } B_2} \end{aligned}\right\} \tag{6.60}$$

whence

$$\frac{\tan \alpha_2}{\tan \alpha_1} = \frac{\text{tangential } B_2}{\text{tangential } B_1} \times \frac{\text{normal } B_1}{\text{normal } B_2}, \tag{6.61}$$

and by the use of eqns. (6.53) and (6.58),

$$\frac{\tan \alpha_2}{\tan \alpha_1} = \frac{1}{\mu_r} \,. \tag{6.62}$$

If $\mu_r \gg 1$, α_2 will always be a small angle. As $\mu_r \to \infty$ $\alpha_2 \to 0$. Often the *external* field can be calculated on the assumption that flux emerges at right angles from an iron surface. In terms of scalar potential the surface of the iron is an equipotential surface as far as the external region is concerned.

6.4.3. *The Magnetic Field Inside Iron*

The discussion of what is meant by the field *inside* a magnetic material is similar to the dielectric case with which we dealt in § 4.2.3. In a macroscopic theory it is meaningless to talk of being inside a material. There has to be a hole before any measurement can be made. The shape of the hole will be of great importance.

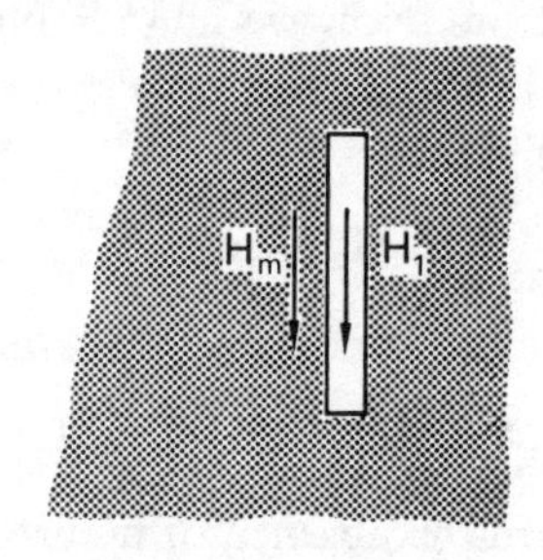

FIG. 6.11. Worm-hole force in magnetic material.

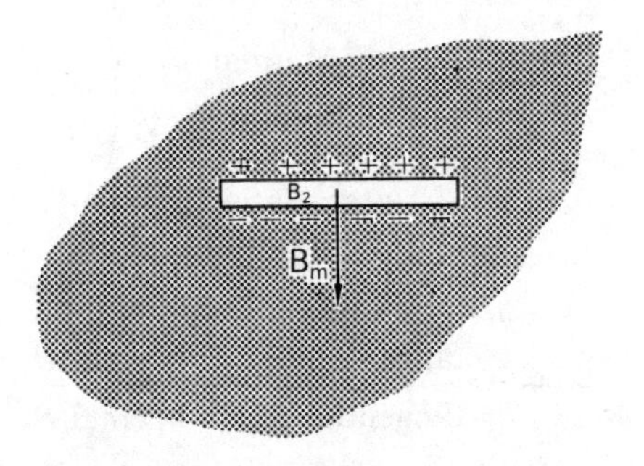

FIG. 6.12. Disc-hole force in magnetic material.

As in electrostatics the only shapes which will not affect the outside field are the long worm-hole (Fig. 6.11) along the direction of the field and the flat disc-cavity perpendicular to the direction of the field (Fig. 6.12). If we define the magnetic field in the material by H_m and observe the magnetic field in the worm-hole as H_1, we have, by eqn. (6.58),

$$H_m = H_1 . \tag{6.63}$$

In the disc-cavity of Fig. 6.12 surface polarity will be present. We can make use of eqn. (6.53) to deduce that

$$B_m = B_2, \tag{6.64}$$

where B_2 is the magnetic flux density in the cavity. Since $B_m = \mu_0\mu_r H_m$ by definition of μ_r, we have

$$B_2 = \mu_0\mu_r H_1. \tag{6.65}$$

Hence in discussing **B** inside a material we have in mind a disc-cavity and in discussing **H** we think of a worm-hole.

This discussion does not give any information about the field that would act on a high-speed charged particle passing through a magnetic material. This problem is discussed by Panofsky and Phillips† who state that experiments show that very high-speed particles experience a force corresponding to **B**, while at lower speeds the force corresponds more nearly to **H**. At high speed the material therefore acts as if it consisted of current loops and at low speeds as if it were composed of impenetrable dipoles.

6.4.4. *B–H Curves*

The domain structure and magnetic characteristics of different materials vary greatly. A useful way of presenting information about the magnetic behaviour is by means of *B–H* curves.

Figure 6.13 shows what happens when a magnetic material is subjected to an increasing magnetic field. At the origin $B = 0$, $H = 0$, $P^* = 0$ so that the effect of the domains cancel on the macroscopic scale. As the applied field is increased there is at first a reversible movement of domain walls in a direction to increase the size of the domains which point more or less in the direction of the applied field (Fig. 6.14). Above the point *a* the *B–H* curve becomes steeper and the region up to point *b* corresponds to further extensive domain wall movement coupled with sudden changes in the direction of some domains. This portion of the curve is irreversible. Above *b* there is a further reversible movement as the domains are aligned more and more closely to the direction of the applied field. The region

† W.K.H.Panofsky and M.Phillips, *Classical Electricity and Magnetism*, Adison Wesley, chapter 8.

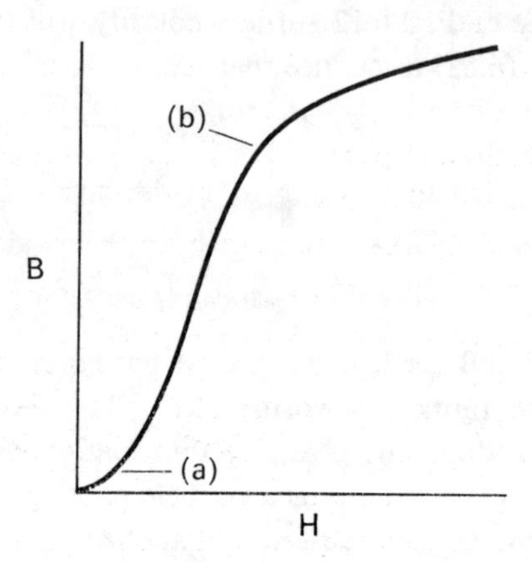

FIG. 6.13. Saturation curve (different regions).

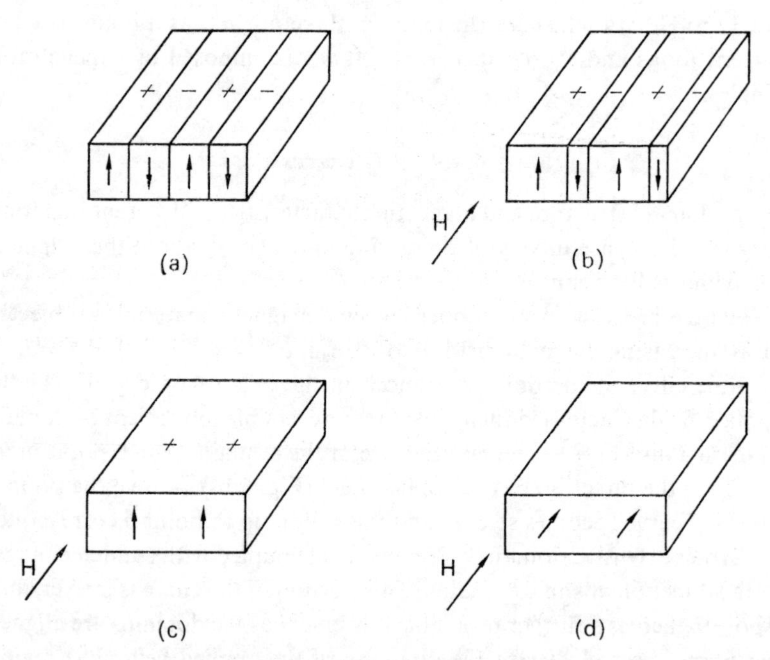

FIG. 6.14. Domains.

near b is called the knee of the B–H curve and above it the material is said to become *saturated.* The whole B–H curve is often called the saturation curve of the material.

The saturation curve of Fig. 6.13 is single-valued. However, if the applied field is decreased, the curve traced out will be as in Fig. 6.15. This is due to hysteresis, a term used to describe the energy loss associated with

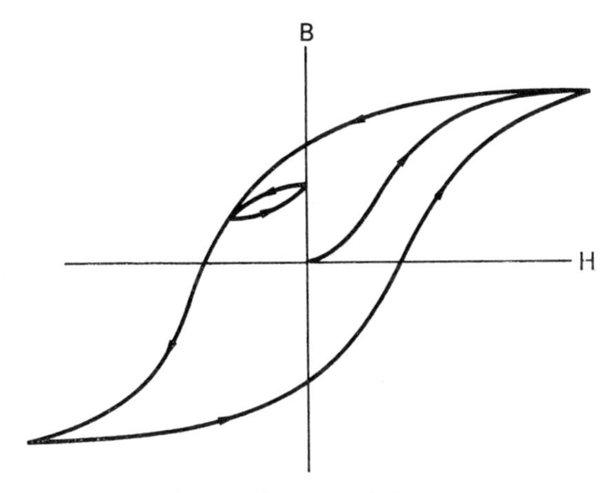

FIG. 6.15. Hysteresis loops.

the magnetization process. This energy loss has various components. The domain movements will cause local eddy-currents to flow in the conducting material. There will also be friction losses caused by impurities in the material which make the domain walls stick to them. Figure 6.15 shows both a minor and a major hysteresis loop. The major loops are obtained by varying the applied field between equal positive and negative values. This has to be done several times until the material is in a cyclic state. Only then does the hysteresis loop become repeatable. In general the position on the B–H loop depends on the previous magnetic history of the material. The curve obtained by connecting the tips of major or cyclic hysteresis loops is often known as the reversal curve. It is generally very close to the saturation curve described in Fig. 6.13. The value of B when $H = 0$ is known as the remanent magnetic flux density, and the value of H when $B = 0$ is known as the coercive magnetic field strength.

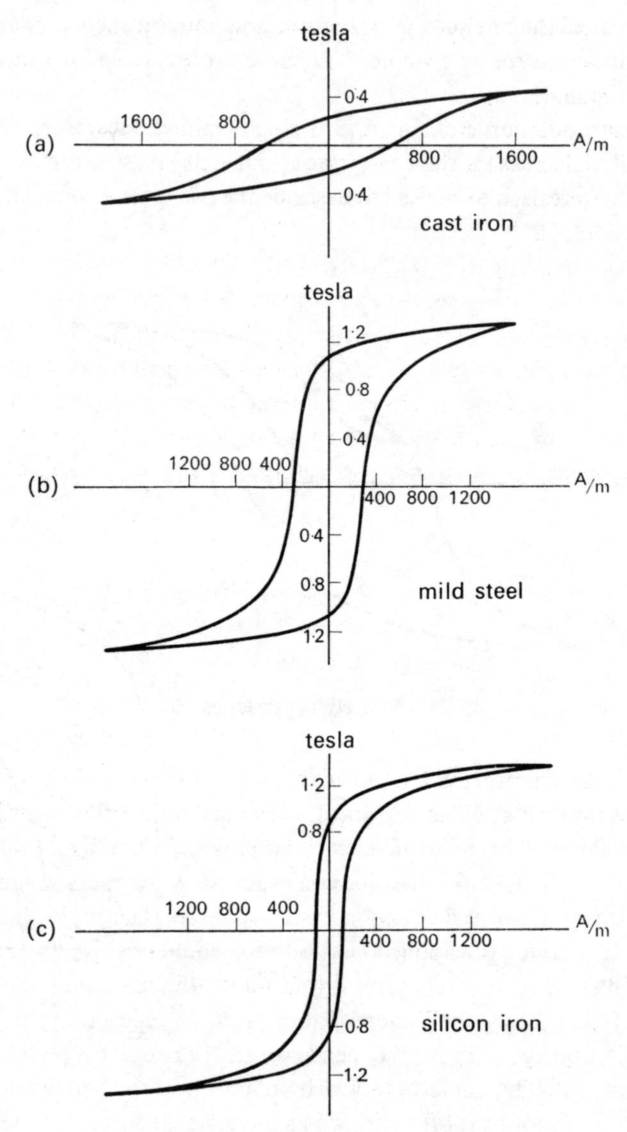

FIG. 6.16. Typical hysteresis loops.

The hysteresis behaviour so far discussed is common to all ferromagnetic materials. Nevertheless, the magnitude of the effect varies enormously as is illustrated in Figs. 6.16 and 6.17, which give results for some typical materials. It is useful to distinguish between two classes: the hard magnetic materials and the soft ones. Hard magnetic materials are used in permanent magnets. For such an application the material itself provides the external field and a high value of B (for example the remanent B) is clearly desirable. More important still there should be little change of B if the material is subjected to external fields which might demagnetize it. This implies a high value of the coercive field strength and will be discussed in more detail in the next chapter. In permanent magnetic materials hysteresis is, therefore, a very desirable quality. Maximum hysteresis is achieved by alloying and by heat treatment. Magnetic hardness is accompanied by mechanical hardness, so that the materials have to be cast and

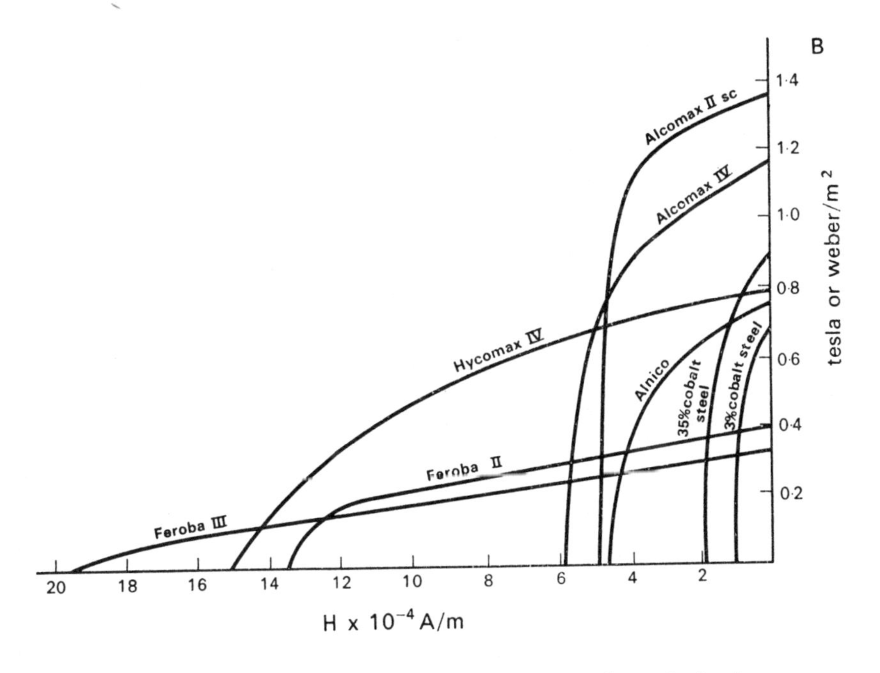

(By Courtesy of The Permanent Magnet Association)

FIG. 6.17. Typical hysteresis curves of permanent magnet materials.

are finished by grinding. Figure 6.17 shows a part of some typical hysteresis curves of permanent magnet materials.

Soft magnetic materials are used in alternating current applications, where the domains have to follow the applied field of the exciting current. Hysteresis causes the polarization to lag behind the applied field and it

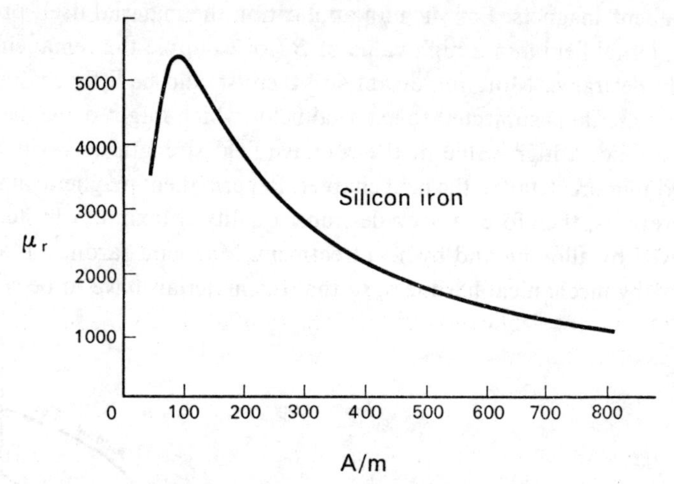

FIG. 6.18. Variation of μ_r of soft magnetic material.

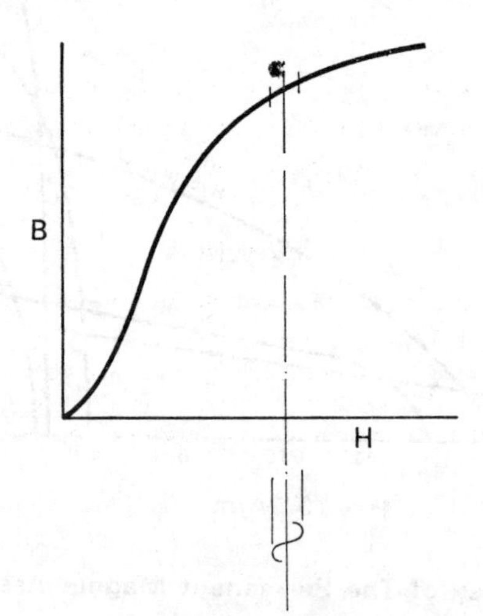

FIG. 6.19. A.c. variation with d.c. bias.

introduces an energy loss. It is therefore very desirable to have negligible hysteresis and at the same time to have large values of flux density. The addition of silicon is found to be very beneficial. A typical hysteresis curve is shown in Fig. 6.16c. The loop is so thin that it can often be replaced by a single-valued saturation curve. This makes it reasonable to talk about the permeability of the material. The permeability will, of course, vary as shown in Fig. 6.18, but at least it will not range between minus and plus infinity as it does with a hysteresis loop, where the idea of permeability is virtually useless as a tool in calculation. If in a soft material there is a biasing field with a superimposed alternating field as illustrated in Fig. 6.19, it is convenient to use an incremental permeability given by the slope of the saturation curve.

In between the hard and the soft magnetic materials there are ferromagnetics of high flux density and moderate hysteresis. These are used in direct current applications, where the iron reinforces the field of a constant exciting current.

Summary

In this chapter we have assembled a set of useful results giving the magnetic field of various distributions of poles and currents.

We have briefly discussed the use of the vector potential in magnetic field calculations.

Lastly we have elucidated the meaning of magnetic permeability and have shown that iron can be represented either by magnetic pole distributions or by magnetization currents.

Exercises

6.1. Show that the torque on a magnetic dipole $\mathbf{P}^*$ in a field $\mathbf{H}$ is $\mathbf{P}^* \times \mathbf{H}$.

6.2. Two co-planar dipoles have their centres at a fixed distance from each other. The angles between the dipoles and the line joining their centres are θ and θ' respectively. Show that if θ is held fixed, the equilibrium condition is given by $\tan\theta = -\frac{1}{2}\tan\theta'$. Explain why in general θ has to be held fixed. Are there any exceptions? [Ans.: $\theta = 0$ or $\theta = \pi$.]

6.3. Show that the magnetic scalar potential of a current loop is given by $V^* = I\Omega/4\pi$, where I is the current and Ω the solid angle at the field point. Determine the potential at a distance x vertically above the centre of a horizontal current loop of radius R. $\left[\text{Ans.: } \frac{I}{2}\left[1 - \frac{x}{(x^2 + R^2)^{1/2}}\right].\right]$

6.4. Show that the force on a dipole of strength $\mathbf{P}^*$ in a magnetic field $\mathbf{H}$ is given by $\mathbf{F} = (\mathbf{P}^* \, . \, \nabla) \, \mathbf{H}$. Calculate the force between two co-axial circular currents I and i of radius R and r respectively at a distance x from each other. Assume that r is much smaller than R. [Ans.: $\frac{3}{2}\mu_0 \pi i I r^2 R^2 x \, (R^2 + r^2)^{-5/2}$.]

6.5. A long circular cylinder of iron is magnetized to saturation by a magnetic field transverse to its axis. The dipole moment per unit volume at saturation is $\mathbf{P}^*$. Show that the cylinder has on its surface a pole strength of density $\mathbf{P}^* \cos \theta$, where θ is the angle between the direction of the magnetic field and the radius of the cylinder at the point at which the pole strength is measured. If the radius of the cylinder is a, show that the field outside the cylinder due to $\mathbf{P}^*$ is the same as that of a line doublet of strength $\pi a^2 \mathbf{P}^*$ per unit length.

6.6. Calculate the contribution to the self-inductance which arises from the internal magnetic field of a steady current flowing in a long straight conductor of circular cross-section. [Ans.: $\mu_0/8\pi$ H/m.]

6.7. In a two-dimensional magnetic field of coordinates x, y the current flow is entirely in the z-direction. Show that a line of magnetic force is a line along which the vector potential is constant.

6.8. Show that the vector potential at a distance r from the axis of a long conductor of circular cross-section and radius a carrying a steady current I is

$$A = \frac{\mu_0 I}{2\pi} \ln r, \text{ where } r > a.$$

6.9. The current in an electrical machine is represented by a current sheet of density $J_z = J \cos \beta x$ A/m. Find the vector potential at a distance y from the sheet. [*Hint:* Away from the current distribution A has to fit Laplace's equation $\partial^2 A/\partial x^2 + \partial^2 A/\partial y^2 = 0$. Separation of variables suggests solutions of the form $A = e^{j\beta x} e^{\pm \beta y}$. The exact expression can be obtained from the magnetic field at the surface of the current sheet.

$$\left[\text{Ans.: } A = \frac{\mu_0 J}{2\beta} \cos \beta x \, e^{-\beta y}.\right]$$

6.10. A conductor of high relative permeability has an elliptical cross-section, so that its boundary is given by the equation $x^2/a^2 + y^2/b^2 = 1$. Assuming that no magnetic flux emerges from the conductor determine the vector potential inside the conductor when it carries a current of uniform density J A/m^2.

$$\left[\text{Ans.: } A = -\frac{\mu_0 \mu_r a^2 b^2 J}{2\,(a^2 + b^2)} \left(\frac{x^2}{a^2} + \frac{y^2}{b^2}\right).\right]$$

6.11. Show that vector potential has the dimensions of momentum per unit charge. Discuss this in terms of the sources of the vector potential.

6.12. A sample of iron has a relative permeability of 12 when the flux density is 2·25 T. What external magnetic field is required to produce this flux density if the iron is in the form of (a) a cylinder with its axis perpendicular to the field, and (b) a flat plate perpendicular to the direction of the field? [Ans.: (a) $9{\cdot}7 \times 10^5$ A/m; (b) $1{\cdot}79 \times 10^6$ A/m.]

6.13. The demagnetizing factor N_D of an iron specimen is defined by the equation

$$H = H_0 - N_D P^*,$$

where H is the magnetic field strength in the specimen and P^* is the polarization. Determine the demagnetization factor (a) for a cylinder placed transversely to a field H_0, and (b) for a plate placed transversely to H_0. [Ans.: (a) $1/2\mu_0$; (b) $1/\mu_0$.]

CHAPTER 7

MAGNETOSTATICS II

7.1. Magnetic Circuits

7.1.1. A Ring Without an Air Gap

The vital parameter which engineers consider in designing magnetic fields is the flux linkage with an electric circuit. The greater the flux linkage, the more energy can be transferred and the greater are the forces which can be transmitted. For two reasons ferromagnetics are of enormous usefulness in increasing the flux linkage. The first is obvious and is described by the concept of relative permeability. If this is of the order 1000 then we know that the applied field is strengthened a thousand times by the additional magnetic sources provided by the material.

The second property of iron is less obvious, although it is equally important. It is the facility which iron has to conduct flux from one place to another. If the exciting current is, for example, threaded on an iron core then that core can be made to produce a strong field at a place far from the current which is causing the field.

Consider this with reference to Fig. 7.1, where an iron ring is uniformly wound by a coil of N turns carrying a current I. If the current

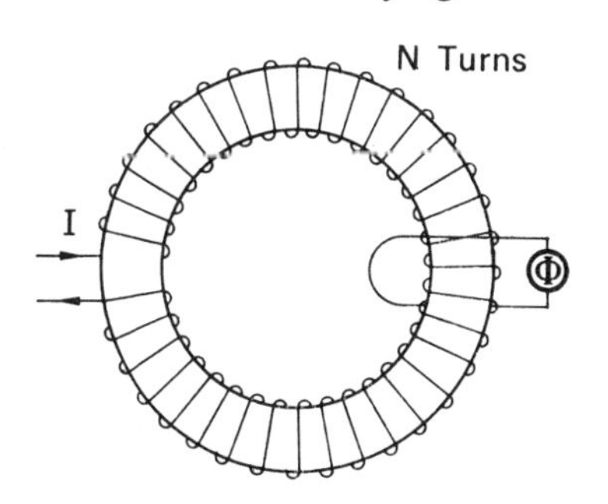

FIG. 7.1. Iron ring with uniform winding.

were applied by means of a current sheet all the magnetic field would be confined to the ring. With a coil there is some slight external field because there is effectively a single turn of the winding in the plane of the ring. However, we shall disregard this small field. The magnetic field strength H in the ring is therefore uniform around the ring and the flux density B is also uniform. By symmetry there is no surface polarity on the iron of the ring. The flux in the ring is measured by a fluxmeter as indicated in the figure. The m.m.f.† of the winding is NI, the mean diameter d, the permeability of the ring is μ, and the cross-sectional area s.

$$\oint \mathbf{H} \,.\, d\mathbf{l} = H\pi d = NI, \tag{7.1}$$

$$\Phi = Bs = \mu Hs, \tag{7.2}$$

where we have assumed that the flux density is uniform over the small area s. Hence

$$\Phi = \frac{\mu NIs}{\pi d}. \tag{7.3}$$

Now consider the altered arrangement of Fig. 7.2, where the exciting winding has been concentrated at one cross-section of the ring. The magnetic field strength H due to the current at the point P at the opposite end of a diameter, i.e. at the place where the flux meter is, can be calculated by the use of eqn. (6.4). Let θ be the angle between the radius vector and

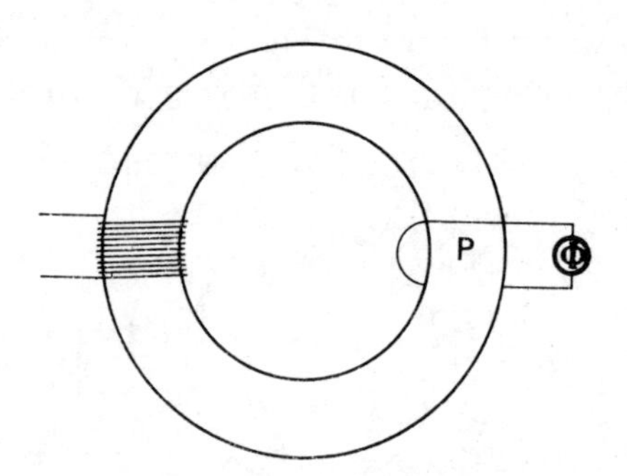

FIG. 7.2. Iron ring with concentrated winding.

† For a definition refer to eqn. (2.3.3). Note that the total current linkage is NI rather than I.

the vector representing the area of the exciting coil. Then

$$V^* = \frac{NIs \cos \theta}{4\pi r^2} \tag{7.4}$$

and

$$H_6 = -\frac{1}{r}\frac{\partial V^*}{\partial \theta} = \frac{NIs \sin \theta}{4\pi r^3}. \tag{7.5}$$

Hence at $r = d$, $\theta = \pi/2$, i.e. at the point P in Fig. 7.2,

$$H_\theta = \frac{NIs}{4\pi d^3}, \tag{7.6}$$

$$B_\theta = \frac{\mu NIs}{4\pi d^3}, \tag{7.7}$$

$$\Phi = \frac{\mu NIs^2}{4\pi d^3}. \tag{7.8}$$

Hence the ratio of fluxes is calculated as

$$\frac{\Phi_2}{\Phi_1} = \frac{s}{4d^2}, \tag{7.9}$$

where Φ_1 is the flux due to a uniform winding and Φ_2 due to a concentrated winding. As an example suppose that s is 1 $\text{cm}^2 = 10^{-4}\ \text{m}^2$ and $d = 10^{-1}$ m. Then $\Phi_2/\Phi_1 = 1/400$, i.e. the concentration of the exciting coil has reduced the flux in the ratio of 400 : 1.

If this calculation is put to the test of an experiment it is found to be correct if the ring is non-magnetic, but if the ring is magnetic, Φ_2 and Φ_1 are virtually the same and it does not seem to matter where the exciting winding is. The iron "conducts" the flux around the ring irrespective of the position of the m.m.f. How does it do so?

The reader will rightly be suspicious of the term "conducts", since there is no flow of pole strength. The domains do not move through the material. The correct electrical analogy of magnetic flux is not with electric current, but with electric flux, which means with electric charge. The bound dipoles in iron behave like the bound charges in a dielectric. But the shaping of electric fields by dielectrics is not so important as the shaping of magnetic fields by iron. The reason lies in the large values of permeability which iron attains as against the smaller values of permittivity possessed by dielectric materials.

The reader will have noticed that in the calculation for Φ_2 we ignored the sources of the magnetic field provided by the iron ring. If we assume a fairly constant permeability throughout the ring we can use the result of § 6.2.5 to predict that the extra sources must be on the surface of

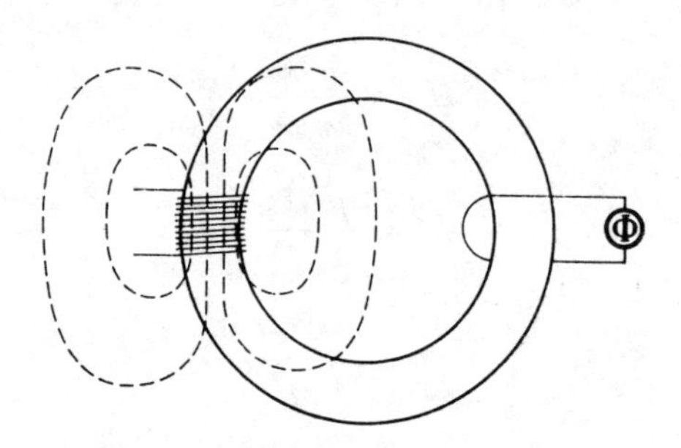

FIG. 7.3. Field of current by itself.

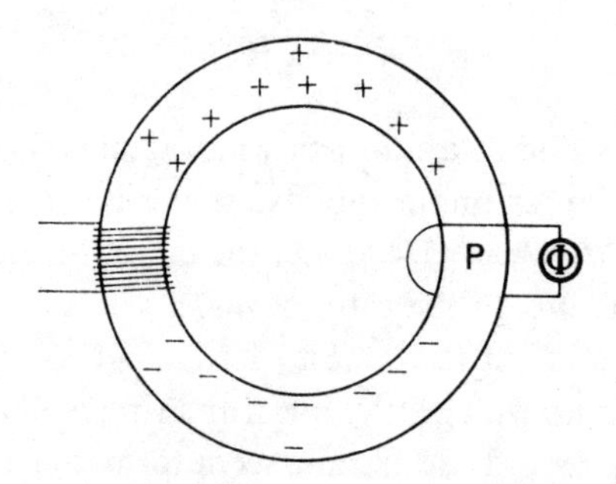

FIG. 7.4. Surface polarity on ring.

the ring. The field of the current alone is as shown in Fig. 7.3. This field cuts the surface of the iron and surface polarity is induced as shown in Fig. 7.4.

Let us make a rough estimate of the surface polarity required to give a constant magnetic field around the ring and let us estimate the leakage flux associated with this surface polarity.

The extra magnetic field strength needed at P is

$$H = \frac{399}{400}\frac{NI}{\pi d} \doteqdot \frac{NI}{\pi d}. \tag{7.10}$$

Consider the surface polarity near P as shown in Fig. 7.4. Assume a band of uniform surface polarity $+q^*$ at a distance a from P and a similar band of polarity $-q^*$ distant a on the other side of P. Let the cross-section of the ring have diameter b and let the width of the bands of polarity be c.

The pole strength of each band is $Q^* = q^*\pi bc$. The magnetic field strength at P due to the two bands is

$$H = \frac{2Q^*}{4\pi\mu_0 a^2} = \frac{q^*bc}{2\mu_0 a^2}, \tag{7.11}$$

where Q^* has been treated as a point source.

From eqn. (7.10) we then have

$$\frac{NI}{\pi d} = \frac{q^*bc}{2\mu_0 a^2}. \tag{7.12}$$

The leakage flux is equal to the pole strength Q^* and as a fraction of total flux it is, by use of eqns. (7.3) and (7.12),

$$\frac{\Phi_l}{\Phi} = \frac{q^*\pi bc}{(\mu_0\mu_r NIs)/\pi d} = \frac{2a^2\pi}{\mu_r s}. \tag{7.13}$$

Suppose $a = 10^{-2}$ m and $\mu_r = 3000$,

$$\frac{\Phi_l}{\Phi} = \frac{2\pi}{3000} = \frac{1}{4}\%. \tag{7.14}$$

Thus a leakage flux of $\frac{1}{4}\%$ of the "mutual" flux in the ring is sufficient to give the correct field strength so that the flux remains constant. Of course, this is a rough estimate only. All the surface will have polarity and therefore leakage flux. But the total leakage flux will not be increased much because all the surface polarity will contribute to the magnetic field in the ring. A more accurate estimate could be made by means of a digital computer study. Our estimate is, however, sufficient to show the large effect which can be brought about by a small amount of leakage flux. In many calculations leakage flux is neglected and it is assumed that the iron conducts the flux without leakage. We can now see how this method can give such good results. The reader should note, however, that there must always be *some* leakage flux even if the amount of it is small.

7.1.2. An Iron Ring with an Air Gap

In transformers we can use closed iron paths, but if there is to be motion, as in machines, then the moving part has to be separated from the stationary part of the device. In such apparatus the iron circuit will have one or more air gaps and it is important to discuss the behaviour of such circuits. For simplicity we shall discuss a ring with a single air gap as shown in Fig. 7.5.

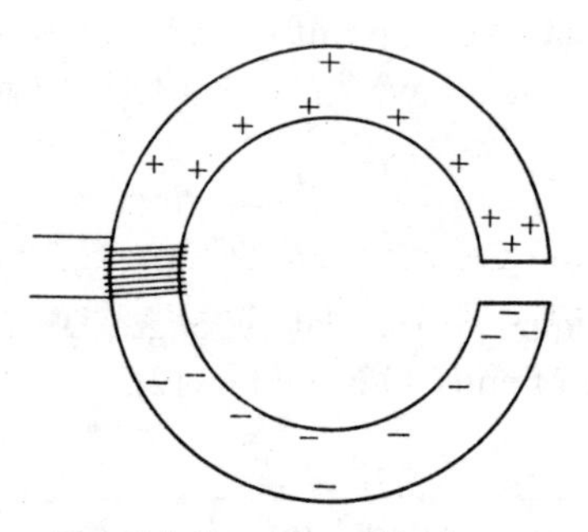

FIG. 7.5. Iron ring with air gap.

The previous section has shown that the leakage flux is small and as a first approximation we shall neglect it. This means that B and H will be assumed constant around the iron ring. Hence

$$NI = \oint \mathbf{H} \,.\, d\mathbf{l} = H_{iron} l_{iron} + H_{gap} l_{gap}. \tag{7.15}$$

Since B is continuous across the gap, we have

$$B_{iron} = B_{gap}. \tag{7.16}$$

Hence

$$NI = \frac{B_{gap} l_{gap}}{\mu_0} \left[1 + \frac{l_{iron}}{\mu_r l_{gap}} \right]. \tag{7.17}$$

If $\mu_r = 3000$ and $l_{iron} = 30 l_{gap}$,

$$NI = \frac{B_{gap} l_{gap}}{\mu_0} [1 + 0{\cdot}01]. \tag{7.18}$$

In spite of the greater length of the iron path, the magnetic flux is determined by the air gap. The chief use of the iron is to allow the designer to move the exciting coil away from the gap. Where there are air gaps the iron does not increase the working flux, but it conducts it to the place at which it can be used.

Reverting to the discussion of the previous section we note, by comparing eqns. (7.1) and (7.18), that the cutting of the air gap has reduced the flux in the ratio of 100 : 1. This means that H in the iron has been reduced in this ratio. How has this come about? We remember that H was dominated by the effect of leakage flux and this is still true when there is an air gap.

The pole strength on the faces of the gap has little effect because the positive and negative polarities are close together. This makes the air gap behave like a dipole and the field of a dipole decays with the cube of the distance. Consider the field at the centre of the exciting coil. When the coil was distributed uniformly we had

$$H = \frac{NI}{\pi d}. \tag{7.19}$$

When the coil is concentrated, use of eqn. (2.37) gives the field due to the current alone as

$$H_c = \frac{NI}{b}, \tag{7.20}$$

where b is the diameter of the cross-section of the ring. Also since $s = \pi b^2/4 = 10^{-4}$ m^2, $b = 1{\cdot}13$ cm.
Thus

$$\frac{H_c}{H} = \frac{\pi d}{b} = \frac{\pi \times 0{\cdot}1}{0{\cdot}0113} = 27{\cdot}8. \tag{7.21}$$

The field of the coil by itself is too strong to give the correct uniform flux density. Thus when there is no gap the leakage flux has to reduce this field from 27·8 H to H by providing a negative force of 26·8 H. We have seen that when the gap is cut in the ring the field is reduced a hundredfold. We now need 0·01 H, but H_c is still 27·8 H. Hence the leakage flux must give $-27{\cdot}79\ H$ as against the previous $-26{\cdot}8\ H$. This means that the leakage flux has hardly changed at all. Of course the approximation is again very rough, but it does show how small variations in leakage flux can distribute the magnetic field. It should also be noted that the magnitude of the leakage flux will remain the same but its relative proportion will have increased from an estimated $\frac{1}{4}\%$ to around 25% of the working flux.

The chief point to remember from all this is that iron "conducts" flux by virtue of surface polarity and this surface polarity is necessarily associated with leakage flux.

7.1.3. Permanent-magnetic Calculations

In the example of § 7.1.2 the primary source of the magnetic field was the current in the exciting winding. This current magnetized the iron and produced additional sources of polarity on the surface of the iron. In the context of this example we assumed that the iron was magnetically "soft" and that it responded to an applied magnetic field with the relationship $B = \mu_0 \mu_r H$. The relative permeability was not necessarily constant, but it was single-valued. In other words, we had neglected hysteresis.

In considering a "permanent" or "hard" magnet such an assumption would give a nonsensical answer. Since there is no exciting current there would apparently be neither H nor B. The action of a permanent magnet must be explained with reference to the action of magnetic polarization rather than the more artificial notion of permeability. We have from § 6.4.1,

$$\mathbf{B} = \mu_0 \mathbf{H} + \mathbf{P}^*. \tag{6.36}$$

Apply this equation to the consideration of Fig. 7.6. Outside the magnet the lines of **B** and **H** run in the same direction, because $\mathbf{B} = \mu_0 \mathbf{H}$. But the internal field, as defined in § 6.4.3, is very different. Inside the magnet the direction of **H** is *opposite* to the direction of **B**. The reason for this is

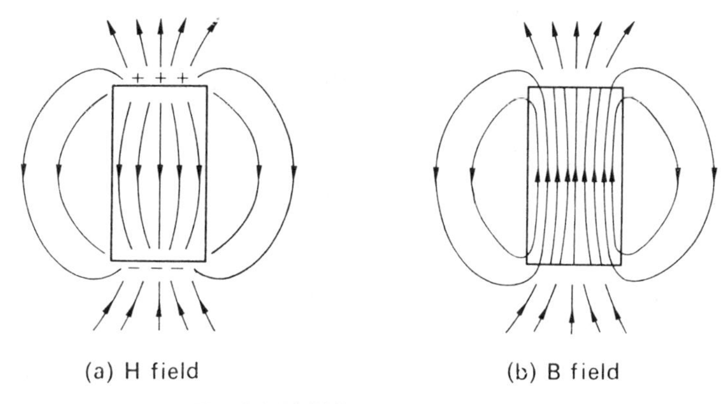

FIG. 7.6. Field in permanent magnet.

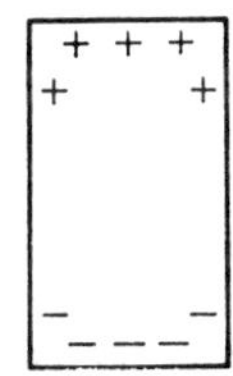

FIG. 7.7. Surface polarity on permanent magnet.

to be found in the action of the surface polarity which produces the external flux and also causes a negative magnetic field strength inside the iron (see Fig. 7.7). How is it possible to have positive flux density **B** and negative **H**? The answer lies in the action of **P*** which is sufficient to overcome **H** and provide **B**. If we refer to the hysteresis curves (Fig. 6.17), we see that the magnetic state of a permanent magnet is represented by positive **B** and negative **H**.

Consider a permanent magnet with an air gap as illustrated in Fig. 7.8. As a first approximation we shall assume that the flux density in the magnet and air gap is uniform. This implies that leakage flux is being neglected. As pointed out in § 7.1.2, we do not imply that there is no leakage flux, but merely that the leakage flux is a small fraction of the working flux. We are, of course, aware that there must be some leakage flux to produce the necessary surface polarity which guides the working flux around the magnetic path. From consideration of magnetic scalar potential we have

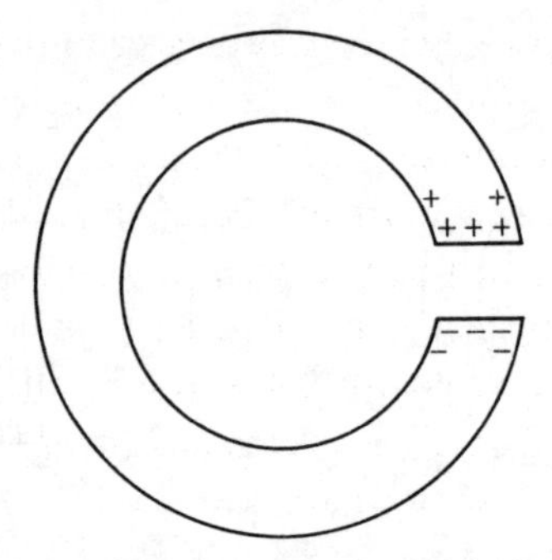

FIG. 7.8. Permanent magnet with air gap.

$\oint \mathbf{H} \, . \, d\mathbf{l} = 0$, so that

$$H_{\text{iron}} l_{\text{iron}} + H_{\text{gap}} l_{\text{gap}} = 0, \tag{7.22}$$

also

$$B_{\text{iron}} = B_{\text{gap}} = \mu_0 H_{\text{gap}}, \tag{7.23}$$

hence

$$B_{\text{iron}} = -\mu_0 H_{\text{iron}} \frac{l_{\text{iron}}}{l_{\text{gap}}}. \tag{7.24}$$

This is a straight-line relationship between B and H and is plotted on Fig. 7.9. The working point is given by P, the intersection of the hysteresis loop and the line of eqn. (7.24). The steepness of the hysteresis curve has

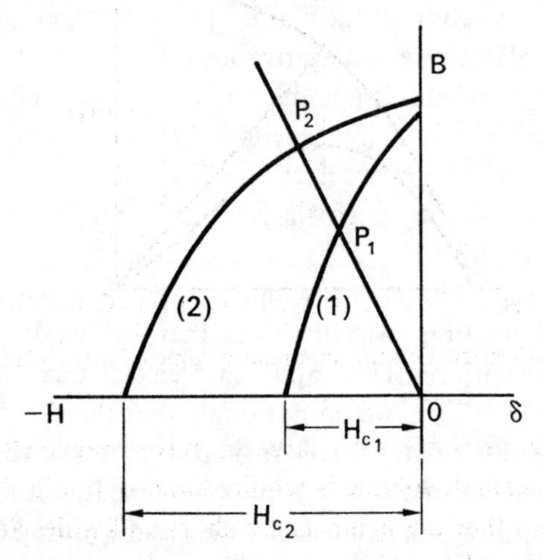

FIG. 7.9. Working point for permanent magnet calculations.

a marked effect on the value of the flux density at the working point and this is illustrated in Fig. 7.9. Thus it is not sufficient that the material should have a high value of remanent flux density. A large value of coercive field strength H_c is also important. Figure 6.17 shows some typical hysteresis curves for modern permanent magnet materials.

A well-designed magnetic circuit will use as little of the expensive permanent magnet material as possible. The volume of the permanent magnet is given by $v_{\text{iron}} = l_{\text{iron}} s_{\text{iron}}$, where s is the cross-sectional area. If we assume that the gap has the same area, we have

$$v_{\text{iron}} = l_{\text{iron}} s = -\frac{H_{\text{gap}} l_{\text{gap}} s}{H_{\text{iron}}} = \frac{H_{\text{gap}} B_{\text{gap}} v_{\text{gap}}}{-H_{\text{iron}} B_{\text{iron}}}. \tag{7.25}$$

The numerator of this expression is proportional to the stored magnetic energy in the gap as may be checked by dimensional analysis. If this is specified, the minimum volume of permanent magnet will be achieved by choosing the maximum value for $-H_{\text{iron}} B_{\text{iron}}$. It is found in practice that the construction of Fig. 7.10 gives a point P for which this condition holds.

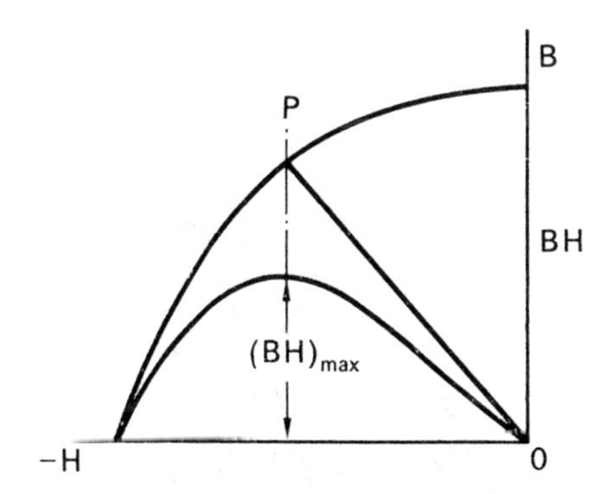

FIG. 7.10. Construction for minimum volume of permanent magnet.

Modern materials have such large coercive magnetic field strengths that often only a small volume is required and that a large area s can be used. This means that the magnets are short and wide. Soft iron is used to guide the flux to the working section.

7.1.4. Magnetic Equivalent Circuits

Although an understanding of the effect of surface polarity and leakage flux is essential to an engineer who is closely concerned with magnetic fields, it is sometimes justifiable to forget about such matters. One can then regard the iron as a conductor of flux in the same way as copper conducts current. Indeed, many people feel very much more at home with electric circuits and they value the analogy. Let us consider the example of the soft iron ring and air gap magnetized by a current in a winding, as in § 7.1.2. We have

$$NI = \frac{B_{\text{gap}} l_{\text{gap}}}{\mu_0}\left[1 + \frac{l_{\text{iron}}}{\mu_r l_{\text{gap}}}\right]; \tag{7.17}$$

also the flux is given by

$$\Phi = Bs, \tag{7.2}$$

where s is the cross-sectional area of the ring and gap. Hence

$$NI = \Phi\left[\frac{l_{\text{gap}}}{\mu_0 s} + \frac{l_{\text{iron}}}{\mu_0 \mu_r s}\right]. \tag{7.26}$$

NI, the magnetomotive force, is the magnetic analogue of electromotive force or voltage. If then we regard flux as analogous to current, the expression in brackets can be regarded as analogous to resistance as shown in Fig. 7.11. This expression is called "reluctance"*. The chief practical

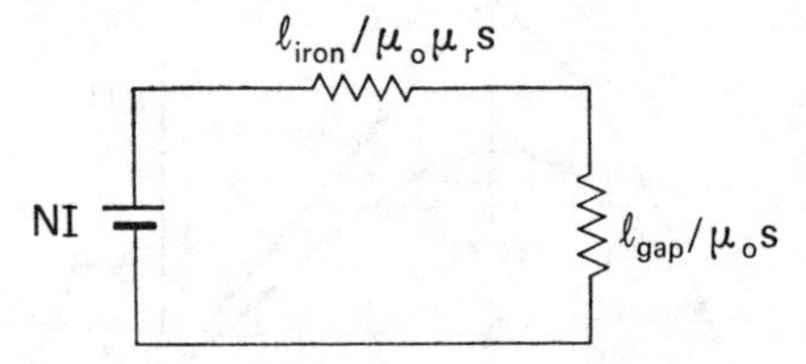

FIG. 7.11. Equivalent magnetic circuit.

difficulty in this procedure is that μ_r depends on B, whereas Ohm's law gives a resistivity independent of current density. This restricts the usefulness of eqn. (7.26) severely.

Moreover the circuit can only be used for steady magnetic fields. Attempts have recently been made to widen the application of magnetic

* The reciprocal of reluctance is called permeance.

circuits,† and the reader may like to refer to them. Caution is needed, especially as the correct physical analogue of magnetic flux is electric flux and not electric current. Reluctance should therefore be taken as the inverse of "magnetic capacitance" rather than "magnetic resistance", if a *physical* analogue is sought. This restriction does not apply if the analogue is only a *mathematical* equivalence. The papers mentioned in the footnote stress the usefulness of magnetic equivalent circuits in the analysis of devices where the flux paths are well defined and the current paths are complicated.

7.2. Application of Laplace's Equation to Magnetic Problems

7.2.1. The Field of a Long Current Filament

We touched on this problem in § 6.2.8, where we found the magnetic field of such a current placed along the z-axis of a cylindrical coordinate system. We shall now find the field of a current filament of strength I along a line defined by the coordinates c, ψ as shown in Fig. 7.12. We shall make use of the results in § 4.3.2, where we found the electric field of a filament of charge.

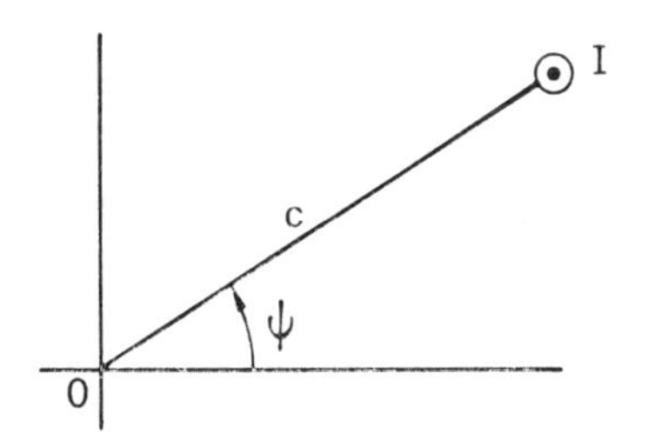

FIG. 7.12. Field of a current filament at point C, ψ.

Consider first the simpler case where $\psi = 0$ as shown in Fig. 7.13 and take $r > c$. At the point $(r, 0)$ the magnetic field is known by eqn. (6.16) as

$$H_\theta = \frac{I}{2\pi(r - c)} = \frac{I}{2\pi r}\left(1 - \frac{c}{r}\right)^{-1}. \tag{7.27}$$

† E. R. Laithwaite, Magnetic equivalent circuits for electrical machines, *Proc. I.E.E.*, 1967, Vol. 114 (11), and C. J. Carpenter, Magnetic equivalent circuits, *Proc. I.E.E.*, 1968, Vol. 115 (10).

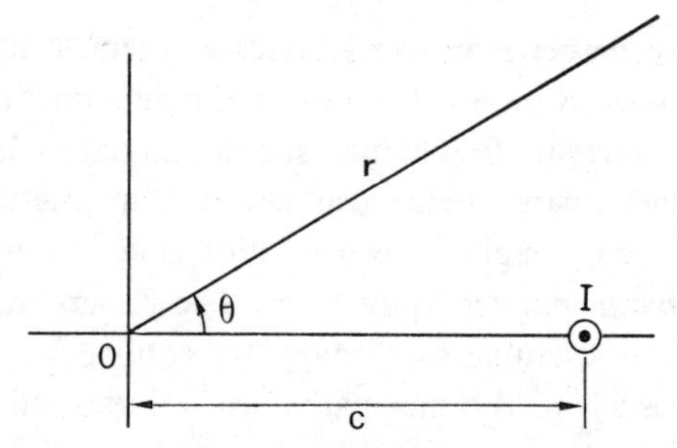

FIG. 7.13. Field of a current filament at point C, 0.

Expanding this expression we have

$$H_\theta = \frac{I}{2\pi r}\left(1 + \frac{c}{r} + \frac{c^2}{r^2} + \cdots + \frac{c^n}{r^n} + \cdots\right). \tag{7.28}$$

In terms of a scalar magnetic potential V^* we have, by the use of eqn. (6.18),

$$H_\theta = -\frac{1}{r}\frac{\partial V^*}{\partial \theta} = \frac{I}{2\pi r}\sum_{n=0}^{\infty}\frac{c^n}{r^n} \tag{7.29}$$

and

$$\frac{\partial V^*}{\partial \theta} = -\frac{I}{2\pi}\sum_{n=0}^{\infty}\frac{c^n}{r^n}. \tag{7.30}$$

We know from eqn. (4.69) that at the point r, θ the circular functions $\sin n\theta$, $\cos n\theta$ are associated with $r^{\pm n}$, if Laplace's equation is to be satisfied.

Hence, at the point (r, θ),

$$\frac{\partial V^*}{\partial \theta} = -\frac{I}{2\pi}\sum_{0}^{\infty}\frac{c^n}{r^n}\cos n\theta \tag{7.31}$$

and

$$V^* = -\frac{I}{2\pi}\left[\theta + \sum_{1}^{\infty}\frac{c^n}{nr^n}\sin n\theta\right]. \tag{7.32}$$

In order to examine whether this is the complete solution we refer to eqns. (4.69) and (4.70). Since H_θ does not change its sign as θ changes sign, we cannot admit $\sin n\theta$ in eqn. (7.31) and hence $\cos n\theta$ cannot appear in eqn. (7.32). Since H_r is zero at $\theta = 0$ and $\theta = \pi$, we cannot admit either $\ln r$ or $\theta \ln r$. Hence eqn. (7.32) is the complete solution. If now the filament is placed at c, ψ as in Fig. 7.12, the potential can be found by replacing θ by $(\theta - \psi)$ in eqn. (7.32). The case of $c > r$ is left as an exercise for the reader. The potential of several currents can be obtained by superposition.

7.2.2. A Magnetic Cylinder in a Uniform Transverse Magnetic Field

Let the applied uniform field be H_0 and let us remember that the field *was* uniform before the cylinder was introduced. Let the relative permeability of the cylinder be μ_r and its radius be a. Equations (4.77)–(4.83) apply if we substitute H_0 for E_0 and μ_r for ε_r. The outside potential is

$$V_1^* = H_0 r \cos\theta \left(\frac{a^2}{r^2}\frac{\mu_r - 1}{\mu_r + 1} - 1\right). \tag{7.33}$$

The first term gives the action of the cylinder and the second the action of the applied field.

The inside potential is

$$V_2^* = -\frac{2H_0 r \cos\theta}{(\mu_r + 1)}. \tag{7.34}$$

The surface polarity has a density

$$q^* = 2\mu_0 H_0 \cos\theta \left(\frac{\mu_r - 1}{\mu_r + 1}\right). \tag{7.35}$$

The flux through the cylinder per unit length can be found from eqn. (7.33). The radial external field is

$$H_r = -\frac{\partial V_1^*}{\partial r} = H_0 \cos\theta \left(\frac{a^2}{r^2}\frac{\mu_r - 1}{\mu_r + 1} + 1\right) \tag{7.36}$$

and the flux density is given by $B_r = \mu_0 H_r$. This flux density is continuous through the surface of the cylinder, so that

$$\Phi = \int_{-(\pi/2)}^{+(\pi/2)} B_r r \, d\theta \tag{7.37}$$

at $r = a$.

Hence

$$\Phi = 2\mu_0 H_0 a \frac{2\mu_r}{\mu_r + 1}. \tag{7.38}$$

The value of $2\mu_r/(\mu_r + 1)$ will be very close to two so that

$$\Phi \doteqdot 4\mu_0 H_0 a, \tag{7.39}$$

but this is twice the flux which would have passed through a non-magnetic cylinder of radius a. We therefore have the important result that a cylinder of magnetic material doubles the flux. The actual value of μ_r does not matter, nor need it be constant at all points. What does matter is that $\mu_r/(\mu_r + 1) \doteqdot 1$. Figure 7.14 is a flux plot showing **B** both outside and

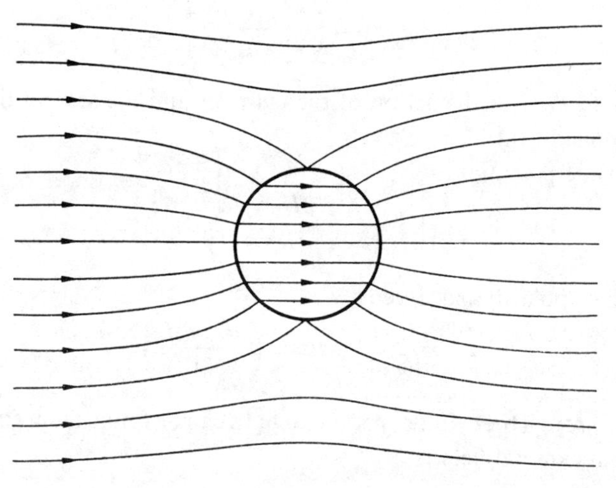

FIG. 7.14. Flux plot of a magnetic cylinder in a uniform field.

inside such a cylinder with $\mu_r = 1000$. It will be seen that the cylinder "gathers flux to itself" from a region equivalent to twice its diameter. The doubling of the flux is not dependent on the fact that the applied field was uniform, but depends only on $\mu_r/(\mu_r + 1) \doteqdot 1$.†

† A sphere in a *uniform* field trebles the flux.

7.2.3. *Magnetic Screening by Means of a Tube*

Consider a tube of external radius a, internal radius b, and constant relative permeability μ_r, which is inserted into an applied uniform field H_0. Three regions need to be considered as in Fig. 7.15. Since the applied

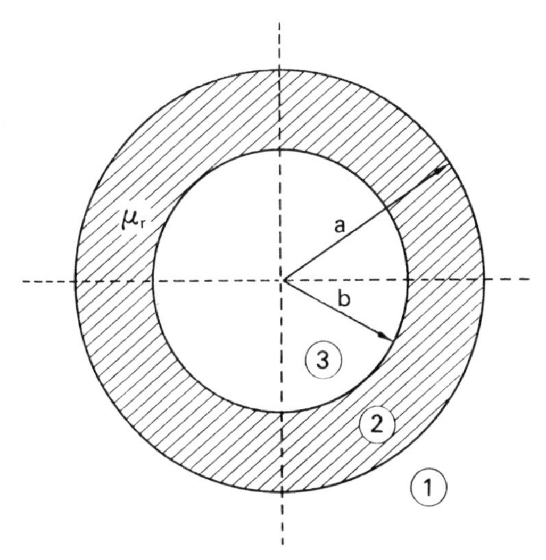

FIG. 7.15. Screening effect of an iron tube.

field varies as $\cos\theta$ and since μ_r is constant, the field of the tube will also vary as $\cos\theta$. A knowledge of possible solutions of Laplace's equation [eqn. (4.69)] tells us that the potential due to the tube must vary as $r\cos\theta$ and $(1/r)\cos\theta$. Outside the tube, where r becomes large, we cannot use $r\cos\theta$ because the effect of the tube will become smaller as we move away from it. In region 3, the "screened" region, we must discard $(1/r)\cos\theta$ which would become infinite at $r = 0$. We therefore have

$$\left.\begin{aligned} V_1^* &= -H_0 r\cos\theta + \frac{A}{r}\cos\theta, \\ V_2^* &= -H_0 r\cos\theta + \frac{C}{r}\cos\theta + Dr\cos\theta, \\ V_3^* &= -H_0 r\cos\theta + Er\cos\theta. \end{aligned}\right\} \quad (7.40)$$

The constants A, C, D, and E can be found by using the boundary conditions of eqns. (6.52) and (6.58) at the outer and inner radius.

After some reduction,

$$V_3^* = -\frac{4\mu_r H_0 r \cos\theta}{(\mu_r + 1)^2 - (b^2/a^2)(\mu_r - 1)^2}. \tag{7.41}$$

If $\mu_r \gg 1$,

$$V_3^* = -\frac{4}{\mu_r}\frac{H_0 r \cos\theta}{1 - (b^2/a^2)}, \tag{7.42}$$

and if the inside radius is small compared with the outside one,

$$V_3^* = -\frac{4}{\mu_r} H_0 r \cos\theta, \tag{7.43}$$

so that the field in the screened region is then $4/\mu_r$ times the original applied field.

If the tube is thin,

$$V_3^* = -\frac{H_0}{\mu_r}\frac{d}{t} r \cos\theta, \tag{7.44}$$

where d is the mean diameter and t the thickness. The walls of the tube carry most of the flux, and this corresponds to the fact that V_3^* is much smaller than the potential of the applied field. But this implies that the tube must be thick enough so that the iron does not saturate. Otherwise the effective value of μ_r will drop rapidly and the screening action will be impaired. For a detailed study the reader may like to refer to a paper by the author.†

7.3. Magnetostatic Images

We briefly discussed the method of images in § 4.3.4 in the context of electrostatics and the reader may like to refer to that section. We seek to solve a problem of current sources surrounded by magnetic boundaries by replacing these boundaries by equivalent currents.

† P. Hammond, The screening effect of iron tubes, *Proc. I.E.E.*, Vol. 163, part C, 1955, Monograph No. 144.

Consider Fig. 7.16 which shows a line current I near a surface of magnetic material of relative permeability μ_r. Since $\mu_r \gg 1$ in many cases, let us first solve the simpler problem of $\mu_r = \infty$. There must be no magnetic field strength along the surface of the material. The current by itself does produce such a tangential field strength, so the surface polarity must produce an equal and opposite field. Clearly a second current filament placed as shown in Fig. 7.17 also gives the correct opposing field. Hence we can replace the action of the magnetic material by the equivalent "image" current. The uniqueness theorem discussed in §§ 3.2.7 and 4.3.1 ensures that if the field is correct on the boundary it will be correct everywhere outside the magnetic material.

It should be noted explicitly that the image of a charge filament in a conductor is of opposite sign, whereas the image of a current filament is of the same sign.

If we now assume a constant finite value for μ_r we realize that the pole strength on the material will be as shown in § 6.4.2.

$$\frac{q^*}{2\mu_0} = \frac{\mu_r - 1}{\mu_r + 1}\,\hat{\mathbf{n}}\,.\,\mathbf{H}. \tag{7.45}$$

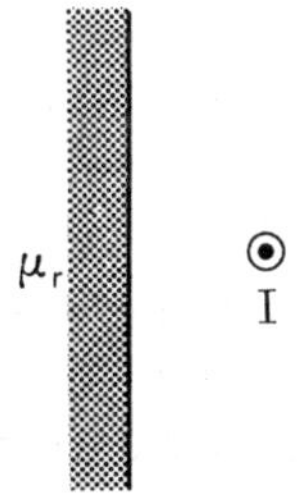

FIG. 7.16. Line current in front of magnetic slab.

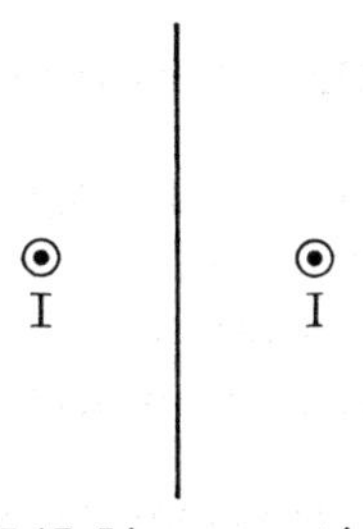

FIG. 7.17. Line current image.

Thus the field of this pole strength is proportional to $(\mu_r - 1)/(\mu_r + 1)$ and the image current will also have to be reduced in this ratio. Similar reasoning shows that the field inside the material can be represented by an image current outside. The image distribution must always be outside the region considered because we need the effect at a distance but not the source itself. Images of coils, current elements, and various shapes of conductor are discussed in a paper by the author.†

The method of magnetic images is useful if the boundaries are planes and cylinders. With such boundaries the shape of the image source is often the same as the true source. Unfortunately, the method of images is of little use unless we can guess the shape of the image source. In general we are driven back to the calculation of the surface polarity.

7.4. Energy in the Magnetostatic Field

7.4.1. Magnetic Field Energy in Terms of H and B

Energy has to be expended to cause electric current to flow. Some of this energy is recoverable and can, therefore, be thought of as having been stored in the magnetic field which is associated with the electric current. In circuit language this phenomenon is described by the term inductance. The analogous mechanical phenomenon is inertia, which describes the kinetic energy of moving mass, while inductance is associated with moving electric charge. The magnetic field energy is affected by the presence of magnetic materials, as might be expected when it is remembered that these materials contain additional moving charges.

Energy may also have to be expended to assemble currents and magnets, because work has to be done against the forces acting between them.

Both the inductance and the assembly work energies are associated with the magnetic field. When there is no field there is no inductance nor any interaction. It is therefore reasonable to seek an expression for the stored energy in terms of the field vectors. In § 5.1 we carried out a similar investigation for the electrostatic field and arrived at expressions for the

† P.Hammond, Electric and magnetic images, *Proc. I.E.E.*, Vol. 107, Part C, 1960, Monograph No. 319.

field energy from a consideration of the work needed to assemble charges. But having done so we noticed in § 5.2 that the same expressions could have been more quickly obtained by considering the tubes of electric flux. We argued that since the energy was associated with the flux it must be possible to find the energy by considering the flux. We found that this was so and that the field contains all the information to determine the energy.

In considering the magnetic field energy we shall use the same technique straight away, without first calculating the assembly work. Consider the tube of flux in Fig. 7.18. In the electric field we replaced the tube by a succession of small capacitors (see Fig. 5.2). In the magnetic field we shall instead replace each tube of flux by a number of small solenoids placed side by side as in Fig. 7.19. The current on the solenoids will be

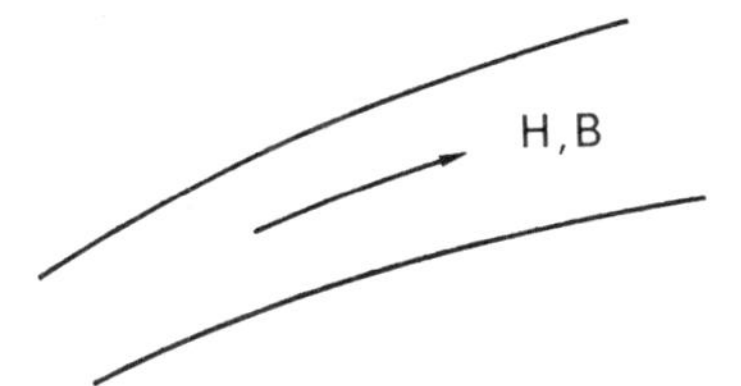

FIG. 7.18. Tube of magnetic flux.

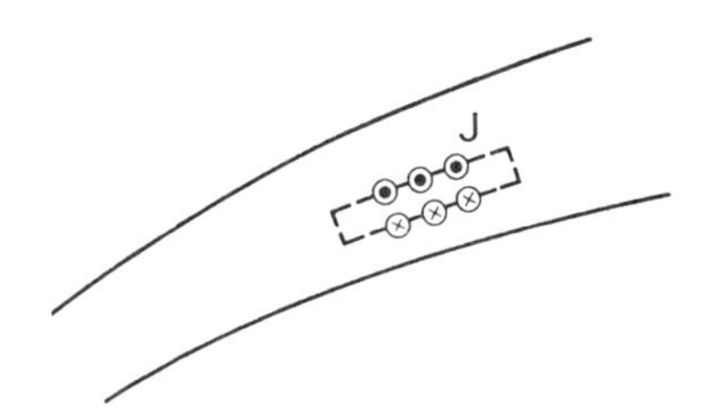

FIG. 7.19. Replacement of magnetic flux by solenoid.

J A/m and in order to give the correct magnetic field strength inside the solenoid we have $J = H$. If the solenoid has length δl and cross-section δs, the stored energy associated with it is given by

$$\delta U^* = \int iV\,dt, \tag{7.46}$$

where i is the current and V the voltage. Hence

$$\delta U^* = \int J\,\delta l\,\frac{\partial B}{\partial t}\,\delta s\,dt = \int_0^B \mathbf{H}\,.\,d\mathbf{B} \qquad \text{per unit volume}\,. \tag{7.47}$$

Hence the field energy is

$$U^* = \iiint \left[\int_0^B \mathbf{H} \, . \, d\mathbf{B} \right] dv \, . \tag{7.48}$$

If **B** is proportional to **H** this can be written

$$U^* = \tfrac{1}{2} \iiint \mathbf{H} \, . \, \mathbf{B} \, dv \, . \tag{7.49}$$

Notice that in eqns. (7.47) and (7.48) **H** and **B** are the values during the process of magnetization, but in eqn. (7.49) they are the final values as measured in a flux plot.

If **B** is not proportional to **H** but is a single-valued function of **H**, as for instance in Fig. 6.13, eqn. (7.48) gives a unique answer. But if there is hysteresis as shown in Fig. 6.15 the previous magnetic state will decide the value of the integral. The hysteresis loss in a cyclic process is given by the cyclic integral

$$\Delta U^* = \iiint [\oint \mathbf{H} \, . \, d\mathbf{B}] \, dv \tag{7.50}$$

which is equal to the area of the hysteresis curve integrated throughout the volume of the material.

7.4.2. Magnetic Field Energy in Terms of Current and Flux

When dealing with current circuits it is often convenient to know the field energy in terms of current and flux linkage. We can derive this expression from the results of the previous section. Let us take eqn. (7.49) as the starting point

$$U^* = \tfrac{1}{2} \iiint \mathbf{H} \, . \, \mathbf{B} \, dv \, . \tag{7.49}$$

We can make use of the vector potential **A**, where

$$\mathbf{B} = \text{curl } \mathbf{A} \tag{6.25}$$

and of the vector identity

$$\text{div } (\mathbf{A} \times \mathbf{H}) = \mathbf{H} \, . \, \text{curl } \mathbf{A} - \mathbf{A} \, . \, \text{curl } \mathbf{H} \, . \tag{7.51}$$

Hence

$$U^* = \tfrac{1}{2} \iiint \mathbf{H} \, . \, \text{curl } \mathbf{A} \, dv = \tfrac{1}{2} \iiint \mathbf{A} \, . \, \text{curl } \mathbf{H} \, dv + \iint (\mathbf{A} \times \mathbf{H}) \, . \, d\mathbf{s} \, . \tag{7.52}$$

The surface integral is zero if the integration is carried out over all space, because **A** decays at least as $1/r$ and **H** as $1/r^2$, whereas ds increases as r^2. Since curl $\mathbf{H} = \mathbf{J}$ we have

$$U^* = \tfrac{1}{2} \iiint \mathbf{A} \,.\, \mathbf{J} \, dv \,. \tag{7.53}$$

In this equation the field energy is described in terms of the current source of the field instead of the field itself. Compare the similar expression for the electrostatic field in eqn. (5.4) which uses the integral $\frac{1}{2} \iiint V\varrho \, dv$. We shall make use of this source formulation again in the next chapter, when we come to discuss electromagnetic radiation.

Let us now apply eqn. (7.53) to a system of n filamentary circuits carrying currents I_k.

$$U^* = \tfrac{1}{2} \sum_{k=1}^{n} I_k \oint \mathbf{A} \,.\, d\mathbf{l}_k \,. \tag{7.54}$$

By the use of eqn. (6.27) we can write

$$\Phi_k = \oint \mathbf{A} \,.\, d\mathbf{l}_k \tag{7.55}$$

so that

$$U^* = \tfrac{1}{2} \sum_{k=1}^{n} I_k \Phi_k \,. \tag{7.56}$$

Having obtained the magnetic field energy let us now consider other possible energy terms, so that later we can correctly apply the principle of conservation of energy and also find the forces and torques on electric currents and magnetized matter.

7.4.3. *Assembly Work of Magnetic Poles and Dipoles*

If we regard the magnetostatic field as arising from poles, we can make use of the results in § 5.1 and write the assembly work in free space as

$$W^* = \frac{\mu_0}{2} \iiint H^2 dv - \frac{\mu_0}{2} \iiint H_i^2 dv \tag{7.57}$$

and, in terms of field energy,

$$W^* - U^* - U_s^* \,, \tag{7.58}$$

where U^* is the field energy and U_s^* the self-energy of the system of poles. Notice that we have been able to equate assembly work and field energy, because a system of constant poles is conservative so that energy cannot be exchanged with it in any cyclic process. It behaves as if its stored energy were potential energy, although our discussion in the previous section led us to consider magnetic energy as being kinetic in origin. We conclude that poles of constant strength do not constitute an adequate model for real magnets.

A slightly more realistic model is furnished by an assembly of dipoles of constant strength. Consider the assembly work of two dipoles of moment $\mathbf{P}_1^*$ and $\mathbf{P}_2^*$. The second dipole is inserted into the field of the first,

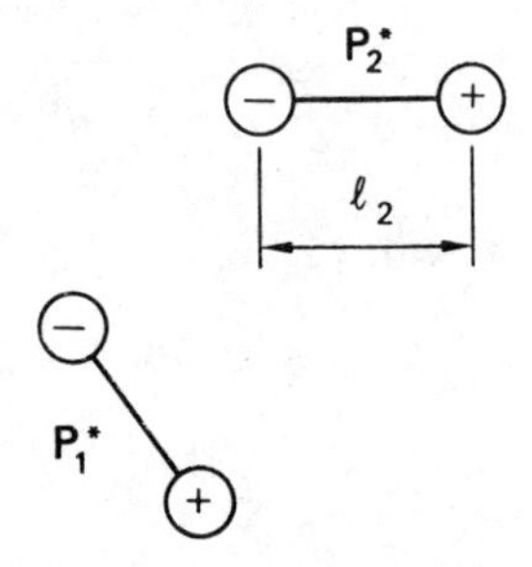

FIG. 7.20. Interaction of two dipoles.

as shown in Fig. 7.20. We can envisage this insertion as having been brought about by the movement of the positive end of the dipole through the distance $\mathbf{l}_2$. The assembly work is

$$W_{12}^* = -\mathbf{H}_{12} \cdot \mathbf{P}_2^*, \tag{7.59}$$

where $\mathbf{H}_{12}$ is the field due to the first dipole at the second dipole. It cannot matter which dipole is inserted first, so that

$$W_{12}^* = W_{21}^* = -\tfrac{1}{2}(\mathbf{H}_{12} \cdot \mathbf{P}_2^* + \mathbf{H}_{21} \cdot \mathbf{P}_1^*). \tag{7.60}$$

Hence, for a set of n dipoles,

$$W^* = -\tfrac{1}{2}\sum_{i=1}^{n} \mathbf{H}_i \cdot \mathbf{P}_i^*, \tag{7.61}$$

where $\mathbf{H}_i$ is the field at the ith dipole due to all the other dipoles. If we now take $\mathbf{P}^*$ as the dipole moment per unit volume the summation of eqn. (7.55) becomes an integral

$$W^* = -\tfrac{1}{2}\iiint \mathbf{H}\,.\,\mathbf{P}^*\,dv, \tag{7.62}$$

and, since $\mathbf{B} = \mu_0\mathbf{H} + \mathbf{P}^*$, we can write

$$W^* = \frac{\mu_0}{2}\iiint H^2 dv - \tfrac{1}{2}\iiint \mathbf{H}\,.\,\mathbf{B}\,dv. \tag{7.63}$$

We shall show in a moment that the second integral is zero, so that

$$W^* = \frac{\mu_0}{2}\iiint H^2\,dv = U^*.$$

Hence the only difference between the expression for poles and dipoles is the different self-energy, which does not enter into calculations involving changes of energy.

We now have to show that $\iiint \mathbf{H}\,.\,\mathbf{B}\,dv = 0$. Consider the vector identity

$$\operatorname{div}(V^*\mathbf{B}) = V^* \operatorname{div}\mathbf{B} + \mathbf{B}\,.\,\operatorname{grad} V^*. \tag{7.64}$$

For the magnetic field $\operatorname{div}\mathbf{B} = 0$ and for a field due to poles or dipoles only, $\operatorname{curl}\mathbf{H} = 0$ and hence $\mathbf{H} = -\operatorname{grad} V^*$. Hence

$$\iiint \mathbf{H}\,.\,\mathbf{B}\,dv = -\iiint \mathbf{B}\,.\,\operatorname{grad} V^* dv = -\iiint \operatorname{div} V^*\mathbf{B}\,dv = -\oiint V^*\mathbf{B}\,.\,d\mathbf{s}. \tag{7.65}$$

Now V^* decays at least as $1/r$ for poles and $1/r^2$ for dipoles and $\mathbf{B}$ decays at least as $1/r^2$ for poles and $1/r^3$ for dipoles, whereas $d\mathbf{s}$ increases as r^2. Taken over a large surface the integral tends to zero and so the volume integral over all space is zero.

At first this is a surprising result, because in eqn. (7.49) this integral described the field energy of a system in which $\mathbf{B}$ was proportional to $\mathbf{H}$, and this field energy is by no means zero in general. The answer to the puzzle is revealing. The field energy is different from zero if there are current sources, so that $\operatorname{curl}\mathbf{H} \neq 0$. In a system without current sources, in which the magnetic material is "soft", so that $\mathbf{B}$ is proportional to $\mathbf{H}$, we should have div **H** = 0 because div **B** = 0. Hence **H** would have neither curl nor divergence sources. Hence there would be no field and of course no field energy.

It is interesting to notice that the *B–H* characteristic for a material of constant dipole moment would be a straight line as shown in Fig. 7.21. This characteristic is observed in barium ferrite, a ceramic permanent magnet material.

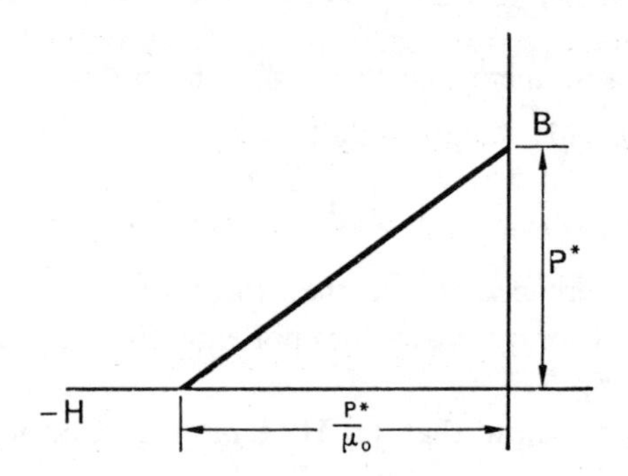

FIG. 7.21. *B–H* curve for material of constant polarization.

We conclude that the equivalence of currents and magnetic sources is of great help in calculating external magnetic fields, but that poles and dipoles are not very useful in the calculation of field energy. The more general method of § 7.4.1 is to be preferred for such calculations.

7.4.4. Energy of Current Circuits

We must now consider the assembly of current circuits, because this will enable us to understand the energy processes far better than the study of poles and dipoles. When currents are moved or changed in magnitude, e.m.f.s are induced and energy can be supplied to the system or taken from it. This is exactly the sort of process which engineers need in order to drive machinery or generate electric power.

Consider two coils carrying steady currents I_1, I_2, and provided with sources of electromotive force ε_1, ε_2, which supply or absorb energy in such a manner as to keep the currents constant. Let the flux linkage with the coils be Φ_1, Φ_2 and the ohmic resistance R_1, R_2. Then

$$\left.\begin{aligned} \varepsilon_1 - \frac{d\Phi_1}{dt} &= R_1 I_1, \\ \varepsilon_2 - \frac{d\Phi_2}{dt} &= R_2 I_2. \end{aligned}\right\} \tag{7.66}$$

Let the two coils be brought near to each other from a great distance at which there is no mutual flux linkage. The energy supplied is

$$\begin{aligned} \int_0^t (\varepsilon_1 I_1 + \varepsilon_2 I_2)\, dt &= I_1 \int_0^t \varepsilon_1 dt + I_2 \int \varepsilon_2\, dt \\ &= (I_1^2 R_1 + I_2^2 R_2)t + I_1(\Phi_1 - \Phi_{01}) + I_2(\Phi_2 - \Phi_{02}), \end{aligned} \tag{7.67}$$

where Φ_1, Φ_2 are the total fluxes and Φ_{01}, Φ_{02} the fluxes of self-inductance. Thus the sources of e.m.f. supply the ohmic losses and an amount of energy which can be written by the use of eqn. (7.56) as $2U^* - 2U_0^*$. This energy must include the change in field energy plus the work done against *external* mechanical forces. But the change in field energy is $U^* - U_0^*$. Hence the mechanical work is also $U^* - U_0^*$. Apart from the ohmic loss half of the energy has been turned into mechanical work and half has been stored as field energy.

In terms of inductance, eqn. (7.67) can be rewritten

$$\int_0^t (\varepsilon I_1 + \varepsilon I_2)\, dt = (I_1^2 R_1 + I_2^2 R_2)\, t + 2MI_1 I_2, \tag{7.68}$$

so that

$$U^* - U_0^* = MI_1 I_2 \tag{7.69}$$

and

$$U^* = \tfrac{1}{2} L_1 I_1^2 + \tfrac{1}{2} L_2 I_2^2 + MI_1 I_2. \tag{7.70}$$

Thus the mechanical work done during the assembly of the coils, their currents being kept constant, is MI_1I_2. It should be noted that the field energy depends only on the final state and not on the process, but the total energy supplied and the mechanical work done depend on the process used.

7.5. Force in the Magnetostatic Field

7.5.1. Forces on Magnetic Poles and Electric Currents

From the definition of the magnetic field strength the force on a pole of strength Q_1^* can be written

$$\mathbf{F}_1 = \mathbf{H}_1 Q_1^* , \tag{7.71}$$

where $\mathbf{H}_1$ can be calculated from the pole and current sources by the use of eqns. (2.22) and (2.37),

$$\mathbf{H}_1 = \sum_{i=2}^{n} \frac{Q_i^*}{4\pi\mu_0 r_{i1}^2} \hat{\mathbf{r}}_{i1} + \sum_{j=2}^{n} \frac{I_j (\delta\mathbf{l}_j \times \hat{\mathbf{r}}_{j1})}{4\pi r_{j1}^2} . \tag{7.72}$$

In terms of pole density ϱ^*, the force per unit volume is

$$\mathbf{f} = \mathbf{H}\varrho^* . \tag{7.73}$$

Force on electric current can be obtained from § 2.3.

$$\mathbf{F}_1 = I_1 \delta\mathbf{l}_1 \times \mathbf{B}_1 . \tag{2.41}$$

The use of **H** as force per unit pole and **B** as force on current is dimensionally correct, but the reader may wonder what would happen if these forces were to be calculated for a magnetic material of relative permeability μ_r. As far as force on pole strength is considered, the answer is straightforward. We showed in § 6.4.1 that μ_r was introduced in order to account for the additional polarity of magnetized material. To introduce μ_r into the force formulae would be to count the polarity twice, because it is merely an alternative description of pole strength. Equations (7.71)–(7.73) are correct as they stand.

The discussion of eqn. (2.42) is slightly more troublesome. Does **B** in this equation signify $\mu_0\mathbf{H}$ or $\mu_0\mu_r\mathbf{H}$? Rather surprisingly the answer is that it does not matter, but that it is simpler to use $\mathbf{B} = \mu_0\mathbf{H}$. Consider an iron wire of circular cross-section and relative permeability μ_r carrying a current. Let this wire be subjected to the magnetic field of external currents and magnets. What is the force on the wire? Let us first consider the wire without the current. The magnetic effect of the iron can be represented by a surface layer of poles and the force can be calculated in the manner typified by eqn. (7.71). Now let the current flow in the wire. This will not affect the pole strength. The outside sources will, however, experience the field of the current, and since action and reaction are equal and opposite there will be an additional force on the wire. This force is the same as would be experienced by a current in a non-magnetic material. We can use $\mathbf{B} = \mu_0 \mathbf{H}$ in our calculation. Why then did we say that it did not matter which expression for $\mathbf{B}$ was used?

The answer lies in the fact that we have avoided the discussion of the forces between the current and the iron. Externally these forces cancel out. But if we want the details we shall have to decide what is the local field in the iron. This immediately leads us back to the discussion of § 6.4.3. The force on the current will depend on the type of cavity in which it is postulated to flow. But if we choose the disc cavity we must remember the additional surface polarity on the walls of the cavity and add the force of the current on these poles. Since the outside world will not be affected by the choice of cavity, the total force on iron and current will remain the same. This discussion is not as academic as it sounds, because in many devices such as electrical machines the engineer deliberately puts his current into slots or cavities. It is, therefore, worth while to discuss a particular example in full.

7.5.2. *Force on a Current Screened by a Magnetic Tube*

In § 7.2.3 we discussed the screening effect of a magnetic tube. Let us extend that discussion by considering the effect of a current I along the axis of the tube as shown in Fig. 7.22. The field in the screened region (region 3) can be derived from eqn. (7.41). It is

$$H_3 = \frac{4\mu_r}{(\mu_r + 1)^2 - (b^2/a^2)(\mu_r - 1)^2} H_0 . \tag{7.74}$$

Suppose $\mu_r = 1000$ and $a = 2b$, then $H_3 = 0{\cdot}0053 H_0$, so that the field in the screened region is 0·53% of the original field. Hence the force on the current is 0·53% of the force which the current would experience in the absence of the tube. How can this be reconciled with the statement in the last section that the force on the tube and the current is the same as if the current were flowing in empty space?

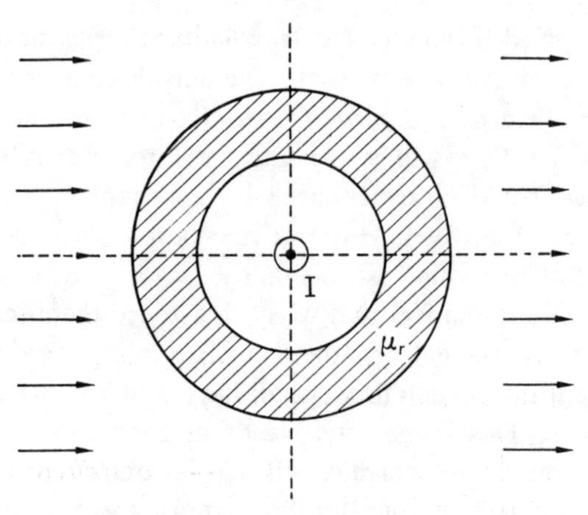

FIG. 7.22. Current screened by magnetic tube.

Consider the pole strength on the surfaces of the tube. This can be obtained by the method of § 4.3.2 and eqn. (4.83). Let q_a^*, q_b^* be the surface densities at radius a and b respectively.

$$\frac{q_a^*}{\mu_0} = (H_{r\,1} - H_{r\,2})_{r=a} = \left(\frac{\partial V_2^*}{\partial r} - \frac{\partial V_1^*}{\partial r}\right)_{r=a}, \tag{7.75}$$

$$\frac{q_b^*}{\mu_0} = \left(\frac{\partial V_3^*}{\partial r} - \frac{\partial V_2^*}{\partial r}\right)_{r=b}. \tag{7.76}$$

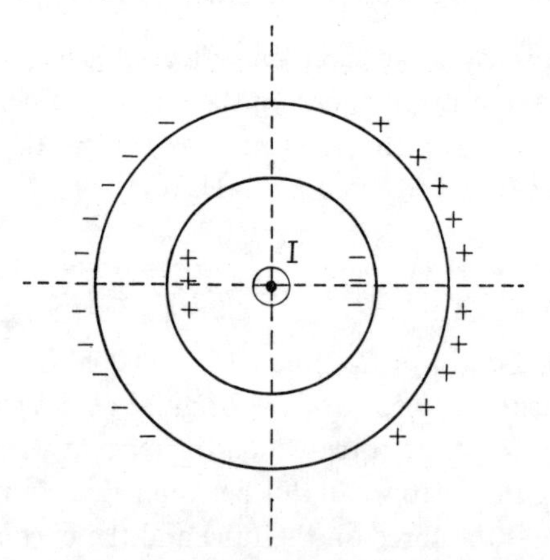

FIG. 7.23. Surface polarity on magnetic tube.

The potentials V_1^*, V_2^*, and V_3^* can be calculated by the use of eqn. (7.40), and after some reduction

$$q_a^* = \frac{2(\mu_r - 1)[(\mu_r + 1) - (\mu_r - 1)(b^2/a^2)]}{(\mu_r + 1)^2 - (\mu_r - 1)^2 (b^2/a^2)} \mu_0 H_0 \cos\theta, \tag{7.77}$$

$$q_b^* = -\frac{2(\mu_r - 1)}{(\mu_r + 1)^2 - (\mu_r - 1)^2 (b^2/a^2)} \mu_0 H_0 \cos\theta. \tag{7.78}$$

These polarities are indicated in sign and position, but not in magnitude, on Fig. 7.23. Consider now the vertical force due to the action of the magnetic field of the current on the polarity.

$$\begin{aligned} F &= \int_0^{2\pi} \left[\frac{I}{2\pi a} q_a^* a \, d\theta \cos\theta + \frac{I}{2\pi b} q_b^* b \, d\theta \cos\theta \right] \\ &= \frac{(\mu_r - 1)^2 [1 - (b^2/a^2)]}{(\mu_r + 1)^2 - (b^2/a^2)(\mu_r - 1)^2} IH_0. \end{aligned} \tag{7.79}$$

If we add this force to the force on the current IH_3, we obtain a force

$$F = \frac{(\mu_r - 1)^2 [1 - (b^2/a^2)] + 4\mu_r}{(\mu_r + 1)^2 - (b^2/a^2)(\mu_r - 1)^2} IH_0 = IH_0 \tag{7.80}$$

which is, of course, the force on the current in the absence of the tube. Thus in our example $99\frac{1}{2}\%$ of the force has been transferred from the current to the tube, but the total force remains unchanged. In electrical machines it is often very desirable for mechanical reasons that the forces should act on the iron rather than on the conductors. These are, therefore, generally put into slots or tunnels in the iron.

7.5.3. Force on Current Circuits

Equation (2.41) can be applied to the force on current circuits. Consider two such circuits. The force on the second due to the first is given by

$$\begin{aligned} \mathbf{F}_2 &= I_2 \oint d\mathbf{l}_2 \times \mathbf{B} \\ &= \frac{\mu_0 I_2 I_1}{4\pi} \oint_2 d\mathbf{l}_2 \times \oint_1 \frac{d\mathbf{l}_1 \times \hat{\mathbf{r}}_{12}}{r_{12}^2} \\ &= \frac{\mu_0 I_1 I_2}{4\pi} \oint_2 \oint_1 \frac{d\mathbf{l}_2 \times (d\mathbf{l}_1 \times \hat{\mathbf{r}}_{12})}{r_{12}^2}. \end{aligned} \tag{7.81}$$

This expression can be simplified by the use of the vector identity

$$\mathbf{A} \times (\mathbf{B} \times \mathbf{C}) = (\mathbf{A} \,.\, \mathbf{C})\,\mathbf{B} - (\mathbf{A} \,.\, \mathbf{B})\,\mathbf{C}, \tag{7.82}$$

$$\mathbf{F}_2 = \frac{\mu_0 I_1 I_2}{4\pi} \oint\oint \left[\frac{(d\mathbf{l}_2 \,.\, \hat{\mathbf{r}}_{12})}{r_{12}^2} \, d\mathbf{l}_1 - \frac{(d\mathbf{l}_2 \,.\, d\mathbf{l}_1)}{r_{12}^2} \, \hat{\mathbf{r}}_{12} \right]. \tag{7.83}$$

The first term in this expression is zero because the integral with respect to $d\mathbf{l}_2$ can be written $\oint \nabla (1/r) \,.\, d\mathbf{l}_2$ and the cyclic line integral of a gradient is zero. Hence

$$\mathbf{F}_2 = -\frac{\mu_0 I_1 I_2}{4\pi} \oint_1 \oint_2 \frac{(d\mathbf{l}_1 \,.\, d\mathbf{l}_2)}{r_{12}^2} \, \hat{\mathbf{r}}_{12}. \tag{7.84}$$

A useful alternative form in terms of flux linkage and current can be obtained from this expression by writing

$$\mathbf{F}_2 = \frac{\mu_0 I_1 I_2}{4\pi} \nabla \oint_1 \oint_2 \frac{d\mathbf{l}_1 \,.\, d\mathbf{l}_2}{r_{12}} = I_2 \nabla \Phi, \tag{7.85}$$

making use of eqn. (6.29).

Thus

$$F_x = I \frac{\partial \Phi}{\partial x} \tag{7.86}$$

and similarly the torque is given by

$$T_\theta = I \frac{\partial \Phi}{\partial \theta}. \tag{7.87}$$

Equations (7.86) and (7.87) are quite general and apply to any system of current circuits, as can be seen by retracing the steps of the argument of the section. Moreover, if the current links magnetic material, this material will increase the flux. For a complete magnetic circuit the increase will be in the ratio μ_r. The correct value for the flux in the equations is therefore the surface integral of $\mathbf{B} = \mu_0 \mu_r \mathbf{H}$.

In terms of mutual inductance eqns. (7.86) and (7.87) can be written

$$F_x = I_1 I_2 \frac{\partial M}{\partial x}, \tag{7.88}$$

$$T_\theta = I_1 I_2 \frac{\partial M}{\partial \theta}. \tag{7.89}$$

7.5.4. Magnetic Field Stress

The most powerful, and in many cases the simplest, method for calculating the force due to magnetic interaction is the method of field stresses. If the magnetic field surrounding a body is known, for instance by means of a flux plot, we can find the force on the body by postulating a magnetic field stress at every point on the enclosing surface. The parallel electric case was discussed in great detail in § 5.4 and the reader should refer to that section. As in the electric case, so in the magnetic one we have to terminate the tubes of magnetic flux by suitable surface layers. The force on these surfaces then gives the required field stress.

Figure 7.24 shows a transverse cut in a typical tube of flux and Fig. 7.25

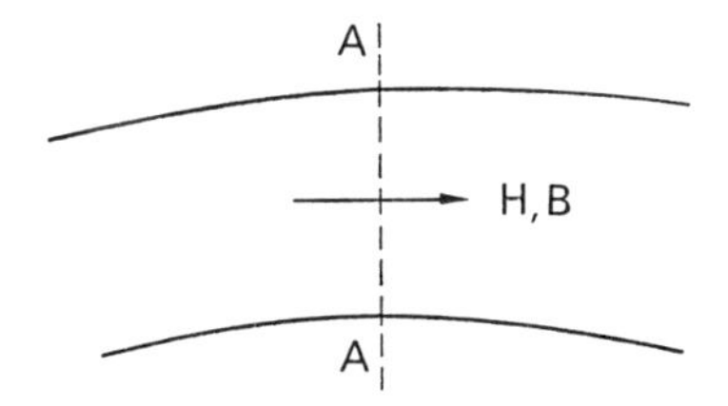

FIG. 7.24. Transverse cut in tube of magnetic flux.

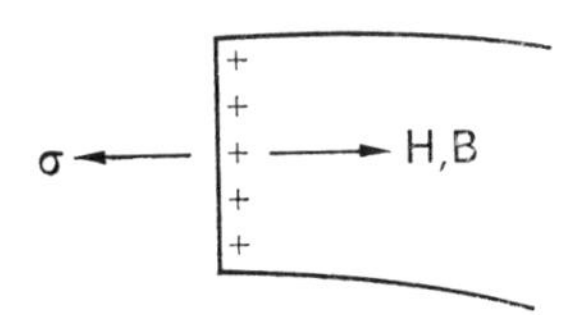

FIG. 7.25. Tensile stress along tube of flux.

shows the layer of surface polarity and the required stress to hold this layer in equilibrium. By comparison with eqn. (5.63) there is a tensile stress

$$\sigma = \tfrac{1}{2}\,\mathbf{H}\,.\,\mathbf{B} \tag{7.90}$$

Figure 7.26 illustrates the manner of making tangential cuts along a tube of flux. This requires the insertion of a surface current and a compressive stress is needed to keep the current in equilibrium, as illustrated in Fig. 7.27. The compressive stress is

$$\sigma = \tfrac{1}{2}\,\mathbf{H}\,.\,\mathbf{B}. \tag{7.91}$$

These two stresses are the principal stresses. If cuts are made on a surface at an arbitrary angle to the tubes of flux, the required stress system consists of a direct stress

$$\sigma = \tfrac{1}{2}(H_n B_n - H_t B_t) \tag{7.92}$$

and a shear stress

$$\tau = H_t B_n \tag{7.93}$$

where n denotes the direction normal to the surface and t the direction along the surface (see Fig. 5.19).

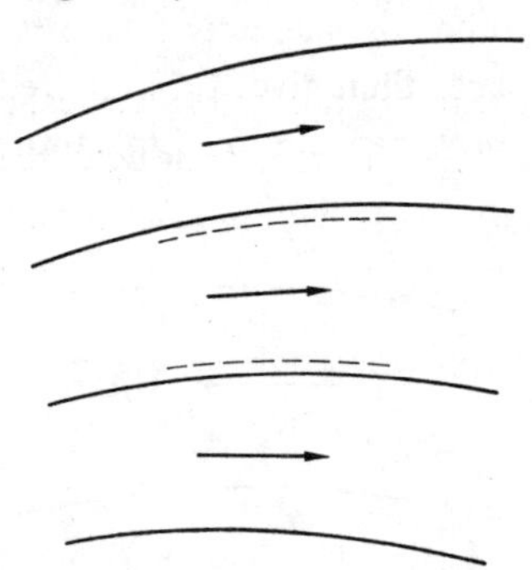

FIG. 7.26. Tangential cut along tube of magnetic flux (the same as 5.15).

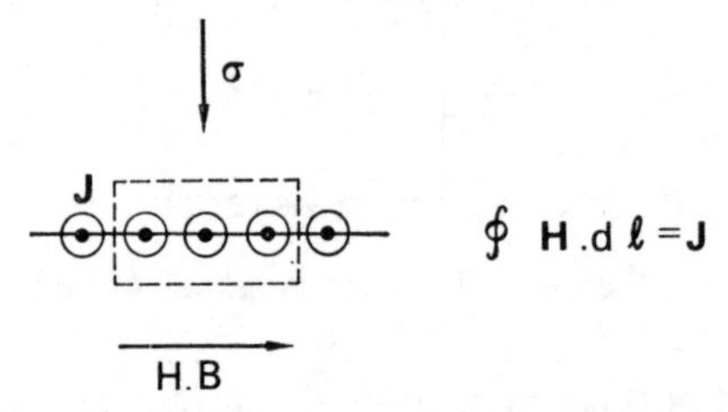

FIG. 7.27. Compressive stress perpendicular to tube of flux (as 5.16).

The method of field stresses relies solely on a knowledge of the field surrounding a body and not on the sources of the field. It can thus deal equally well with pole and current sources or with a mixture of the two. To obtain a unique answer for the force, the surface surrounding the body must lie in free space. If the surface cuts through matter there will in general be additional forces of *mechanical* origin. If the material cut is also magnetized, the usual difficulties of defining the internal field arise.

These were discussed in § 6.4.3. It is not possible to define a unique magnetic force per unit volume. Various expressions for the volume force are possible, depending on the choice of fundamental particle from which the material is thought to be constructed. None of the expressions are very useful because the domain structure makes magnetized bodies highly anisotropic. Only where the magnetic flux through the material is uniform over a region is there likely to be any success in predicting the surface layers, and hence the stress of magnetic origin when a cut is made through the material. In general it is best to calculate the field stress in free space.

As an illustration of the method of field stresses, consider the following simple example. Figure 7.28 shows a conductor carrying current I in a slot of width d cut in iron of very high relative permeability. A typical flux

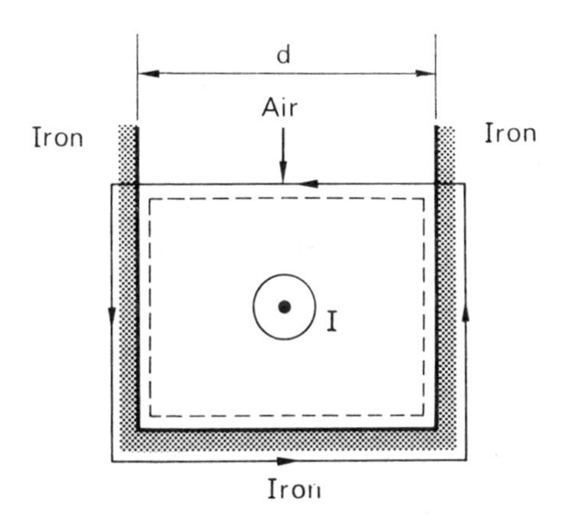

FIG. 7.28. Force on current in a slot in iron.

line is shown. To find the force on the conductor by means of the field stress we must choose a surface over which the field stress is to be integrated. The dotted line indicates a suitable surface. The only part of the surface in the field contributes a downwards force of $\frac{1}{2}$ HBd per unit length into the paper. Using $\oint \mathbf{H}\,.\,d\mathbf{l} = I$, we have $H = I/d$. Hence the force is $F = \frac{1}{2}\mu_0 I^2/d$.

7.5.5. Force in Terms of Energy

The force expressions in §§ 7.5.1–7.5.4 have been derived without any reference to energy. A careful reading of those sections shows that a knowledge of the sources or of the magnetic field is entirely adequate to find the force. Nevertheless it is occasionally convenient to derive the force from a consideration of energy.

Let us consider first a system of coils carrying current. In § 7.4.4 we saw that the work done by such a system against external forces is equal to the mutual field energy $U^* - U_0^*$, if the currents are kept constant. Hence the force is given by the expression

$$F_x = \frac{\partial}{\partial x}(U^* - U_0^*) = \left.\frac{\partial U^*}{\partial x}\right|_{I \text{ constant}} \tag{7.94}$$

and the torque

$$T_\theta = \left.\frac{\partial U^*}{\partial \theta}\right|_{I \text{ constant}} \tag{7.95}$$

The work is done *against* the external forces. Hence the force applied to each coil by the rest of the system is in a direction to *increase* the field energy. Equations (7.94) and (7.95) give the direction for the forces and torques on the coils. Comparing these equations with eqns. (5.35) and (5.36) we note a reversal in sign between the magnetic and electric forces. The same reversal is experienced in mechanical systems where the forces act to *increase* the kinetic energy and *decrease* the potential energy. This reminds us that the magnetic field energy U^* is kinetic in origin, being due to moving charges, whereas the electric field energy is the potential energy of stationary charges. Thus like currents attract to increase the field energy and like charges repel each other to reduce the field energy.

So far we have thought of currents without magnetic material. If there is such material present we can account for its effect by using $B = \mu_0\mu_r H$ in the expression for the field energy U^* given in eqn. (7.49). But unless μ_r is substantially constant, it will be very difficult to obtain the force by the differentiation process of eqn. (7.94). In devices which use saturated iron or rely on hysteresis effects the energy method cannot be recommended. It is much better to use the method of field stresses.

Let us consider next a system of permanent magnets. As with all magnetic materials we have the choice of describing their action in terms of dipoles or magnetization currents (§ 6.4.1). If the magnets are truly permanent we can assign to the material either a constant dipole moment or a constant magnetization current density. The B–H characteristic for such material is shown in Fig. 7.21. The use of the dipole description implies an equivalent potential energy. This is physically unreal, because all magnetic energy is kinetic, but it is convenient for calculation. Reference to § 7.4.3 shows that the force can be obtained from the expression

$$F_x = -\frac{\partial U^*}{\partial x} \tag{7.96}$$

where $B = \mu_0 H$ in the expression for U^* in eqn. (7.49). The alternative choice of magnetization current as a model for a permanent magnet leads to the formulation of the field energy in terms of the vector potential $\mathbf{A}$. This is the kinetic energy description and we have

$$F_x = +\frac{\partial U^*}{\partial x} \tag{7.97}$$

where the expression for U^* is that given in eqn. (7.53).

In devices which use both soft and hard magnetic materials as well as electric currents it is not possible to attribute part of the energy to dipoles and part to currents. The magnetic material must be regarded as a distributed current and the vector potential must be used. But the calculation of energy inside the material is likely to be complicated and unreliable. To try to derive force from a change of energy then becomes an almost impossible undertaking.

The chief value of the energy method is that by defining kinetic and potential energy it gives general guidance about the direction of force and torque. It is useful to remember that magnetic systems defined in terms of current sources always move in a direction to increase the flux linkage.

Summary

We have carried out a detailed investigation into the manner by which iron "conducts" flux. This has drawn our attention to the dominant role played by the pole strength on the surface of the iron.

We have discussed the application of Laplace's equation to the study of magnetic problems and looked at the possibility of representing magnetic boundaries by image currents.

Finally, we have obtained expressions for the energy stored in magnetic fields and for the forces on currents and magnets in magnetic fields.

Exercises

7.1. Explain what is meant by the *reluctance* of an iron circuit. Discuss the effect on the flux linkage when the following iron cores are inserted into a current loop: (a) a flat disc in the plane of the loop, (b) a cylinder with its axis in the plane of the loop, (c) a sphere, (d) a cylinder co-axial with the loop.

7.2. The *permeance* of a magnetic circuit is the inverse of the reluctance. Show that in a circuit of highly permeable iron with an air gap the permeance can be written as $\Lambda = \mu_0 k A_g l_g$, where A_g is the area of the gap, l_g is the length of the gap, and k is a constant. Explain the physical significance of k and suggest how it may be estimated.

7.3. A permanent magnet has a uniform pole strength P^* on its circular face of radius r. Show that the flux density of a distance z from the centre of the pole face is

$$B = \frac{P^*}{2}\left[1 - \frac{z/r}{(1+z^2/r^2)^{1/2}}\right].$$

7.4. A permanent magnet is to be made with Alcomax IV (see Fig. 6.17). The working point on the $B-H$ curve is to be $B = 0.8$ T, $H = -5\times10^4$ A/m. The air gap is to be 0·0015 m and the m.m.f. drop in the iron of the magnetic circuit except for the Alcomax is 20% of the total m.m.f. Estimate the length of the Alcomax block. [Ans.: 0·023 m.]

7.5. Figure 7.24 shows an iron circuit energized by a winding at A. The mean path lengths are: DAB, 0·5 m; $BEFD$, 0·4 m; $BC = CD$, 0·08 m; air-gap at C, 0·6 mm. The cross-sectional areas are DAB, 25 cm²; $BEFD$, 5 cm²; BCD, 12 cm². Estimate

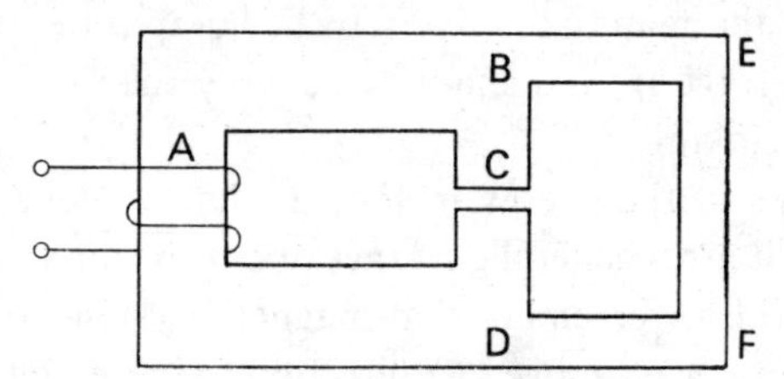

FIG. 7.29. Iron circuit.

the m.m.f. of the winding and the flux density in EF if the flux density in the air gap is 0·5 T. The magnetization curve for the iron is as follows:

B (T)	0·2	0·4	0·6	0·8	1·0	1·2
H (A/m)	250	450	700	760	1040	1750

[Ans.: 520; 0·82.]

7.6. A non-magnetic cylinder of radius a carries on its surface an axial current distribution of line density $I \cos 2\theta$ A/m, where the angle θ is measured around the circumference of the cylinder. Find the magnetic field both inside and outside the cylinder. Generalize the result for any arbitrary current distribution and discuss the design of multi-pole machines in the light of your analysis. [Ans.: Inside potential $Ir^2 \sin 2\theta/4a$; outside potential $-Ia^3 \sin 2\theta/4r^2$.]

7.7. Show that the magnetic field of a pair of equal and unlike current filaments separated by a distance $2c$, referred to the point midway between them as origin is given by

$$V^* = -\frac{I}{\pi}\sum_1^\infty \frac{c^n}{nr^n}\sin n\theta \quad (n \text{ is odd})$$

for $r > c$, and

$$V^* = \frac{I}{\pi}\sum_1^\infty \frac{r^n}{nc^n}\sin n\theta \quad (n \text{ is odd})$$

for $r < c$.

7.8. Determine the magnetic field due to a current filament I placed at a distance c from a cylinder of relative permeability μ and radius a, where $c > a$.

$$\left[\text{Ans.: For } a < r < c,\ V^* = \frac{I}{2\pi}\sum_1^\infty \frac{r^n}{nc^n}\left(1 - \frac{\mu-1}{\mu+1}\frac{a^{2n}}{r^{2n}}\right)\sin n\theta\right.$$

$$\left. r > c,\ V^* = -\frac{I}{2\pi}\sum_1^\infty \frac{c^n}{nr^n}\left(1 + \frac{\mu-1}{\mu+1}\frac{a^{2n}}{c^{2n}}\right)\sin n\theta\,.\right]$$

7.9. Use the result of the previous question to determine the force between the current and the cylinder.

$$\left[\text{Ans.: } F = \frac{(\mu-1)\,I^2}{2\pi c\,(\mu+1)}\left(\frac{a^2}{c^2} + \frac{a^4}{c^4} + \cdots\right) = \frac{1}{2\pi}\frac{\mu-1}{\mu+1}\frac{a^2}{c\,(c^2-a^2)}I^2.\right]$$

7.10. Show that the magnetic field of a current I placed outside a semi-infinite block of iron of relative permeability μ_r, and parallel to its surface, can be represented as far as the field outside the iron is concerned by the current I and an image current $I\,(\mu_r - 1)/(\mu_r + 1)$. Show also that the field inside the iron is that of a current $2I/(\mu_r + 1)$ at the position of the current filament.

7.11. Show that the magnetic field of a current I placed inside a semi-infinite block of iron of relative permeability μ_r, and parallel to its surface, can be represented as far as the field outside is concerned by a current $2I\mu_r/(\mu_r + 1)$ instead of the current I. Show also that the field inside the iron is the same as the field of the original current I and an image current outside the iron of strength $-I\,(\mu_r - 1)/(\mu_r + 1)$.

7.12. A cylindrical region of diameter d is to be screened from a magnetostatic field B_0. Various iron tubes are available. The iron saturates at flux density B_m. Defining a criterion of thickness discuss whether it would be better to use (a) a thin tube, (b) a thick tube, (c) two co-axial thin tubes, (d) two co-axial thick tubes, or (e) a thick tube and a thin tube.

7.13. A d.c. generator is to be designed for a fixed output. Discuss the advantages and disadvantages of putting the armature conductors (a) on the surface of a smooth cylindrical armature, (b) in open slots, or (c) in closed tunnels.

7.14. Figure 7.30 shows a current filament of strength I in the right angle between two large iron surfaces. Estimate the force on the current assuming that the iron has a very high permeability.

$$\left[\text{Ans.: } F_x = -\frac{\mu_0 I^2 (2a^2 + b^2)}{4\pi a (a^2 + b^2)}, \quad F_y = -\frac{\mu_0 I^2 (a^2 + 2b^2)}{4\pi b (a^2 + b^2)}.\right]$$

7.15. Figure 7.31 shows a current filament in an air gap of an iron circuit of infinite permeability. Using the method of magnetic field stresses show that the force on the current is given by

$$F_x = \frac{\mu_0 I^2}{2g} \frac{(l - 2x)}{l} \text{ N/m.}$$

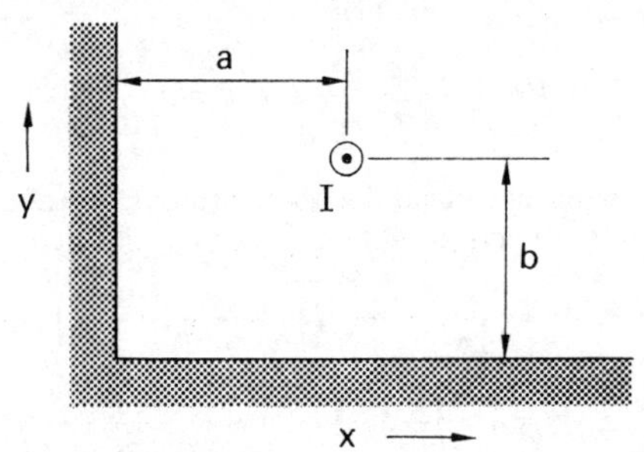

FIG. 7.30. Current near a magnetic block.

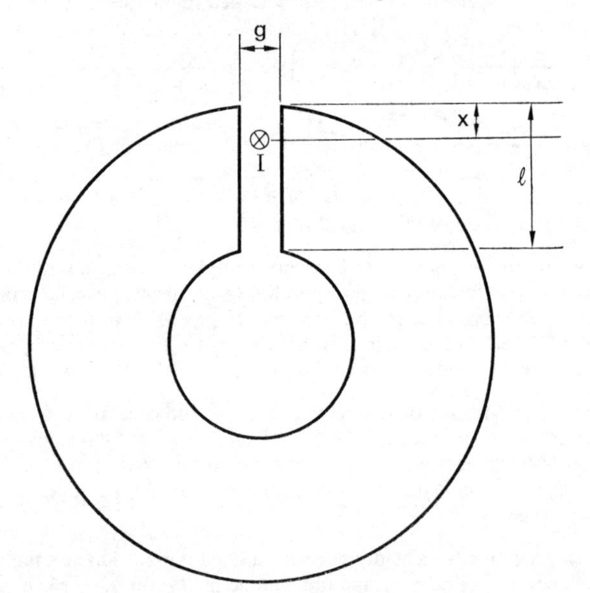

FIG. 7.31. Force on a current in an air gap.

7.16. Figure 7.32 shows a simple torque motor. If the exciting ampere-turns are NI, both air gaps have length g, the iron reluctance is negligible, the rotor diameter is d, and the axial length of the machine is l, estimate the torque using the method of field stresses ($d \gg g$).

$$\left[\text{Ans.}: \frac{1}{8}\mu_0 \frac{N^2 I^2 l d}{g} \text{Nm}.\right]$$

FIG. 7.32. Torque motor.

CHAPTER 8

TIME-VARYING CURRENTS AND FIELDS IN CONDUCTORS

8.1. Current and Charge in a Conductor

How does current distribute itself throughout the volume of a conductor? We avoided this question in the last chapter by assuming the current to flow in filaments rather than in thick conductors. Of course conductors of finite cross-section can be considered as bundles of filaments, but this does not solve the question because we have had to assume that we knew the individual currents. Nor is the question merely of theoretical interest. Conductor material is expensive. It is wasteful to provide thick conductors if the current flows only along the surface. We shall devote this chapter to the discussion of the distribution of currents and fields in conducting material.

Ohm's experimental law provides a useful strating point. It tells us that the current flow is determined by the conductivity of the material and by the geometry of the conductor. Let us first consider constant current. If a potential difference V is applied between the ends of a conductor, we have

$$V = RI. \tag{2.17}$$

If the conductor has constant cross-section s, length l, and conductivity σ,

$$R = \frac{l}{\sigma s}. \tag{2.18}$$

The appearance of the cross-section in eqn. (2.18) tells us that the current distributes itself with uniform density. Consider the conductor as a bundle of filaments of equal cross-section and length. The p.d. and the resistance will be the same for each filament, and hence by Ohm's law the current will be the same. Hence the current density will be uniform. This knowledge enables us to write Ohm's law as a point relationship in the conducting material. Remembering that current flows down the potential gradient,

$$-\delta V = E\,\delta l = \frac{\delta l}{\sigma\,\delta s}\,J\,\delta s$$

and *vectorially*

$$\mathbf{E} = \frac{\mathbf{J}}{\sigma}. \tag{2.20}$$

Thus in a conductor, in which steady current flows, the current density is uniform and the electric field strength is uniform across the section of the conductor.

In the absence of any e.m.f. in the conductor, **E** is the electrostatic field. Such a field can come only from electric charge and the question arises: Where are the charges which cause the current to flow? They must be either in the conducting material or on its surface or in both places. Let us elucidate this matter.

The equation of continuity of charge and current is illustrated in

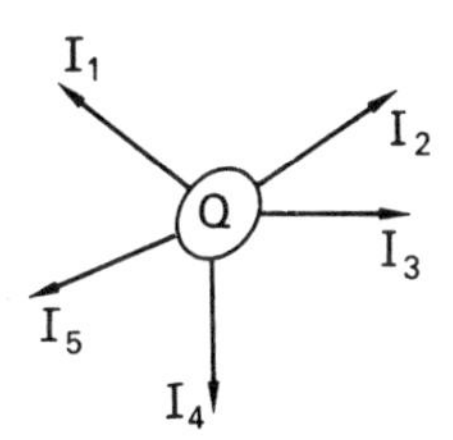

FIG. 8.1. Continuity of current and charge.

Fig. 8.1 and can be written as

$$\sum I_{\text{out}} = -\frac{\partial Q}{\partial t}. \tag{8.1}$$

In terms of current and charge density,

$$\oiint \mathbf{J} \,.\, d\mathbf{s} = -\iiint \frac{\partial \varrho}{\partial t}\,dv, \tag{8.2}$$

and by the divergence theorem, eqn. (3.16),

$$\operatorname{div} \mathbf{J} + \frac{\partial \varrho}{\partial t} = 0. \tag{8.3}$$

We also have the relation between charge and field in the absence of polarization

$$\operatorname{div} \varepsilon_0 \mathbf{E} = \varrho. \tag{8.4}$$

Substituting eqns. (2.20) and (8.4) in eqn. (8.3) we obtain

$$\frac{\sigma}{\varepsilon_0}\varrho + \frac{\partial \varrho}{\partial t} = 0 \tag{8.5}$$

or

$$\varrho = \varrho_0 \, e^{-(\sigma/\varepsilon_0)t}. \tag{8.6}$$

Thus if there is within the conductor a charge density ϱ_0, it will decay exponentially with a time-constant $\tau = \varepsilon_0/\sigma$. A typical conductivity for a metal is of the order 10^7 siemens (mho/m) and ε_0 is of the order 10^{-11}. Hence τ is of the order 10^{-18} sec. Thus the charge will diffuse to the conductor surface with extreme rapidity. We conclude that even at high frequencies there can be no charge inside a conductor. Instead of eqn. (8.3) we can write

$$\operatorname{div} \mathbf{J} = 0. \tag{8.7}$$

The electrostatic field, therefore, arises from *surface* charges. In circuit language surface charge is associated with capacitance. It is interesting to note that even in steady current flow there is always this capacitance effect. Because of the high values of conductivity in metals the capacitance of conductors is very small, but at high frequencies this so-called "stray" capacitance becomes important.

In terms of electric and magnetic fields, the knowledge that inside a conductor there is virtually no charge even under time-varying conditions, enables an important simplification to be made. We noted in § 2.3 that the full m.m.f. equation is

$$\oint \mathbf{H} \,.\, d\mathbf{l} = I + \frac{d\Psi}{dt}, \tag{2.48}$$

which for a point relationship becomes

$$\operatorname{curl} \mathbf{H} = \mathbf{J} + \frac{\partial \mathbf{D}}{\partial t}. \tag{8.8}$$

At an angular frequency ω the magnitudes of the two terms $\mathbf{J}$ and $\partial\mathbf{D}/\partial t$

compare as J and $\omega\varepsilon_0/\sigma$, i.e. as 1 and $\omega\tau$, where τ is of the order 10^{-18}. We can, therefore, adopt the simpler magnetostatic expression

$$\oint \mathbf{H} \,.\, d\mathbf{l} = I, \tag{3.26}$$

or at a point

$$\text{curl } \mathbf{H} = \mathbf{J} \tag{3.32}$$

which is consistent with eqn. (8.7), since the divergence of a curl is zero. We are, therefore, able to use inside a conductor carrying *time-varying* current the expressions for the magnetic field which have been derived on the assumption of *constant* current.

We are now in a position to derive the general equation for the distribution of time-varying current. Let us assemble the results so far:

$$\text{curl } \mathbf{H} = \mathbf{J}, \tag{3.32}$$

$$\text{curl } \mathbf{E} = -\frac{\partial \mathbf{B}}{\partial t} \quad \text{(Faraday's law)}, \tag{3.38}$$

$$\mathbf{J} = \sigma\mathbf{E}, \tag{2.20}$$

$$\mathbf{B} = \mu_0\mu_r\mathbf{H}. \tag{2.24}$$

Eliminating **E**, **H**, and **B** we obtain

$$\text{curl curl } \mathbf{J} = -\sigma\mu_0\mu_r \frac{\partial \mathbf{J}}{\partial t}. \tag{8.9}$$

Using the vector identity

$$\text{curl curl } \mathbf{J} = \text{grad div } \mathbf{J} - \nabla^2\mathbf{J} \tag{8.10}$$

and remembering that the current density has zero divergence [eqn. (8.7)], we obtain

$$\nabla^2\mathbf{J} = \sigma\mu_0\mu_r \frac{\partial \mathbf{J}}{\partial t}. \tag{8.11}$$

In a similar way it can be shown that the form of the equation is the same if instead of **J** we write **E**, **H**, or **B**. This is one of the most useful equations in the whole of science. It is called the *diffusion* equation.

8.2. General Discussion of the Diffusion Equation

The diffusion equation describes a considerable number of physical processes. We have met it in connection with current in conductors and it is helpful to realize that this is a diffusion process. If in Ohm's law we replace the current density **J** by a charge density ϱ flowing at an average velocity **u**,

$$\mathbf{J} = \varrho\mathbf{u}, \tag{8.12}$$

we can write

$$\mathbf{E} = \frac{\varrho}{\sigma}\mathbf{u}. \tag{8.13}$$

Thus the force represented by **E** is proportional to the velocity, whereas in free space we should have force proportional to acceleration. Equation (8.13) is typical for flow through a porous solid.

Other processes described by equations of the form of eqn. (8.11) are the diffusion of compressible fluids through matter. But perhaps the most important example of a diffusion process is the flow of heat through matter, for which we can write

$$\nabla^2\theta = \frac{k}{cd}\frac{\partial\theta}{\partial t}, \tag{8.14}$$

where θ is the temperature, k the thermal conductivity, c the specific heat, and d the density.

Our direct experience of temperature makes eqn. (8.14) a useful starting point for a qualitative discussion. Suppose the temperature on the surface of the ground varies during the year in a sinusoidal manner as shown in Fig. 8.2, curve *A*. How is the temperature below the surface going to vary? It is reasonable to suppose that the variation will be less as the depth increases. Also there must be some time delay as the heat diffuses inward, so that there will be a phase angle varying with depth. This is shown in Fig. 8.2, curves *B* and *C*. If now the frequency of temperature variation is increased as shown in Fig. 8.3, which gives a daily variation, it is likely that both the attenuation and the phase angle will increase. This is borne out by this figure which like Fig. 8.2 has been calculated by use of eqn. (8.14).

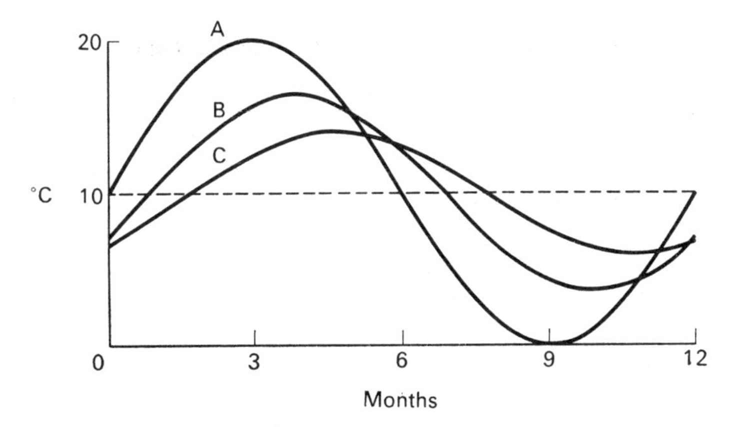

FIG. 8.2. Annual temperature variation of soil at the surface and at depths 1 m and 2 m below the surface.

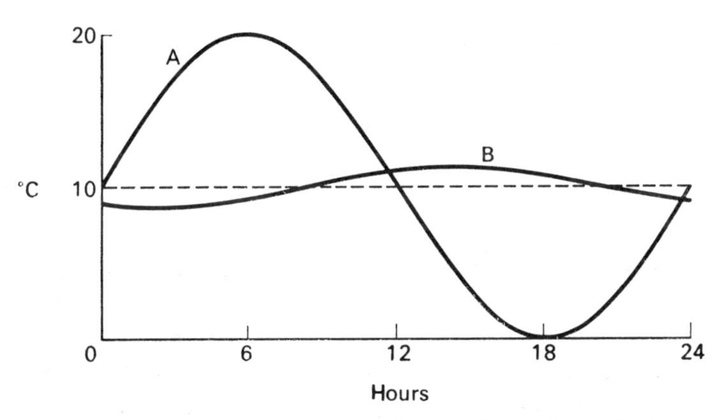

FIG. 8.3. Daily temperature variation of soil at the surface and at a depth of 25 cm below the surface.

8.3. Current and Field in a Semi-infinite Slab

Below a certain depth there will be negligible variation and it is, therefore, sensible to regard a thick material as a semi-infinite slab with only one free surface, as shown in Fig. 8.4. Let us obtain a mathematical solution for the current density near the surface of a very thick conductor. Let

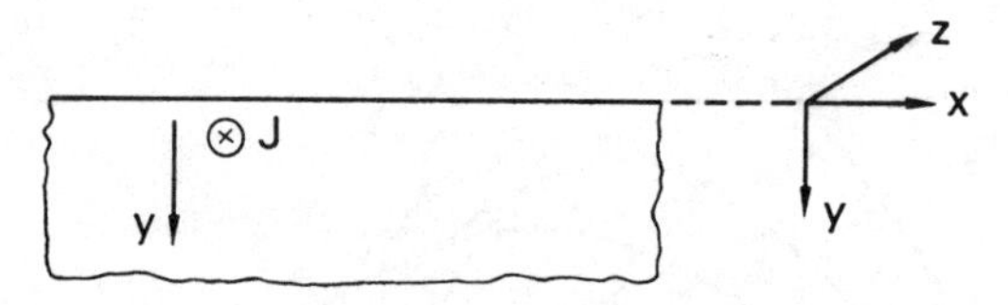

FIG. 8.4. Current flow in a semi-infinite slab.

us choose a right-handed set of axes as shown in Fig. 8.4 and let the current J flow in the z-direction only. From eqn. (8.11)

$$\nabla^2 J = \sigma \mu_0 \mu_r \frac{\partial J}{\partial t}. \tag{8.15}$$

Since there are no variations in the x- and z-directions, this reduces to

$$\frac{\partial^2 J}{\partial y^2} = \sigma \mu_0 \mu_r \frac{\partial J}{\partial t}. \tag{8.16}$$

If the angular frequency is ω, we can express the time variation as $e^{j\omega t}$. Substituting this in eqn. (8.16) we note that the space variation will have to satisfy

$$\frac{\partial^2 \bar{J}}{\partial y^2} = j\omega\sigma\mu_0\mu_r \bar{J}, \tag{8.17}$$

where $\bar{J}$ is a complex (or phasor) quantity.

If we put

$$\delta^2 = 2/\omega\sigma\mu_0\mu_r, \tag{8.18}$$

$$\bar{J} = A\, e^{[(1+j)y]/\delta} + B\, e^{-[(1+j)y]/\delta}. \tag{8.19}$$

The constant A is zero, because J must decrease with depth. B is the surface current density J_s. So that

$$J = J_s\, e^{-[(1+j)y]/\delta}\, e^{j\omega t}$$

$$= J_s\, e^{-y/\delta}\, e^{j(\omega t - y/\delta)}. \tag{8.20}$$

The real part of this expression gives the instantaneous value

$$J_z = J_s\, e^{-y/\delta} \cos(\omega t - y/\delta). \tag{8.21}$$

Equation (8.21) shows the expected decay with depth, and we now see that this decay is exponential and therefore very quick. We also see the phase change with depth.

Suppose now that the total current I per unit width is known. Then

$$I = \int_0^\infty J\,dy = \int_0^\infty J_s\, e^{-[(1+j)y]/\delta} e^{j\omega t} dy$$

$$= \frac{\delta}{1+j} J_s\, e^{j\omega t} = \frac{J_s\delta\,(1-j)}{2}\, e^{j\omega t}. \tag{8.22}$$

The real part of this expression is

$$I_z = \frac{J_s\delta}{2}(\cos \omega t + \sin \omega t) = \frac{J_s\delta}{\sqrt{2}} \cos\left(\omega t - \frac{\pi}{4}\right). \tag{8.23}$$

Thus the total current lags by $\pi/4$ on the surface current density. The ohmic loss P per unit width x and length z, i.e. per unit surface area, is

$$P = \int_0^\infty \frac{|J|^2}{2\sigma} dy = \int_0^\infty \frac{J_s^2}{2\sigma} e^{-2y/\delta} dy = \frac{J_s^2\delta}{4\sigma} = \frac{I_{\text{r.m.s.}}^2}{\sigma\delta}. \tag{8.24}$$

Since from eqn. (8.23) the r.m.s. current is $(J_s\delta)/2$. Hence the ohmic, resistance is given by

$$R = \frac{P}{I_{\text{r.m.s.}}^2} = \frac{1}{\sigma\delta}. \tag{8.25}$$

Hence the energy loss is the same as if a uniformly distributed current flowed in a surface layer of depth δ. This parameter is of the utmost importance in all diffusion processes. It is called the *skin depth* or *penetration depth*. The reader should notice that there is current below the depth δ, although it decreases rapidly. At a depth of 4δ, for example, the current density has fallen to less than 2% of the value at the surface.

The skin depth varies inversely as the square root of the frequency, the conductivity and the permeability. A useful figure to keep in mind is that the skin depth for hot copper at power frequencies is 1 cm. This corresponds to a conductivity of $5{\cdot}06 \times 10^7$ mho/m and a temperature of about 50°C at 50 Hz, and to a temperature of 100°C at 60 Hz. At 50 MHz the skin depth for copper at 50°C is 0·01 mm. Aluminium at 50 Hz and 50°C has a skin depth of $1\frac{1}{4}$ cm. The skin depth in magnetic materials is reduced by the high relative permeability. Thus for a silicon iron (unsaturated) used in transformer cores the skin depth at 50 Hz is of the order of $1\frac{1}{2}$ mm. These figures show that the results for a semi-infinite slab can be applied realistically to materials of quite small thickness.

So far we have solved the diffusion problem in terms of the current flow through a thick conductor. But the results can equally be applied to the problem of flux distribution in a magnetic slab. The diffusion equation for magnetic flux density is

$$\nabla^2 \mathbf{B} = \sigma \mu_0 \mu_r \frac{\partial \mathbf{B}}{\partial t}, \tag{8.26}$$

and in the simple geometry of the thick slab this becomes

$$\frac{\partial^2 B}{\partial y^2} = \sigma \mu_0 \mu_r \frac{\partial B}{\partial t}. \tag{8.27}$$

The total flux can be obtained by the use of eqn. (8.23),

$$\Phi = \frac{B_s \delta}{\sqrt{2}} \cos\left(\omega t - \frac{\pi}{4}\right). \tag{8.28}$$

Thus the flux is largely confined to a surface layer of the thick slab. At a depth greater than 4δ the flux density will have fallen to less than 2% of its value at the surface. If the purpose of the material is to carry the flux, it would clearly be wasteful to use a block much thicker than the skin depth.

The current density in the magnetic slab can be obtained by the use of

$$\mathbf{J} = \text{curl } \mathbf{H}. \tag{3.32}$$

If B is in the z-direction, this simplifies to

$$J_x = \frac{\partial H_z}{\partial y} = \frac{1}{\mu_0 \mu_r} \frac{\partial B_z}{\partial y}. \tag{8.29}$$

Rewriting eqn. (8.20) in terms of B, we have

$$B = B_s e^{-(1+j)y/\delta} e^{j\omega t} \tag{8.30}$$

and

$$J = -\frac{B_s}{\mu_0 \mu_r} \frac{(1+j)}{\delta} e^{-(1+j)y/\delta} e^{j\omega t}. \tag{8.31}$$

The ohmic power loss per unit surface area is given by

$$P = \int_0^\infty \frac{|J|^2}{2\sigma}\,dy = \frac{B_s^2}{\sigma\mu_0^2\mu_r^2}\,\frac{1}{2\delta} = \frac{H_s^2}{2\sigma\delta}. \tag{8.32}$$

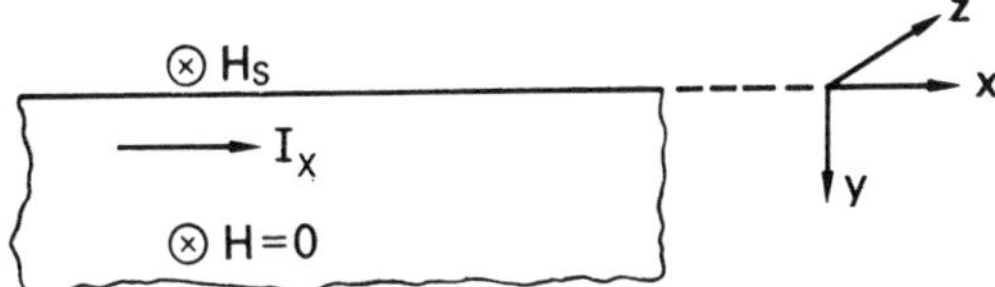

FIG. 8.5. Magnetic field strength in a semi-infinite slab.

The total current flow can be obtained by integrating eqn. (8.31), or more simply by applying $\oint \mathbf{H}\,.\,d\mathbf{l} = I$, as in Fig. 8.5, which gives the current per unit distance in the z-direction as

$$\bar{I} = -\bar{H}_s.$$

Hence

$$P = \frac{I_{\text{r.m.s.}}^2}{\sigma\delta}. \tag{8.33}$$

Thus, as before in eqns. (8.24) and (8.25), the loss is the same as if the total current were uniformly distributed to a depth equal to the skin depth.

8.4. Current Flow in a Flat Sheet

Let us now investigate the flow of alternating current in a sheet of finite thickness $2b$ (Fig. 8.6). We can follow the treatment of § 8.3 up to eqn. (8.19), but we must no longer put A equal to zero. Let $y = 0$ be on the centre line of the sheet and let the current density there be J_0. By symmetry we have in eqn. (8.19) $A = B$ and also $J = A + B$ at $y = 0$.

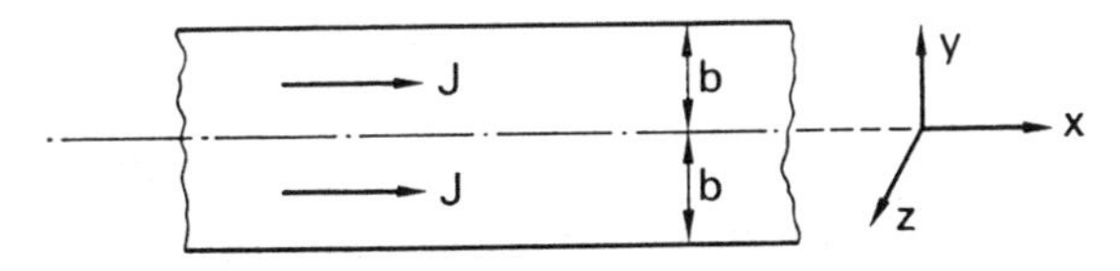

FIG. 8.6. Current flow in a flat sheet.

Thus the space variation of J is given by

$$\bar{J} = \frac{\bar{J}_0}{2}\left(e^{(1+j)y/\delta} + e^{-(1+j)y/\delta}\right) = \bar{J}_0 \cosh(1+j)\,y/\delta. \quad (8.34)$$

At the surface, $\bar{J} = \bar{J}_s$, so that

$$\bar{J}_s = \bar{J}_0 \cosh(1+j)\,b/\delta, \quad (8.35)$$

and J in terms of J_s is given by

$$\bar{J} = \bar{J}_s \frac{\cosh(1+j)\,y/\delta}{\cosh(1+j)\,b/\delta}. \quad (8.36)$$

The total current per unit width of sheet is

$$\bar{I} = \int_{-b}^{+b} \bar{J}\,dy = \frac{2\bar{J}_s\delta}{1+j}\tanh(1+j)\,b/\delta. \quad (8.37)$$

If b/δ is large, $\tanh(1+j)\,b/\delta \to 1$. This is true within 1% if $b/\delta > 2{\cdot}64$. Hence in a sheet of this thickness $\bar{I} = (2\bar{J}_s\delta)/(1+j)$. Comparison with eqn. (8.22) shows that this is the value which would be obtained by thinking of the sheet as two infinite slabs put back to back. The power loss could be obtained in a similar way as in eqn. (8.24), but an interesting alternative is to work out the impedance to the current. The voltage applied to the sheet can be obtained from the surface electric field. Per unit length (in the x-direction)

$$\bar{V} = \frac{\bar{J}_s}{\sigma}. \quad (8.38)$$

Hence the complex impedance per unit area x and z is

$$Z = \frac{\bar{V}}{\bar{I}} = \frac{(1+j)}{2\sigma\delta}\coth(1+j)\,b/\delta. \quad (8.39)$$

Separating real and imaginary terms, we have

$$Z = \frac{1+j}{2\sigma\delta}\,\frac{\cosh b/\delta\cos b/\delta + j\sinh b/\delta\sin b/\delta}{\sinh b/\delta\cos b/\delta + j\cosh b/\delta\sin b/\delta}, \quad (8.40)$$

which after some reduction becomes

$$Z = R + jX = \frac{1}{2\sigma\delta}\,\frac{\sinh 2b/\delta + \sin 2b/\delta}{\cosh 2b/\delta - \cos 2b/\delta} + \frac{j}{2\sigma\delta}\,\frac{\sinh 2b/\delta - \sin 2b/\delta}{\cosh 2b/\delta - \cos 2b/\delta}. \quad (8.41)$$

The reactance X is due to the current inside the sheet and is thus due to the internal part of the inductance. Comparison of eqn. (8.41) with eqn. (8.25) shows that for large values of b/δ the losses in a thick sheet and a double-sided slab are equal.

The expression for R in eqn. (8.41) is the resistance to alternating current. The d.c. resistance is

$$R_{\text{d.c.}} = \frac{1}{2\sigma b}. \tag{8.42}$$

Hence the ratio of a.c. to d.c. resistance is

$$\frac{R_{\text{a.c.}}}{R_{\text{d.c.}}} = \frac{b}{\delta}\,\frac{\sinh 2b/\delta + \sin 2b/\delta}{\cosh 2b/\delta - \cos 2b/\delta}. \tag{8.43}$$

This ratio is plotted in Fig. 8.7. At large values of b/δ it tends to b/δ,

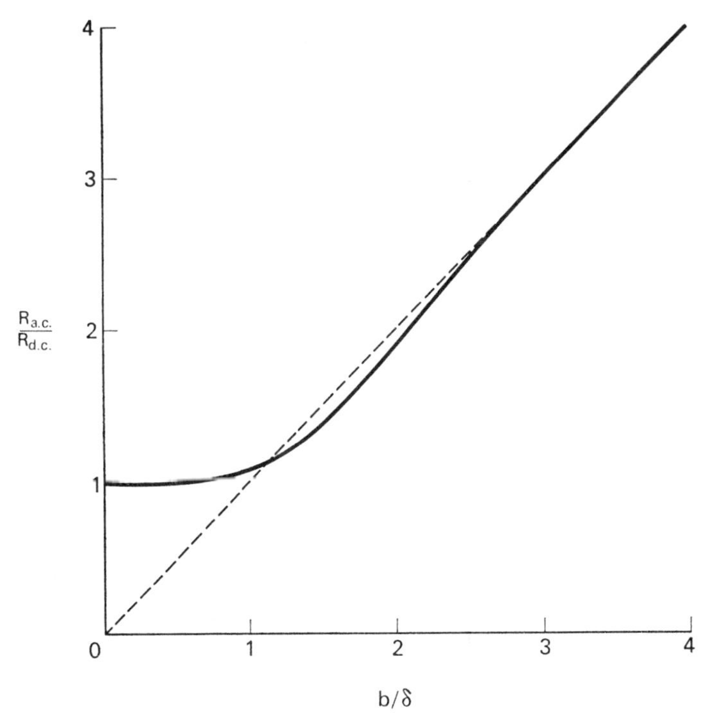

FIG. 8.7. A.c. resistance of a flat sheet.

because then the losses are as if the current were concentrated in two surface layers of depth δ. At low values of b/δ we can expand in ascending powers of b/δ:

$$\frac{R_{\text{a.c.}}}{R_{\text{d.c.}}} \doteqdot 1 + \frac{4}{45}\left(\frac{b}{\delta}\right)^4 . \tag{8.44}$$

Remembering that δ varies inversely as the square root of the frequency, we note that at high frequencies the a.c. resistance varies as $\sqrt{f}$ and at low frequencies as $1 + kf^2$. Figure 8.7 shows that the thickness criterion between the two types of behaviour is very nearly $b > \delta$ or $b < \delta$.

8.5. Eddy Currents in Thin Conductors

The fact that for small values of b/δ the a.c. resistance in § 8.4 has the form $R_{\text{d.c.}}\,(1 + kf^2)$ is very suggestive. It appears the losses can be obtained by adding a term to the direct current loss. Let us consider how this might come about. At very low frequencies the current will be uniformly distributed across the sheet. However, there will be an alternating magnetic field inside the sheet due to this uniform current. This field will induce a non-uniform "eddy" current, which will cause an additional loss. The eddy current will itself have a magnetic field, which in turn will induce a current and so on.

Let the current density and flux density be given by

$$\bar{J} = \bar{J}_0 + \bar{J}_1 + \bar{J}_2 + \cdots \bar{J}_k + \cdots \tag{8.45}$$

and

$$\bar{B} = \bar{B}_0 + \bar{B}_1 + \bar{B}_2 + \cdots \bar{B}_k + \cdots, \tag{8.46}$$

where $\bar{J}_0$ is the uniformly distributed current and $\bar{B}_0$ is the field due to $\bar{J}_0$. From eqn. (8.29) we obtain

$$\bar{J}_k = \frac{1}{\mu_0\mu_r}\frac{\partial \bar{B}_k}{\partial y} \tag{8.47}$$

and from eqns. (3.38) and (2.20)

$$\frac{\partial \bar{J}_k}{\partial y} = j\omega\sigma\bar{B}_{k-1} . \tag{8.48}$$

Hence

$$\bar{B}_0 = \mu_0 \mu_r y \bar{J}_0 . \tag{8.49}$$

There is no constant of integration because $\bar{B}_0 = 0$ at $y = 0$ by symmetry. Also

$$\bar{J}_1 = j\omega\sigma\mu_0\mu_r \frac{y^2}{2} \bar{J}_0 + \bar{C} = j\frac{y^2}{\delta^2} \bar{J}_0 + \bar{C}. \tag{8.50}$$

Since $\bar{J}_1$ is an eddy current there is no net flow of $\bar{J}_1$ across a section of the sheet. Thus

$$\int_{-b}^{+b} \bar{J}_1 \, dy = 0 = \frac{j\bar{J}_0}{\delta^2} \frac{2b^3}{3} + \bar{C}2b, \tag{8.51}$$

whence

$$\bar{C} = -j\frac{b^2}{3\delta^2} \bar{J}_0$$

and

$$\bar{J}_1 = \frac{j\bar{J}_0}{\delta^2}\left(y^2 - \frac{b^2}{3}\right). \tag{8.52}$$

Carrying on with this procedure, we shall find that B_1 will be of the order y^3/δ^3 and J_2 of the order y^4/δ^4. Let us now consider the power loss per unit area which is given by

$$\int_{-b}^{+b} \frac{|\bar{J}|^2}{2\sigma} \, dy.$$

Let us examine the expression

$$\bar{J}^2 = \bar{J}_0^2 + \bar{J}_1^2 + \bar{J}_2^2 + \cdots + 2\bar{J}_0\bar{J}_1 + 2\bar{J}_0\bar{J}_2 + 2\bar{J}_0\bar{J}_3 + \cdots$$
$$+ 2\bar{J}_1\bar{J}_2 + 2\bar{J}_1\bar{J}_3 + \cdots$$

For a thin sheet we can omit terms in higher orders than y^4/δ^4. Thus

$$\bar{J}^2 = \bar{J}_0^2 + \bar{J}_1^2 + 2\bar{J}_0\bar{J}_1 + 2\bar{J}_0\bar{J}_2 .$$

But $\bar{J}_0$ and $\bar{J}_1$ are in phase quadrature and will produce no average power by their interaction. Also the integral of $\bar{J}_0\bar{J}_2$ across the section will be zero, because $\bar{J}_0$ is constant and $\bar{J}_2$ is an even (symmetrical) function of y.

Hence, finally,

$$|\bar{J}|^2 = |\bar{J}_0|^2 + |\bar{J}_1|^2,$$

and the power loss is

$$P = \int_{-b}^{+b} \frac{|\bar{J}_0|^2 + |\bar{J}_1|^2}{2\sigma} dy = \frac{|\bar{J}_0|^2 b}{\sigma} + \frac{4}{45} \frac{b^5}{\delta^4} \frac{|\bar{J}_0|^2}{\sigma}. \quad (8.53)$$

Hence the ratio

$$\frac{R_{\text{a.c.}}}{R_{\text{d.c.}}} = 1 + \frac{4}{45}\left(\frac{b}{\delta}\right)^4, \quad (8.54)$$

which is the same as eqn. (8.44). This important result shows that the a.c. loss for thin sheets can be calculated by leaving out of account the effect of the magnetic field of the eddy currents. If the field of the eddy currents is negligible, their behaviour is said to be *resistance-limited.* For thick conductors in which the current is confined to surface layers the behaviour is said to be *inductance-limited.*

8.6. Current Flow in Cylindrical Conductors

The diffusion equation in cylindrical polar coordinates in which there is no variation along the z-axis of the cylinder can be derived from first principles (Fig. 8.8).

$$(\text{curl } \mathbf{H})_z = \frac{\partial H_\theta}{\partial r} + \frac{H_\theta}{r} = J_z, \quad (8.55)$$

$$(\text{curl } \mathbf{E})_\theta = -\frac{\partial E_z}{\partial r} = -\mu_0 \mu_r \frac{\partial H_\theta}{\partial t}. \quad (8.56)$$

Hence by the use of $\sigma E_z = J_z$, and dropping the suffixes,

$$\frac{\partial^2 J}{\partial r^2} + \frac{1}{r}\frac{\partial J}{\partial r} = \sigma \mu_0 \mu_r \frac{\partial J}{\partial t}. \quad (8.57)$$

The solution of this equation requires the use of Bessel functions and the interested reader is referred to books on these functions.†

† See, for example, N. W. McLachlan, *Bessel Functions for Engineers*, Oxford University Press, 1948.

We can, however, use §§ 8.3 and 8.4 to obtain approximate solutions. Let b be the radius of the conductor. If the conductor is thick, i.e. b/δ is large, we can assume that the effective resistance is that of a thin tube of

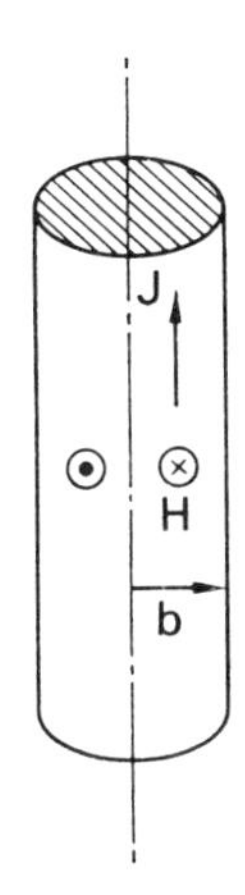

FIG. 8.8. Current flow in a cylindrical conductor.

diameter $2b$ and wall-thickness δ. Hence the a.c. resistance per unit length is

$$R_{\text{a.c.}} = \frac{1}{\sigma \times \text{area}} - \frac{1}{2\pi\sigma b\delta} \tag{8.58}$$

or

$$\frac{R_{\text{a.c.}}}{R_{\text{d.c.}}} = \frac{1}{2}\frac{b}{\delta}. \tag{8.59}$$

This can be compared with $R_{\text{a.c.}}/R_{\text{d.c.}} = b/\delta$ for a thick sheet.

If the conductor is thin, i.e. b/δ is small, we can neglect the magnetic field of the eddy current. Let us investigate this case. Let the total current be $I \cos \omega t$. Then the uniform d.c. current density will be

$$J_0 \cos \omega t = \frac{I}{\pi b^2} \cos \omega t. \tag{8.60}$$

The fields inside the conductor at radius r due to J_0 is given by

$$\oint \mathbf{H} \,.\, d\mathbf{l} = H \times 2\pi r = J_0 \pi r^2 \cos \omega t = I \frac{r^2}{b^2} \cos \omega t. \tag{8.61}$$

Hence

$$H = I\frac{r}{2\pi b^2}\cos\omega t. \tag{8.62}$$

Since the conductor is thin we shall take this as the total magnetic field. Using eqn. (8.56) to find the eddy current density J_1, we have

$$-\frac{\partial E}{\partial r} = -\frac{1}{\sigma}\frac{\partial J_1}{\partial r} = \omega\mu_0\mu_r\frac{Ir}{2\pi b^2}\sin\omega t \tag{8.63}$$

so that

$$J_1 = -\frac{Ir^2}{2\pi b^2\delta^2}\sin\omega t + C. \tag{8.64}$$

The constant C can be obtained by noting that the net flow due to the eddy current density J_1 must be zero

$$\int_0^b J_1 2\pi r\, dr = 0 = -\frac{I\sin\omega t}{4}\frac{b^2}{\delta^2} + \pi b^2 C.$$

Thus

$$C = \frac{I}{4\pi\delta^2}\sin\omega t$$

and

$$J_1 = \frac{I}{2\pi\delta^2}\left(\frac{1}{2} - \frac{r^2}{b^2}\right)\sin\omega t. \tag{8.65}$$

Figure 8.9 shows the normalized current density plotted against r/b, showing that the current flow is in one direction in the core of the conductor and in the opposite direction at a radius greater than $b/\sqrt{2}$.

The power loss per unit length is given by

$$P = \int_0^b\left(\frac{|J_0|^2 + |J_1|^2}{2\sigma}\right)2\pi r\, dr = \frac{I^2}{2\pi b^2\sigma} + \frac{I^2 b^2}{96\pi\sigma\delta^4}, \tag{8.66}$$

whence

$$\frac{R_{\text{a.c.}}}{R_{\text{d.c.}}} = 1 + \frac{1}{48}\left(\frac{b}{\delta}\right)^4. \tag{8.67}$$

This is the approximation for thin conductors. The complete solution for

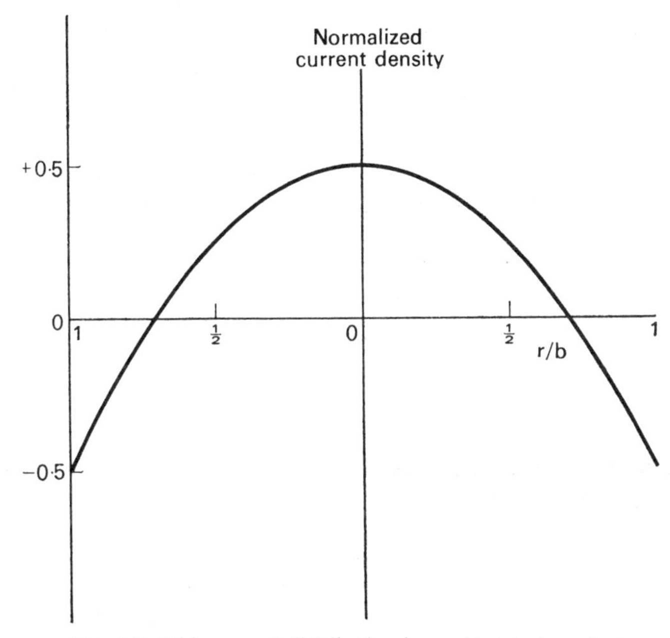

FIG. 8.9. Eddy current distribution in a cylindrical conductor.

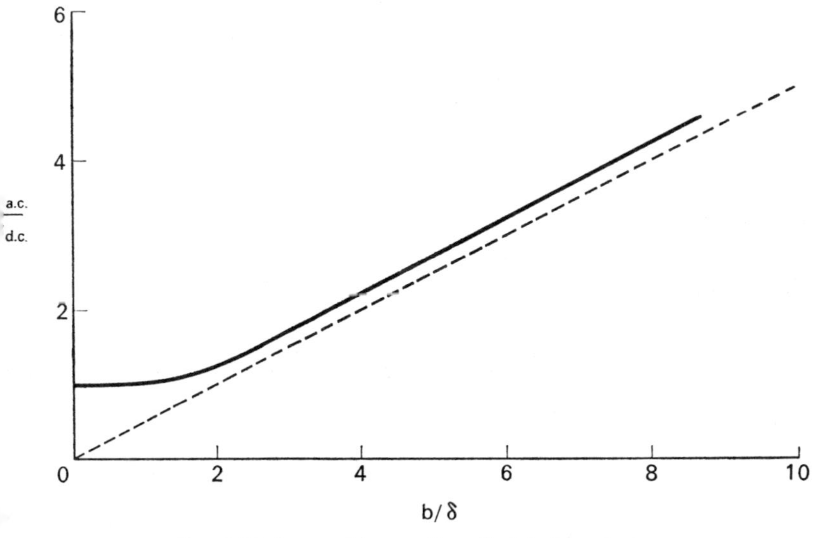

FIG. 8.10. A.c. resistance of a cylindrical conductor.

the ratio $R_{a.c.}/R_{d.c.}$ is shown graphically in Fig. 8.10. For $b/\delta > 2$, $R_{a.c.}/R_{d.c.}$ tends to the value $b/2\delta + \frac{1}{4}$. This value can be obtained by the use of cylindrical (Bessel) functions. It is noteworthy that the approximation employed in eqns. (8.58) and (8.59) is a very close one.

8.7. Induction Heating

The power loss due to eddy currents can sometimes be put to good use. Induction heating is used a great deal in industry, often by means of specially designed induction furnaces. Let us consider the loss in the flat

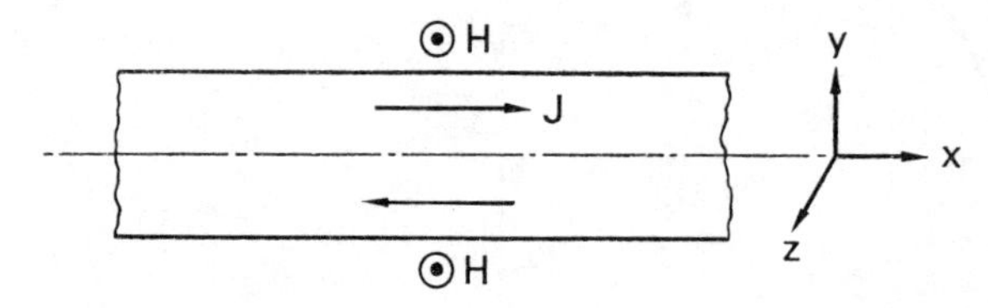

FIG. 8.11. Eddy currents induced in a flat sheet by an applied magnetic field.

sheet discussed in § 8.4, when such a sheet is inserted into an alternating magnetic field parallel to its surfaces, as shown in Fig. 8.11. The magnetic field in the sheet can be obtained by the method used in § 8.4,

$$\bar{H} = \bar{H}_s \frac{\cosh (1 + j)\, y/\delta}{\cosh (1 + j)\, b/\delta}. \tag{8.68}$$

This equation corresponds to eqn. (8.36). Hence the current density is along the x-direction and is given by

$$\bar{J} = \frac{\partial \bar{H}}{\partial y} = \bar{H}_s \frac{1 + j}{\delta} \frac{\sinh (1 + j)\, y/\delta}{\cosh (1 + j)\, b/\delta} \tag{8.69}$$

and the power per unit area of sheet

$$P = \int_{-b}^{+b} \frac{|\bar{J}|^2}{2\sigma}\, dy = \frac{|\bar{H}_s|^2}{\sigma\delta} \frac{\sinh 2b/\delta - \sin 2b/\delta}{\cosh 2b/\delta + \cos 2b/\delta}. \tag{8.70}$$

The power per unit volume can be written

$$P = \frac{|\bar{H}_s|^2}{\sigma\delta^2}\left(\frac{\delta}{2b}\right)\frac{\sinh 2b/\delta - \sin 2b/\delta}{\cosh 2b/\delta + \cos 2b/\delta} \tag{8.71}$$

$$= \frac{|\bar{H}_s|^2}{\sigma\delta^2} F\,(2b/\delta), \tag{8.72}$$

where $F(2b/\delta)$ means the function shown in eqn. (8.71). A knowledge of the behaviour of this function is useful to the user of induction furnaces. At low values of b/δ we can expand in ascending powers of b/δ to obtain the approximation for thin sheets. We find that for low b/δ

$$F \to \frac{2}{3}\left(\frac{b}{\delta}\right)^2. \tag{8.73}$$

At large values of b/δ

$$F \to \frac{1}{2}\left(\frac{\delta}{b}\right). \tag{8.74}$$

The complete function is plotted in Fig. 8.12. We see that for maximum

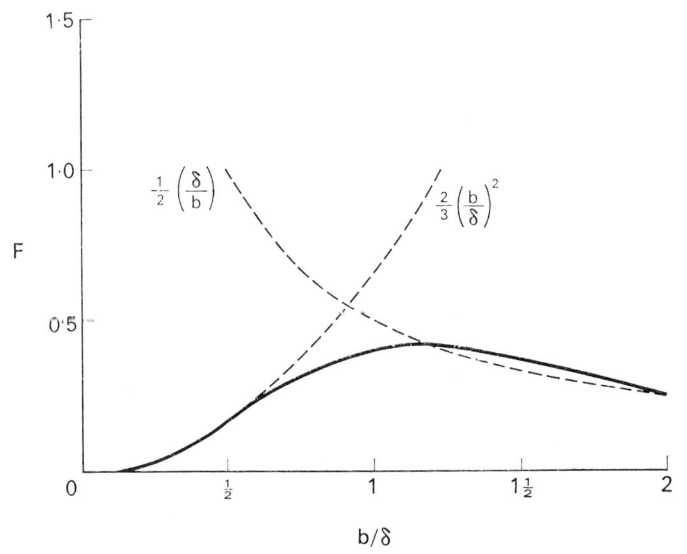

FIG. 8.12. Heating effect in a flat sheet.

heating, δ being fixed, the sheets should have thickness $2b = 2.25\delta$, i.e. b should be slightly bigger than δ. If the sheets are thicker the eddy currents will not penetrate far enough. If they are thinner, the eddy currents will be smaller.

8.8. Proximity Loss

In § 8.4 we calculated the eddy-current loss in an isolated conductor. In general there will be several conductors near each other as in a system of bus-bars, or the conductor may be wound as a coil having several turns side by side. Such grouping of conductors causes additional eddy currents to flow, because each part of the conductor lies in the alternating magnetic field of the rest of the system. The additional loss is called the *proximity loss*.

Consider the simple example, shown in Fig. 8.13, of two sheets forming the go and return conductors of a twin conductor system. The loss per unit area in the top conductor due to the current in that conductor can be obtained from eqn. (8.43),

$$P = \frac{I^2}{2\sigma\delta} \frac{\sinh 2b/\delta + \sin 2b/\delta}{\cosh 2b/\delta - \cos 2b/\delta}, \tag{8.75}$$

where I is the r.m.s. current per unit width.

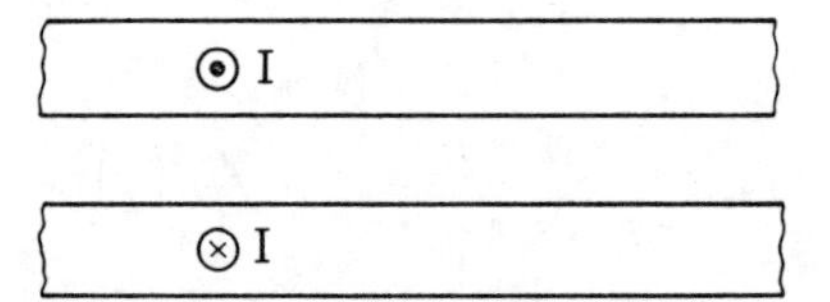

FIG. 8.13. Bus-bar system consisting of a go and a return conductor.

The field due to the bottom conductor at the surfaces of the top conductor is $H_s = I/2$, and by eqn. (8.70) we can find the loss due to this field acting alone as

$$P = \frac{I^2}{2\sigma\delta} \frac{\sinh 2b/\delta - \sin 2b/\delta}{\cosh 2b/\delta + \cos 2b/\delta}. \tag{8.76}$$

Let us now consider the interaction of the two eddy current systems, one due to the current in the conductor and one due to the field of the other conductor. Let us write the instantaneous current density

$$J_{\text{total}} = J = J_0 + J_1 + J_2, \tag{8.77}$$

where J_0 is the uniform current density and J_1 and J_2 are the two sorts of eddy current.

The total instantaneous loss in each conductor is given by

$$P = \int_{-b}^{+b} \frac{J^2}{\sigma}\, dy = \frac{1}{\sigma} \int_{-b}^{+b} (J_0^2 + J_1^2 + J_2^2 + 2J_0J_1 + 2J_0J_2 + 2J_1J_2)\, dy. \tag{8.78}$$

Now

$$\int_{-b}^{+b} J_0J_1\, dy = \int_{-b}^{+b} J_0J_2\, dy = 0 \tag{8.79}$$

because J_0 is uniform and J_1 and J_2 are eddy currents and hence have no net flow across the section (see Fig. 8.9). Also

$$\int_{-b}^{+b} J_1J_2\, dy = 0 \tag{8.80}$$

because J_1, as given in eqn. (8.36), is an even function of y and J_2, as given in eqn. (8.69), is an odd function of y. This leads to the remarkable conclusion that the total loss

$$P = \frac{1}{\sigma} \int_{-b}^{+b} (J_0^2 + J_1^2 + J_2^2)\, dy. \tag{8.81}$$

Hence the losses can be added. Thus from eqns. (8.75) and (8.76) we deduce

$$\begin{aligned} \frac{R_{\text{a.c.}}}{R_{\text{d.c.}}} &= \frac{b}{\delta}\,\frac{\sinh 2b/\delta + \sin 2b/\delta}{\cosh 2b/\delta - \cos 2b/\delta} + \frac{b}{\delta}\,\frac{\sinh 2b/\delta - \sin 2b/\delta}{\cosh 2b/\delta + \cos 2b/\delta} \\ &= \frac{2b}{\delta}\,\frac{\sinh 4b/\delta + \sin 4b/\delta}{\cosh 4b/\delta - \cos 4b/\delta}. \end{aligned} \tag{8.82}$$

The result of eqn. (8.81) also applies to conductors of circular cross-section (and to various other symmetrical shapes). We can write for such conductors:

$$\frac{R_{\text{a.c.}}}{R_{\text{d.c.}}} = 1 + F\left(\frac{b}{\delta}\right) + G\left(\frac{b}{\delta}\right), \tag{8.83}$$

where F describes the eddy-current loss due to the current itself and G describes the proximity loss. Values of F and G are given in electrical handbooks.

8.9. Complex Permeability

Iron cores are used in transformers and similar devices to increase the flux linked with a coil. We have seen that the flux inside conductors is greatly influenced by the eddy currents, so that as b/δ increases there will be less and less flux in the core. To someone taking voltage and current measurements at the ends of the coil surrounding the core it appears as if the permeability of the core is being reduced. If he can measure the phase angle between voltage and current he will notice that the reduction in the magnitude of the permeability is accompanied by a change in phase. Although these effects are not due to a change in permeability, which throughout this chapter we have so far assumed to be constant, and which is constant in this discussion, it is convenient to define an effective permeability due to eddy currents as

$$\mu_e = \frac{\text{average } B}{\text{surface } H} = \frac{\Phi}{2bH_s}. \tag{8.84}$$

The flux can be obtained by the use of eqn. (8.68) so that

$$\overline{\mu_e} = \frac{\mu_r\mu_0}{(1+j)}\frac{\delta}{b}\tanh(1+j)\,b/\delta. \tag{8.85}$$

The magnitude of this effective permeability is

$$|\overline{\mu_e}| = \frac{\mu_r\mu_0}{\sqrt{2}}\frac{\delta}{b}\left(\frac{\cosh 2b/\delta - \cos 2b/\delta}{\cosh 2b/\delta + \cos 2b/\delta}\right)^{1/2} \tag{8.86}$$

and the phase angle is given by

$$\tan\psi = \frac{\sinh 2b/\delta - \sin 2b/\delta}{\sinh 2b/\delta + \sin 2b/\delta}. \tag{8.87}$$

Figure 8.14 illustrates eqn. (8.86). It will be seen that $|\mu_e|$ is close to $\mu_r\mu_0$ for $b/\delta < \frac{1}{2}$ and that for $b/\delta > \pi/4$ it approaches the asymptotic value of $\mu_r\mu_0\ \delta/b\sqrt{2}$. At very high frequencies $|\mu_e|$ tends to zero because δ tends

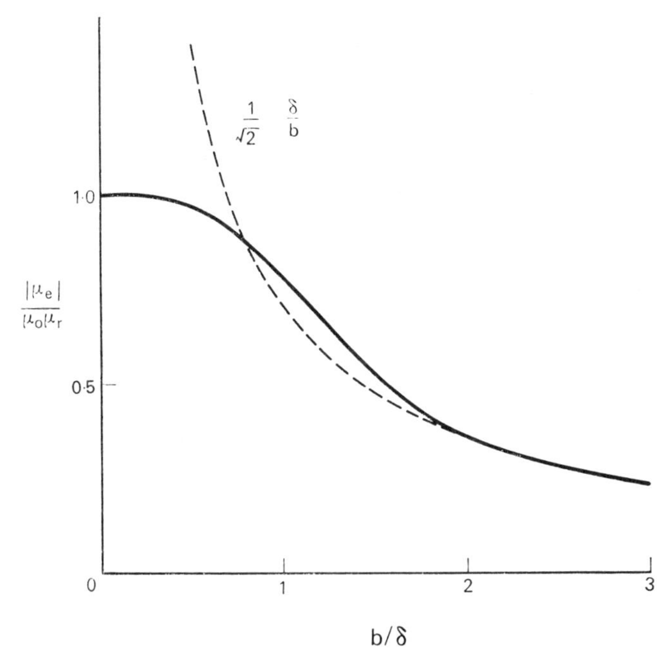

FIG. 8.14. Complex permeability.

to zero. Thus very strong eddy currents produce a permeability which is effectively zero.

So far we have ignored the hysteresis loss. It is in general very difficult to separate hysteresis from eddy current loss, because both are affected by the local relationship between B and H in the material. As we shall discuss in the next section, saturation also plays an important part, and our analysis which has been based on a constant relative permeability μ_r is by no means accurate. Nevertheless, it is useful to indicate the general effect of hysteresis on the total loss.

If we ignore the harmonics which are produced by saturation we can write the instantaneous values of H and B as

$$\left.\begin{aligned} H &= H_m \cos \omega t, \\ B &= B_m \cos(\omega t - \theta), \end{aligned}\right\} \tag{8.88}$$

where θ is called the hysteresis angle.

The equation connecting B and H is an ellipse whose major axis makes an angle of θ with the axis of H. The loss is proportional to

$$\oint H\,dB \quad \text{or} \quad \int H \frac{dB}{dt}\,dt.$$

Hence from eqn. (8.88) the loss is proportional to $\sin\theta$. In terms of exponentials we have $B = B_m e^{j(\omega t - \theta)}$. Inserting this in the diffusion equation, we have

$$\frac{\partial^2 B}{\partial y^2} = j \frac{2}{\delta^2} e^{-j\theta} B. \tag{8.89}$$

If we write

$$\delta' = \delta\, e^{(j\theta/2)} \tag{8.90}$$

the solution proceeds formally as before, and instead of eqn. (8.85) we have the effective permeability due to eddy current and hysteresis

$$\mu_{e+h} = \frac{\mu_r \mu_0}{1+j} e^{-j\theta} \frac{\delta'}{b} \tanh(1+j)\, b/\delta'. \tag{8.91}$$

Hence for large values of b/δ' (or b/δ),

$$\mu_{e+h} = \frac{\mu_0 \mu_r}{\sqrt{2}} \frac{\delta}{b} e^{-j[(\pi/4)+(\theta/2)]}. \tag{8.92}$$

Thus the limiting phase angle has been increased from $\pi/4$ to $[(\pi/4) + (\theta/2)]$. It must, however, be remembered that this treatment gives only general guidance.

8.10. Anomalous Eddy-current Loss and Saturation Effects in Iron

The eddy-current loss predicted by calculations based on a constant relative permeability are found to be much too small when they are compared with measurement. Many workers have sought to elucidate this problem, which is a very complicated one. When the iron is in the form of thin sheets, the domain structure becomes important, particularly if the material has been deliberately made anisotropic by cold rolling. As indicated in Fig. 8.15, the eddy current density will be high near moving domain walls. Since the loss varies as the square of the current density, this bunching of the eddy current increases the loss. Some other possible loss mechanisms are discussed in the book by Brailsford.† The total eddy current loss in cold-rolled, grain-oriented material varies with the direction of magnetization. Brailsford gives eddy current anomaly factors of between 3 and 10, where the anomaly factor has been defined as the measured eddy-current loss divided by the eddy-current loss calculated on the assumption of constant relative permeability. The anomaly factor also varies with domain size, frequency, and maximum flux density.

Domain effects are important in thin sheets, but in massive iron and steel it is the shape of the B–H relationship which really matters. To investigate the eddy currents in a heavily saturated material let us make the drastic assumption that the B–H curve is a step-function as indicated in Fig. 8.16. Under steady-state conditions the flux density can then have

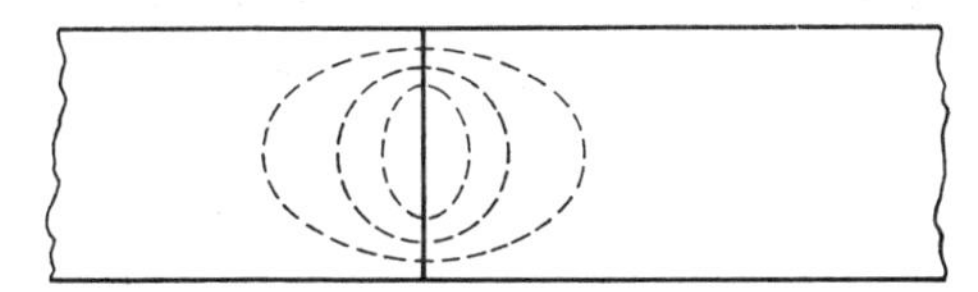

FIG. 8.15. Eddy currents near a moving domain wall.

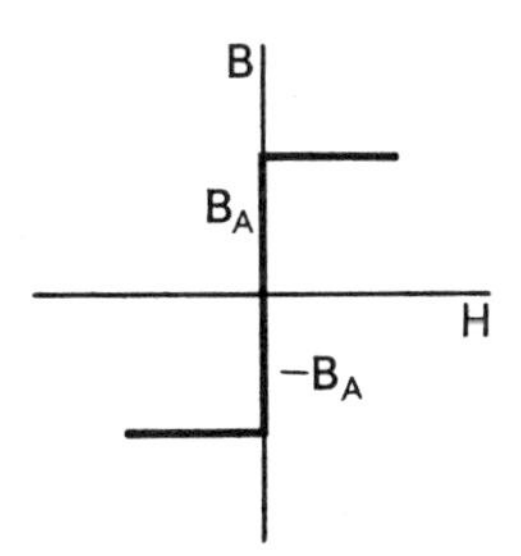

FIG. 8.16. A step-function B–H characteristic.

† F. Brailsford, *Physical Principles of Magnetism*, Van Nostrand, 1966. Interested readers should refer also to a paper by C. R. Boon and J. A. Robey, Effect of domain-wall motion a power loss in grain-oriented silicon-iron sheet, *Proc. I.E.E.*, Vol. 115, No. 10, pp. 1535–40, 1968.

only the values $+B_A$ or $-B_A$. If a magnetic field strength $H_s \sin \omega t$ is applied to the surface of the material then, during a positive half-cycle, the influence of the positive magnetic field will penetrate into the material. Thus a wave of magnetic flux will move more into the material and at the wave front there will be change from $-B_A$ to $+B_A$. This wave will be followed by one of opposite sign.

Let us consider the semi-infinite slab discussed in § 8.3. At any time t let the distance of the wave front from the surface of the material be Y and its velocity dY/dt. Consider Fig. 8.17. The rate of change of flux per unit width x of slab is

$$\frac{d\Phi}{dt} = 2B_A \frac{dY}{dt}. \tag{8.93}$$

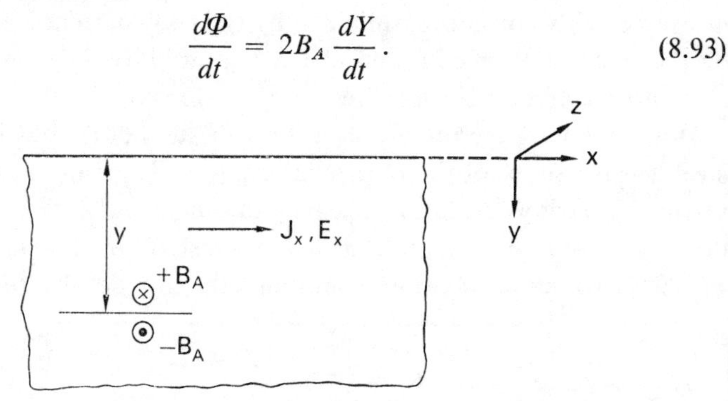

FIG. 8.17. Penetration of flux wave into a solid block.

Beyond the distance Y there are no time-varying effects, because the wave has not penetrated so far. Hence there is no induced electric field for $y > Y$. Applying Faraday's law to any rectangular circuit embracing the wave front, we have for all distance $y < Y$,

$$E_x = -2B_A \frac{dY}{dt} \tag{8.94}$$

and

$$J_x = -2\sigma B_A \frac{dY}{dt}. \tag{8.95}$$

The total current per unit width z of slab can be derived by considering the m.m.f.

$$I = -H_s \sin \omega t = \int_0^Y J_x \, dy = -2\sigma B_A Y \frac{dY}{dt}. \tag{8.96}$$

This gives

$$\sigma B_A \frac{d(Y)^2}{dt} = H_s \sin \omega t, \tag{8.97}$$

whence

$$Y^2 = \frac{H_s}{\omega \sigma B_A}(1 - \cos \omega t) \tag{8.98}$$

and

$$Y = \sqrt{\left(\frac{2H_s}{\omega \sigma B_A}\right)} \sin\left(\frac{1}{2}\omega t\right),\dagger \tag{8.99}$$

where $t = 0$ has been chosen at the start of a positive half-cycle. The maximum depth of penetration occurs when $\sin \frac{1}{2}\omega t = 1$ or $t = \pi/\omega$.

$$\delta_A = Y_{max} = \sqrt{\left(\frac{2H_s}{\omega \sigma B_A}\right)}. \tag{8.100}$$

This should be compared with the skin depth δ of the linear theory

$$\delta = \sqrt{\left(\frac{2}{\omega \sigma \mu_0 \mu_r}\right)}. \tag{8.18}$$

We note that in the non-linear case the permeability $\mu_0\mu_r$ has been replaced by the ratio B_A/H_s. The velocity of the wave front is given by

$$\frac{dY}{dt} = \frac{1}{2}\omega Y_{max} \cos\left(\frac{1}{2}\omega t\right). \tag{8.101}$$

The current density J_x is constant throughout the depth Y. Hence the instantaneous power loss per unit surface area is $J_x^2 Y/\sigma$. The energy lost per half-cycle is

$$U = \frac{1}{\sigma}\int_0^{\pi/\omega} J_x^2 Y \, dt. \tag{8.102}$$

By use of eqns. (8.95), (8.99), and (8.101),

$$U = Y_{max}^3 B_A^2 \sigma \omega^2 \int_0^{\pi/\omega} \cos^2\left(\tfrac{1}{2}\omega t\right) \sin\left(\tfrac{1}{2}\omega t\right) dt = \tfrac{4}{3} Y_{max}^3 B_A^2 \sigma \omega. \tag{8.103}$$

† This is subject to the condition that Y is always positive.

Hence the average power is given by the use of eqn. (8.100) as

$$P = \frac{8}{3\pi}\frac{H_s^2}{\sigma Y_{\max}} = 1{\cdot}7\,\frac{H_s^2}{2\sigma Y_{\max}}. \tag{8.104}$$

Comparing this with eqn. (8.32) we note that if we put $\mu_0\mu_r = B_A/H_s$ the loss in the saturated material is 70% higher than the loss in a material having a linear *B–H* characteristic.

It is found in practice that for saturated iron and steel, eqn. (8.104) gives results more in accord with measurements than eqn. (8.32). There is, however, the difficulty of deciding on the value of B_A, because the actual *B–H* curve does not have a sharply defined saturation flux density. If more accurate results are required, the shape of the *B–H* curve must be taken into consideration. Analytical methods become impossibly difficult if this is done, but it is possible to compute the eddy current numerically.†

8.11. Solution of Complicated Eddy-current Problems by Dimensional Analysis

Throughout this chapter we have assumed that the conducting sheets had a very large surface area and that the applied fields and currents were uniform across this surface. This meant that we could simplify the diffusion equation to include only the dimension perpendicular to the surface. We thus had only one space variable.

Our analysis revealed that the behaviour of the eddy currents depended on the geometrical ratio b/δ, i.e. on the ratio of the linear dimension to the skin-depth. If now there is a variation of the field along the surface, or if the conductor has a cross-section in which the width is of the order of the thickness, it is found that the behaviour depends on other non-dimensional parameters typified by a/δ, where a is another linear dimension, and λ/δ, where λ is the wavelength (twice the pole-pitch) of the applied field. Further complications arise where there are regions of different conductivity or permeability, which will introduce parameters such as $(\mu_r)_1/(\mu_r)_2$ and σ_1/σ_2.

† See the paper by K.K. Lim and P. Hammond, A universal loss chart for the calculation of eddy current losses in thick steel plates, *Proc. I.E.E.*, Vol. 117, No. 4, pp. 857–64, 1970.

A full solution including all the parameters may well be far too complicated. Even in our simple problem of the conducting sheet the full solution, typified by the hyperbolic and circular functions of eqn. (8.43), was cumbersome. But we noticed, for instance in Fig. 8.7, that there were two sharply defined regions for $b < \delta$ and $b > \delta$. The first type of solution we called resistance-limited and the second inductance-limited. A knowledge of these two limiting types of behaviour rendered the full solution almost unnecessary.

In more complicated problems it becomes increasingly useful to divide the solution into types of behaviour in which certain parameters are dominant. A full treatment is beyond the scope of this book, but an example illustrates the method.†

Figure 8.18 shows a wide sheet of conductivity σ and relative permeability μ_r. The thickness of the sheet is $2b$. Parallel to the surface of the sheet and at a distance a from it there is a flat current distribution, in which the

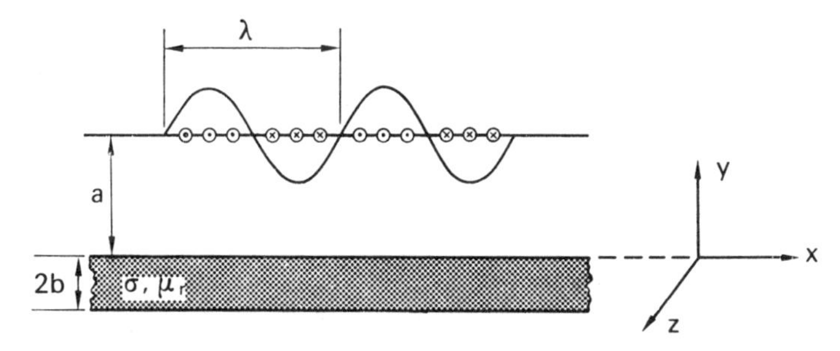

FIG. 8.18. Eddy currents induced in a flat conductor by a current distribution varying sinusoidally in space and time.

current density varies sinusoidally in space with a wavelength λ and in time with angular frequency ω. The current flow is parallel to the sheet.

The parameters governing the behaviour of the problem are $\mu_r : 1$, $\sigma : 0$, $\lambda : \delta$, $\lambda : a$ and $b : \delta$. Let us examine these separately. (1) If the relative permeability μ_r is increased and becomes dominant, we shall expect some magnetostatic behaviour. (2) Since there is only one region with

† See R.L. Stoll and P. Hammond, Calculation of the magnetic field of rotating machines. Part 4: Approximate determination of the field and the losses associated with eddy currents in conducting surfaces, *Proc. I.E.E.*, Vol. 112, No. 11, pp. 2083–94, 1965.

finite conductivity, we shall not expect any special dependence on σ, as long as the ratio $\omega\varepsilon_0/\sigma \ll 1$ as discussed in § 8.1. (3) The ratio λ/δ will decide how far we can disregard the space variation of the applied field in working out the eddy current distribution in the sheet. Moreover for large λ/δ we shall have an inductance-limited behaviour. Since δ varies as $1/\sqrt{\mu_r}$ we shall have to compare the effect of the parameter λ/δ with the parameter μ_r. If λ/δ is small there will be little eddy-current effect because the current will produce very little field. (4) The ratio λ/a will decide the strength of the applied field at the surface of the slab, but cannot by itself affect the type of eddy-current flow. (5) The effect of the ratio b/δ has been fully discussed in this chapter. If $b > \delta$ we can treat the sheet as infinitely thick.

Let us examine a sheet for which $b > \delta$ so that we can disregard b and focus attention on the two parameters μ_r and λ/δ. Close examination reveals three types of eddy current loss:

$$\text{(a)} \quad \lambda \ll \delta \qquad P \propto \sigma\lambda^3\omega^2; \tag{8.105}$$

$$\text{(b)} \quad \lambda \gg \delta \quad \text{and} \quad \lambda\delta \ll \mu_r \quad P \propto \lambda^2\omega^{3/2}\sigma^{1/2}\mu_r^{-1/2}; \tag{8.106}$$

$$\text{(c)} \quad \lambda \gg \delta \quad \text{and} \quad \lambda\delta \gg \mu_r \quad P \propto \omega^{1/2}\mu_r^{1/2}\sigma^{-1/2}. \tag{8.107}$$

In case (a) there is little eddy current effect. If the loss is to be reduced the conductivity of the sheet should be reduced. The loss is independent of the permeability of the sheet. In case (b) the skin effect is appreciable but the permeability is dominant. The loss will be reduced by reducing the conductivity and increasing the permeability. In case (c) the skin effect is dominant. The loss will be reduced if the conductivity is increased.

This example should convince the reader that a method based on dimensional analysis is very well suited to the elucidation of complicated problems involving eddy currents.

Summary

In this chapter we have discussed the flow of alternating current, and the distribution of magnetic flux, in conducting material. We have found that an important parameter is the skin-depth which depends on conductivity, permeability, and frequency. If the thickness of the conductor is large compared with the skin-depth, then the current and flux will be confined to layers close to the surface. If, however, the thickness of the conductor is small compared with the skin-depth, then the current and flux will penetrate throughout the conductor. In such a "thin" conductor the current and flux distribution can be calculated by neglecting the magnetic field of the eddy currents.

Exercises

8.1. Examine possible solutions for the diffusion equation $\nabla^2\phi = (1/h^2)(\partial\phi/\partial t)$. What form of solution is appropriate to the problem of flux penetration into a conducting sheet, when a constant magnetic field is applied to the surfaces of the sheet? What form is appropriate if the applied magnetic field alternates? [Ans.: $\cos\alpha x\, e^{-\beta t}$, $e^{\pm\alpha x}e^{j(\omega t\pm\alpha x)}$.]

8.2. Define the term *skin-depth* applied (a) to the flow of alternating current in a conductor, and (b) to the flux distribution in a conductor. Discuss what is meant by *thin* and *thick* conductors.

8.3. Show that the a.c. resistance of a thin conductor of circular cross-section is given by

$$R_{\text{a.c.}} = R_{\text{d.c.}}\left(1 + \frac{1}{48}\frac{a^4}{\delta^4}\right).$$

Show that the a.c. resistance of copper wire of 1 cm diameter at 100 Hz is approximately $\frac{1}{2}$% higher than the d.c. resistance. What is the increase at 400 Hz? [Ans.: 8%.]

8.4. Discuss the terms *inductance-limited* and *resistance-limited* eddy currents. Show that under certain conditions the a.c. resistance of a circular conductor of radius a is given by the expression $R_{\text{a.c.}} = R_{\text{d.c.}}\, a/2\delta$. What are these conditions? In a conductor to which these conditions apply it is proposed to reduce the loss by cooling. If the resistivity is reduced by 99%, by how much will the losses be reduced? [Ans.: By 90%.]

8.5. A long non-magnetic conductor of square cross-section and side a is placed in a uniform magnetic field $H\cos\omega t$. The conductivity of the conductor is σ and the direction of the field is perpendicular to the length of the conductor and parallel to one of the diagonals of the cross-section. Determine the eddy-current loss per unit length of the conductor, if the magnetic field of the eddy currents can be neglected. [Ans.: $(H^2/6\sigma)(a^4/\delta^4)$ W/m.]

8.6. Two long conductors of rectangular cross-section and sides $2a$ and $2b$, where $a \gg b$, form the go and return paths of a single phase a.c. supply carrying r.m.s. current I. The long sides of the conductors are horizontal and the conductors are arranged with one immediately above the other, the space between them being c, where $a \gg c$. Estimate the peak magnetic field strength on both the long sides of each conductor and obtain in complex form an expression for the magnetic field in one conductor. [Ans.: $I/\sqrt{2}a, 0$; $I/\sqrt{2}a\,[\sinh\alpha(b-y)]/(\sinh 2\alpha b)$, where $\alpha = (1+j)/\delta$].

8.7. Show that the complex magnetic field strength H inside a steel lamination of permeability $\mu_0\mu_r$ and conductivity σ, may be described by the equation

$$\frac{d^2H}{dy^2} = j\omega\sigma\mu_0\mu_r H.$$

Derive an expression for the complex value of the flux Φ per unit width in a lamination of thickness $2b$, subjected to an alternating magnetic field of peak value H_s. [Ans.: $(2\mu_0\mu_r H_s/\alpha)\tanh \alpha b$, where $\alpha = (1+j)/\delta$.]

8.8. Explain what is meant by *complex permeability*. Determine the complex permeability for the lamination described in Exercise 8.7. Can a complex permeability also be associated with hysteresis? If so, what is the essential difference between the two uses of the concept? [Ans.: $(\mu_0\mu_r/\alpha b)\tanh ab$; yes, see § 8.9, the phase angle due to the hysteresis loss applies only at a particular depth in the material because it depends on the hysteresis loop at that depth whereas the phase angle due to eddy-current loss applies at the supply terminals.]

8.9. Alternating current flows parallel to the surface of a semi-infinite slab of conducting material. At what depth is the current density reduced to (a) 10%, (b) 1% of the surface value? At what depth is the phase of the current density (c) 90°, (d) 180° lagging on the surface value? [Ans.: (a) $2{\cdot}3\delta$; (b) $4{\cdot}6\delta$; (c) $1{\cdot}57\delta$; (d) $3{\cdot}14\delta$.]

8.10. Show that under certain assumptions the limiting phase angle taking account of eddy current and hysteresis loss is given by $\phi = \pi/4 + \theta/2$, where θ is the hysteresis angle. Show also that the eddy-current loss associated with the hysteresis loss is less than the eddy-current loss in the absence of hysteresis.

8.11. Show that in a material which has a rectangular B–H characteristic the surface electric field strength is given by

$$E = B_A\omega\delta_A \cos(\tfrac{1}{2}\omega t)$$

during the positive half-cycle of magnetization and

$$E = -B_A\omega\delta_A \sin(\tfrac{1}{2}\omega t)$$

during the negative half-cycle. Sketch this curve and show that its fundamental component has a power factor of 0·894 compared with the power factor of 0·707 for the linear magnetization curve.

8.12. A useful representation of the magnetization curve is given by the expression

$$B = \frac{H}{a+bH}.$$

Discuss the physical significance of a and b and sketch curves for different values of $\xi = a/b$. Examine the dimensionless parameters governing the eddy currents in a sheet of thickness $2d$ and conductivity σ subjected to a surface magnetic field strength $H_0 \sin \omega t$. Show that the diffusion equation can be written in non-dimensional form as

$$\frac{\partial^2 h}{\partial \bar{x}^2} = \frac{\eta}{(1+h)^2}\frac{\partial h}{\partial \tau},$$

where $\bar{x} = x/d$, $\tau = t\omega/2\pi$, $h = H/\xi$, $\eta = \sigma\omega d^2/2\pi\xi b$.

CHAPTER 9

MAXWELL'S EQUATIONS

9.1. Faraday's Law of Electromagnetic Induction

In electrostatics we defined an electric field strength in terms of the magnitude and position of stationary electric charges. Because the law of force between such charges is the inverse square law, we found it possible to postulate a scalar potential and to derive the electric field by taking the gradient of that potential. The electrostatic field is therefore a conservative field and curl $\mathbf{E} = 0$. No energy can be interchanged with such a conservative system by means of a cyclic process. Electrostatic devices are useful only for the storage of electrical energy unless, as in electrostatic generators, the charges are moved by mechanical forces as well as electrical ones. By itself an electrostatic field cannot produce a flow of steady current.

The discovery that there are also non-conservative electric forces was made in 1831 by Michael Faraday, and his law of electromagnetic induction is one of the two foundation stones of electrical engineering. The other one is Maxwell's reformulation of Ampère's law which we shall discuss in § 9.7. Before Faraday made his discovery, Ampère had demonstrated the equivalence of magnets and currents, which led to the idea that a current loop could be replaced by a magnetic shell, which we discussed in § 2.2. This led to the realization that the magnetic field of a steady current is given by

$$\oint \mathbf{H} \, . \, d\mathbf{l} = I \tag{3.26}$$

and that, therefore, the magnetic field of a current is non-conservative. By considering the forces between a current and a magnet [see eqns. (2.39)–(2.45)], it could be deduced that a moving charge in a magnetic field would experience an electric field

$$\mathbf{E} = \mathbf{u} \times \mathbf{B}, \tag{2.45}$$

and that this electric field strength was not necessarily conservative because it was easy to devise circuits in which either u and B were so arranged as to give an electromotive force

$$\oint \mathbf{E} \,.\, d\mathbf{l} = \oint (\mathbf{u} \times \mathbf{B}) \,.\, d\mathbf{l}. \tag{2.46}$$

A typical example of such a circuit was given in Fig. 2.12. Equation (2.46) is a direct consequence of Ampère's experimental discovery of the equivalence of currents and magnets. Faraday's experiments showed that eqn. (2.46) is a special case of a very much more general law which was formulated mathematically by Maxwell as

$$\oint \mathbf{E} \,.\, d\mathbf{l} = -\frac{d\Phi}{dt}, \tag{2.47}$$

where the magnetic flux is given by

$$\Phi = \iint \mathbf{B} \,.\, d\mathbf{s}. \tag{2.26}$$

Comparison of eqns. (2.47) and (2.46) shows that the change of flux in eqn. (2.46) is due to the motion of the circuit across the steady flux density B. In order to obtain the total rate of change of the flux we must add to this *motional* e.m.f. the *transformer* e.m.f. caused by a change of B with time. Equation (2.47) can then be written

$$\oint \mathbf{E}' \,.\, d\mathbf{l} = -\iint \frac{\partial \mathbf{B}}{\partial t} \,.\, d\mathbf{s} + \oint (\mathbf{u} \times \mathbf{B}) \,.\, d\mathbf{l}. \tag{9.1}$$

The use of Stokes's theorem, eqn. (3.34), leads to

$$\text{curl}\, \mathbf{E}' = -\frac{\partial \mathbf{B}}{\partial t} + \text{curl}\, (\mathbf{u} \times \mathbf{B}). \tag{9.2}$$

It should be noticed that the electric field $\mathbf{E}$ in eqns. (9.1) and (9.2) could either be the non-conservative part of the total electric field or the total field itself, because for the electrostatic part of the field there would be no curl. Thus Faraday's law does not by itself specify the electric field completely. This is, of course, in agreement with the argument of § 3.2.7. A unique specification must include the divergence sources of the field as well as the curl sources.

The velocity u in eqns. (9.1) and (9.2) is the velocity of the circuit across the field **B** and the electric field is the field in the frame of reference of this moving circuit. The electric field measured by a stationary observer differs from that of the moving observer, because the latter measures the additional effect defined in eqn. (2.45). Let **E**′ be the field in the moving frame and **E** the stationary field. Then

$$\mathbf{E} = \mathbf{E}' - \mathbf{u} \times \mathbf{B} \tag{9.3}$$

and eqn. (9.2) can be written

$$\text{curl}\,(\mathbf{E}' - \mathbf{u} \times \mathbf{B}) = -\frac{\partial \mathbf{B}}{\partial t}, \tag{9.4}$$

whence

$$\text{curl}\,\mathbf{E} = -\frac{\partial \mathbf{B}}{\partial t}. \tag{9.5}$$

This important result shows that the differential formulation of Faraday's law is independent of the motion of the material through the field. Equations (9.2) and (9.5) are very useful in enabling us to transfer from a moving reference frame to a stationary one and vice versa.

9.2. Some Features of Faraday's Law

Engineers often quote Faraday's law in the form "the induced e.m.f. equals the negative rate of change of the flux turns". This is parallel to the statement that the m.m.f. is equal to the current turns. We did not include the "turns" in our statement of the m.m.f. law, eqn. (3.26), because our discussion showed that the source of the m.m.f. was the total current. It makes no difference whether this current flows in many turns of a thin wire, or in one thick conductor, or whether we have in view the current flow in a plasma, where there is no metallic conductor at all.

In our statement of the e.m.f. law, eqn. (2.47), there is again no reference to the turns, but there the reason is slightly different. On the right-hand side of eqn. (2.47) it is again the total magnetic flux that is under consideration. But when engineers talk of flux turns, they do not mean, as in the m.m.f. law, that the flux is contained in turns, but that turns of a metallic conductor link the flux. Thus the turns are to do with the electric field around the flux. We now see that the left-hand side of the e.m.f. law is in view. The e.m.f. is the line integral of the electric field. Clearly if a wire is wrapped around the flux, then the e.m.f. induced in the wire will depend on its total length and therefore on the turns. But we also note that there is no need for the wire to consist of an integral number of turns.

It is really misleading to introduce the notion of flux turns. One can, for instance, readily construct an apparatus in which a coil of wire can be wrapped around a magnetic flux. If the flux is held constant and the wire does not cross the magnetic field, no e.m.f. is induced by changing the turns.

Another feature of Faraday's law which needs elucidation is the negative sign. This arises in the following manner. We have arbitrarily related positive current flow and positive magnetic field strength by the m.m.f. law. The negative sign in eqn. (2.47) states that the e.m.f. would drive the current in a direction to oppose the change of the magnetic flux. If this were not so, the e.m.f. would set up a current to strengthen the flux, which would then increase the e.m.f. Such a process would be unstable. Of course it would be possible to write eqn. (2.47) with a positive sign, but then we should have to write $\oint \mathbf{H} \cdot d\mathbf{l} = -I$.

It is not always obvious how eqn. (2.47) is to be interpreted when the path of integration for the electric field is not clearly defined. Consider, for instance, Fig. 9.1, which illustrates the motion of a circuit as it surrounds a magnetic flux of constant strength. There is a change of flux linking the circuit, but the voltmeter does not record an e.m.f. Reference to eqn. (9.1) shows why there is no e.m.f. Since $\partial \mathbf{B}/\partial t$ is zero there is no transformer action, and so the first term in eqn. (9.1) is zero. The second term is also zero because the motion of the circuit takes place outside the magnetic field, so that $\mathbf{u} \times \mathbf{B}$ is zero everywhere.

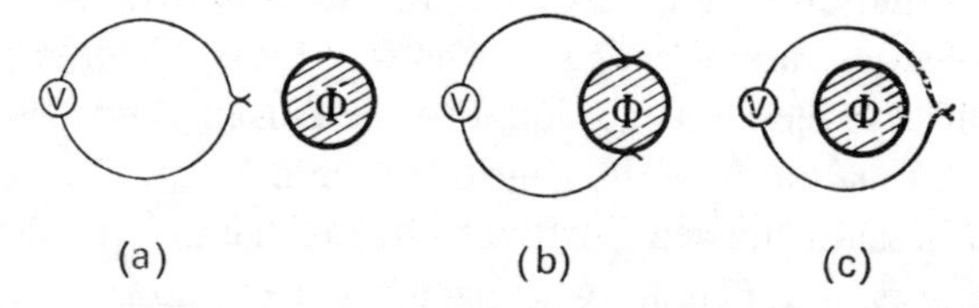

FIG. 9.1. A constant flux surrounded by a moving conductor.

In Fig. 9.2 the converse difficulty is encountered. Here a conducting disc spins in a constant magnetic field. The flux linked with a circuit made up of the disc and a stationary wire touching the centre of the disc and its circumference is apparently constant. Nevertheless, the voltmeter registers an e.m.f. Reference to eqn. (9.1) shows that in this case there is a motional term, because the charges in the disc cross the magnetic field.

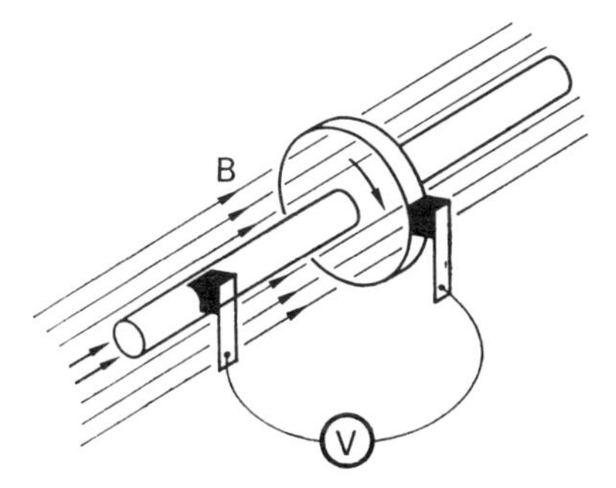

FIG. 9.2. A conducting disc rotating in a constant flux.

9.3. Conducting Fluids in Magnetic Fields

In Chapter 8 we discussed at some length the diffusion of currents and magnetic fields into stationary conductors. If motional e.m.f.s occur as well, the behaviour becomes more complicated, particularly so where the conductor is a fluid. A wholly new set of phenomena are encountered, and these are given the general title "magnetohydrodynamics". A full treatment is impossible in a book of this size,† but a few remarks are justified.

Let us combine eqn. (9.2) with Ohm's law:

$$\text{curl } \mathbf{E} = \text{curl } \frac{\mathbf{J}^{\ddagger}}{\sigma} = -\frac{\partial \mathbf{B}}{\partial t} + \text{curl } (\mathbf{u} \times \mathbf{B}), \tag{9.6}$$

which can be written

$$\frac{\partial \mathbf{B}}{\partial t} = \text{curl} \left(\mathbf{u} \times \mathbf{B} - \frac{\mathbf{J}}{\sigma} \right). \tag{9.7}$$

† Readers whose professional work demands a detailed understanding of the subject are referred to a companion volume in this series, J. A. Shercliffe, *A Textbook of Magnetohydrodynamics*, Pergamon Press, 1965.

‡ The Hall effect current has been neglected.

If the conduction term is negligible compared with the motional term,

$$\frac{\partial \mathbf{B}}{\partial t} = \text{curl } \mathbf{u} \times \mathbf{B} \tag{9.8}$$

and, referring to eqn. (9.1), we have

$$\frac{d}{dt} \iint \mathbf{B} \,.\, d\mathbf{s} = 0 . \tag{9.9}$$

This implies that the magnetic flux through any surface moving with the material remains constant. The flux is "frozen" into the material.

The condition for this to occur is that

$$\left| \frac{\mathbf{J}}{\sigma} \right| \ll \left| \mathbf{u} \times \mathbf{B} \right| . \tag{9.10}$$

Using the m.m.f. relation we can write

$$\text{curl } \mathbf{B} = \mu \mathbf{J} , \tag{9.11}$$

so that the magnitudes of **B** and **J** are related by

$$\left| \mathbf{B} \right| = \mu L \left| \mathbf{J} \right| , \tag{9.12}$$

where L is the typical length governing the flow. Hence the inequality (9.10) becomes

$$\sigma \mu L u \gg 1 . \tag{9.13}$$

The dimensionless group $\sigma \mu L u$ is called the magnetic Reynolds number R_m and should be compared with the ordinary Reynolds number

$$R = \frac{Lu}{\nu} , \tag{9.14}$$

where ν is the kinematic viscosity.

The ordinary Reynolds number is obtained by considering the ratio of inertia forces to viscous forces, and it therefore defines the relative importance of convection and diffusion effects. The magnetic Reynolds number does the same. If it is large, convection is dominant; if it is small, the behaviour is that of the diffusion process discussed in Chapter 8. The magnetic field is then governed by

$$\nabla^2 B = \sigma\mu \frac{\partial B}{\partial t}. \tag{9.15}$$

In electrical machines R_m is large because the high conductivity of copper or aluminium is combined with the high relative permeability of iron. Diffusion effects are secondary and the flux distribution is determined by the exciting currents and the relative motion.† If, on the other hand, R_m is small, the field is scarcely affected by the motion and there are no motionally induced currents. It may then be possible to ignore the induced magnetic field and work entirely in terms of a known applied field.

Consider now some typical values of R_m for conducting fluids. In the application of magnetohydrodynamics to the generation of electric power it has been proposed to pass conducting gases through magnetic fields. For such gases a possible figure of $\sigma = 100$ siemens gives a value for the product $\mu\sigma$ of the order of 10^{-4}. This leads to a small value for R_m and the field may be calculated as though the gas were stationary. On the other hand, if magnetic effects in astrophysics are studied, the characteristic length L is so enormous that R_m is large and the magnetic field is frozen into the stellar material. Finally, if the pumping of liquid metals by electromagnetic induction is being studied, R_m is of the order unity, so that it is not possible to ignore convection or diffusion.

In § 7.5.4 we discussed the stresses set up by a magnetic field. These stresses become very important in plasma physics. The generation of electric power from controlled nuclear fusion processes hold out considerable economic promise. One of the most difficult problems to be solved is how to confine the hot plasma. It would be essential that such a plasma should be kept away from the walls of the container, and this might be possible by the action of the magnetic field stresses. Consider this with reference to Fig. 9.3 which shows a plasma of cylindrical shape in which there is uniform axial current density J. The field stress on the surface of the cylinder is $\frac{1}{2}HB = B^2/2\mu_0$ inward. If the difference between the plasma pressure and the external pressure of the circumference is

† This is discussed in detail in chapter 2 of *Induction Machines for Special Purposes* by E. R. Laithwaite, Newnes, 1966. Professor Laithwaite uses a dimensionless group called the "goodness" factor which is related to R_m.

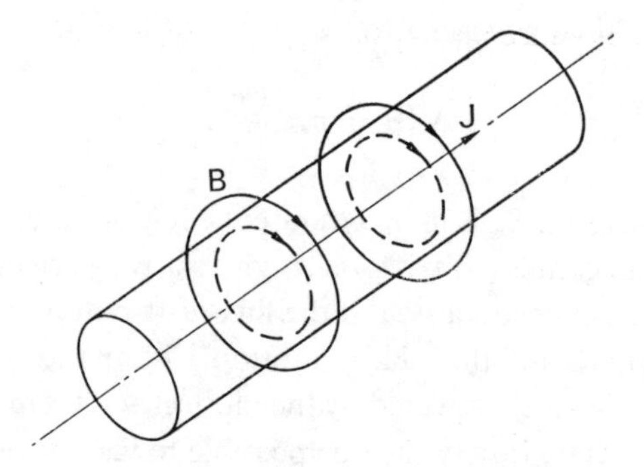

FIG. 9.3. Magnetic field of an axial current of cylindrical cross-section.

equal to this magnetic pressure, equilibrium is possible. Inside the cylinder the field is given by the m.m.f. law, whence

$$B = \frac{\mu_0 r J}{2}. \tag{9.16}$$

Hence the inward pressure at radius r is

$$\frac{B^2}{2\mu_0} = \frac{\mu_0 r^2 J^2}{4}. \tag{9.17}$$

This can be balanced by a radial outward pressure

$$p = \frac{\mu_0 r^2 J^2}{4} + \text{constant}. \tag{9.18}$$

If the pressure of the external radius b is zero we have

$$p = -\frac{\mu_0 J^2}{4}(b^2 - r^2). \tag{9.19}$$

Unfortunately this arrangement is unstable. If for any reason the radius b decreases, there will be a stronger inward magnetic force and the radius will decrease still further.

The study of pressures in conducting fluids is called magnetohydrostatics. Because it is closely allied to magnetohydrodynamics we have mentioned it in this section, although it might well have been included in Chapter 7. The magnetic stresses are of course important in dynamic problems as well as static ones. The equation of motion is

$$\mathbf{J}\times\mathbf{B} - \nabla p + \mathbf{f} = \varrho_m \frac{d\mathbf{u}}{dt}, \tag{9.20}$$

where ∇p is the pressure gradient, $\mathbf{f}$ is the force per unit volume due to such effects as gravity and viscosity and ϱ_m is the mass per unit volume. The total acceleration $d\mathbf{u}/dt$ allows for the motion of the particle under consideration. Thus

$$\frac{d\mathbf{u}}{dt} = \frac{\partial\mathbf{u}}{\partial t} + \frac{\partial\mathbf{u}}{\partial x}\frac{dx}{dt} + \frac{\partial\mathbf{u}}{\partial y}\frac{dy}{dt} + \frac{\partial\mathbf{u}}{\partial z}\frac{dz}{dt} = \frac{\partial\mathbf{u}}{\partial t} + (\mathbf{u}\,.\,\nabla)\,\mathbf{u}. \tag{9.21}$$

To introduce the magnetic stress into eqn. (9.20) we can write

$$\mathbf{J}\times\mathbf{B} = \frac{1}{\mu}(\text{curl}\,\mathbf{B})\times\mathbf{B}. \tag{9.22}$$

By the use of the vector identity

$$\tfrac{1}{2}\,\text{grad}\,B^2 = (\mathbf{B}\,.\,\nabla)\,\mathbf{B} + \mathbf{B}\times\text{curl}\,\mathbf{B} \tag{9.23}$$

we can write eqn. (9.20) as

$$\frac{1}{\mu}(\mathbf{B}\,.\,\nabla)\,\mathbf{B} - \frac{1}{2\mu}\,\text{grad}\,B^2 - \nabla p + \mathbf{f} = \varrho_m\left|\frac{\partial\mathbf{u}}{\partial t} + (\mathbf{u}\,.\,\nabla)\,\mathbf{u}\right|. \tag{9.24}$$

The first two terms of this equation give the force per unit volume due to the magnetic field stress.

9.4. Electromotive Force and Electric Field Strength

The law of electromagnetic induction, eqn. (2.47), gives the line integral of the electric field strength around a closed loop. As long as all the apparatus consists of such "circuits", there is no need to investigate how the total electric field is distributed. But when it is necessary to find the forces on charged particles it is not in general possible to do so from Faraday's law by itself. The exceptions are cases of symmetry. The reader will remember that we used arguments from symmetry in our discussion of the m.m.f. law in Chapter 6. For instance, we derived the magnetic field of a long steady current by choosing a circular path around the current, so that

$$\oint \mathbf{H} \,.\, d\mathbf{l} = H \times 2\pi r = I, \tag{6.20}$$

whence

$$H = \frac{I}{2\pi r}. \tag{6.16}$$

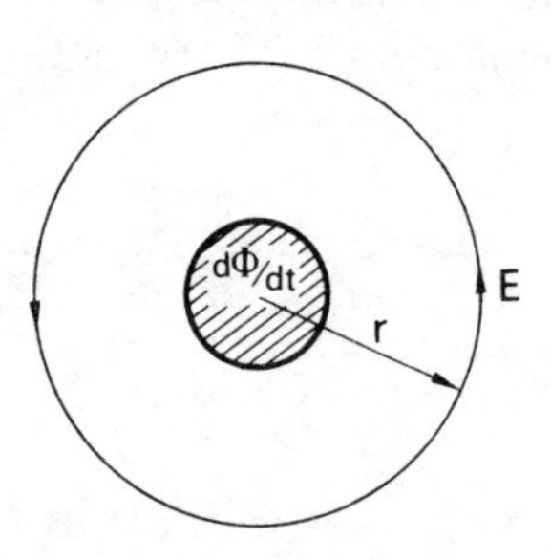

FIG. 9.4. Electric field surrounding a changing flux.

The analogous electrical case is shown in Fig. 9.4, where a long column of flux of cylindrical cross-section is surrounded by an electric field **E**. Here the e.m.f. is

$$e = \oint \mathbf{E} \,.\, d\mathbf{l} = E \times 2\pi r = -\frac{d\Phi}{dt}, \tag{9.25}$$

so that

$$E = -\frac{1}{2\pi r}\frac{d\Phi}{dt}. \tag{9.26}$$

If a conducting ring of resistance R is placed at this radius r, as shown in Fig. 9.5, a current $I = e/R$ will circulate. Consider now Fig. 9.6 where

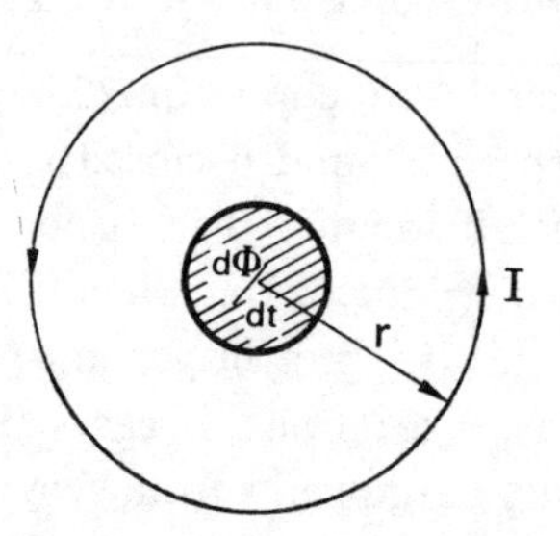

FIG. 9.5. Current in a conductor surrounding a changing flux.

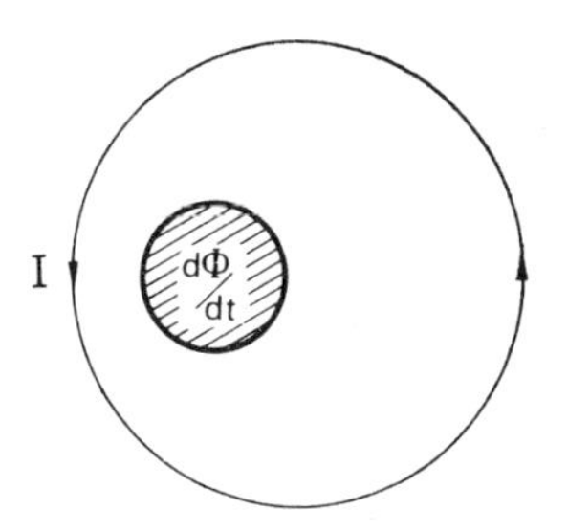

FIG. 9.6. Current flow in a displaced conductor.

the conductor has been displaced. The e.m.f. is unchanged and hence the current is also unchanged. Now at every point in the conductor we have by Ohm's law

$$\mathbf{E} = \mathbf{J}/\sigma, \tag{2.20}$$

where $\mathbf{J}$ is the current density and μ the conductivity. Since the displacement of the ring has not altered I, the current density $\mathbf{J}$ will also be unchanged. Hence $\mathbf{E}$ must be unchanged. But we have found that the electric field strength due to the changing flux is given by eqn. (9.26) and varies inversely as r. We conclude that the electric field strength in Ohm's law cannot be the induced electric field strength by itself. The ohmic field is the *total* field necessary to drive the charge through the material at a point. Since this field is different from the field due to the changing flux we must look for additional sources.

The problem is similar to that of Fig. 9.7 which shows a battery circulating a current through a wire. How is the effect of the battery conveyed along the wire? It can only be conveyed by electric charges. We saw in § 8.1 that a volume charge distribution is impossible in a conductor. Hence there will have to be a surface distribution of electric charge giving the electric field required by Ohm's law.

In the problem of the ring linking a changing flux the e.m.f. is distributed instead of being concentrated as in the battery. In the symmetrical case of Fig. 9.4, the e.m.f. is uniformly distributed so that the average electric field strength is the actual electric field strength. When the ring is displaced a charge distribution as in Fig. 9.8 maintains the uniform average value along the ring by strengthening the induced field where it is too weak and reducing it where it is too strong. The line integral of this electrostatic part of the field is zero, so that it makes no contribution to the e.m.f. around the ring. The charge distribution can be measured. One simple way is to measure the potential difference between different parts of the ring.

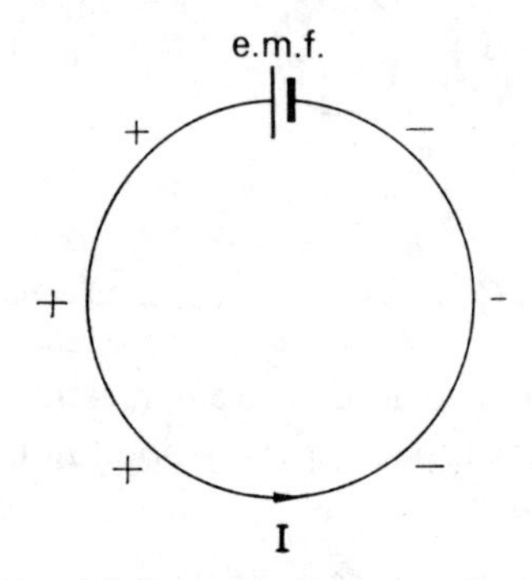

FIG. 9.7. Battery and current circuit.

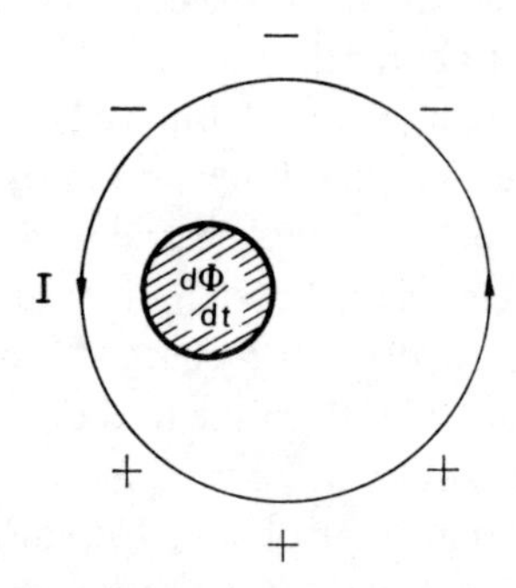

FIG. 9.8. Surface charges on a conductor surrounding a changing flux.

It will have been noticed that this problem of current flow around a ring is very similar to the magnetic field distribution around a magnetic circuit which was discussed in § 7.1. The e.m.f. corresponds to the m.m.f., and surface charge to surface pole strength. The capacitance associated with the surface charges corresponds to the leakage inductance of the magnetic circuit.

Let us now seek to derive an expression for the total electric field strength. We combine eqn. (9.5) with the statement

$$\mathbf{B} = \text{curl}\,\mathbf{A}, \tag{6.25}$$

whence

$$\text{curl}\,\mathbf{E} = -\text{curl}\,\frac{\partial \mathbf{A}}{\partial t}. \tag{9.27}$$

To find **E** we need to "uncurl" this equation. Since curl implies a differentiation, we need to integrate and there will be a constant of integration, which we shall write as a gradient, because the curl of a gradient is zero. We have

$$\mathbf{E} = -\frac{\partial \mathbf{A}}{\partial t} - \text{grad}\,V. \tag{9.28}$$

This equation shows two kinds of electric field. The first term describes a non-conservative (e.m.f. type) field and the second term a conservative (p.d. type) field. In the example illustrated in Fig. 9.6, $\partial \mathbf{A}/\partial t$ is due to the changing flux and grad V to the charge distribution.

9.5. The Vector Potential or Electrokinetic Momentum

In order to be able to calculate **A** and V we need to find how these quantities depend on the sources of the field, i.e. on current and charge. The vector **A** was discussed in § 3.2.5 where, by assuming curl $\mathbf{H} = \mathbf{J}$ and div $\mathbf{A} = 0$, we found for non-magnetic material

$$\mathbf{A} = \frac{\mu_0}{4\pi} \iiint \frac{\mathbf{J}}{r}\, dv', \tag{3.44}$$

where **J** is the current densityper unit area. Similarly, we found in § 3.2.4 that for non-polarizable material

$$V = \frac{1}{4\pi\varepsilon_0} \iiint \frac{\varrho}{r}\, dv', \tag{3.22}$$

where ϱ is the charge density per unit volume.

We have met the vector **A** under the curious name "vector potential". It was given this name because of the similarity between eqns. (3.22) and (3.44), which makes **A** analogous to V the ordinary electrostatic potential. In physical terms this is an unfortunate nomenclature. The name potential is a short way of writing potential energy, and energy is essentially a scalar function. It is meaningless to talk of vector energy, because energy cannot have direction.† Indeed, the whole usefulness of the potential derives from the fact that it is scalar. Instead of adding forces vectorially, we find it easier to add the potentials and derive the total force by taking the gradient of the total potential.

In his famous treatise, Maxwell‡ suggests an alternative name for **A** which shows its nature more clearly. He calls it the "electrokinetic momentum". Let us consider eqns. (3.44) and (9.28) in this light. The current density **J** sets up in space a momentum in a direction parallel to the current flow. The magnitude of the momentum varies inversely as the distance from each particle of moving charge, which constitutes the current. If the momentum changes, a force will be experienced by a charge at the place at which the momentum is being observed. In accordance with the laws of mechanical inertia, the force will oppose the change of momentum.

Once again we notice that mechanics holds the key to the understanding of the physical process. Up to now the vector **A** has been regarded merely as a mathematical tool to "uncurl" Faraday's law. We now see that we could well have postulated that a current or moving charge possesses not only inertia (inductance or magnetic energy) but also momentum. The magnetic field could then have been derived from this momentum. The reader should notice that the electrokinetic momentum is an alternative way of describing the magnetic field. It should also be noticed that **A** describes the *mutual* momentum effect between the current sources and the charge which measures the electric field of eqn. (9.28), just as V describes the *mutual* potential energy between the charge sources and the exploring test charge in the electric field.

9.6. Motional Effects and the Lorentz Force

In the previous section we have discussed the field observed by a stationary charge. Thus the force experienced by such a charge Q is given by

$$\mathbf{F} = Q\mathbf{E}, \tag{9.29}$$

† Where energy flows through a surface enclosing a volume it may be convenient to define an energy surface density which has direction.

‡ J. C. Maxwell, *Electricity and Magnetism*, para. 590.

where $\mathbf{E}$ is defined by eqn. (9.29). If the charge is moving it experiences the force

$$\mathbf{F} = Q(\mathbf{E} + \mathbf{u} \times \mathbf{B}), \tag{9.30}$$

where $\mathbf{u}$ is its velocity across the magnetic field $\mathbf{B}$. The force given by eqn. (9.30) is called the *Lorentz force* after its originator the Dutch physicist H. A. Lorentz. It is instructive to express the Lorentz force in terms of the potentials. We have

$$\mathbf{F} = Q\left(-\frac{\partial \mathbf{A}}{\partial t} - \operatorname{grad} V + \mathbf{u} \times \operatorname{curl} \mathbf{A}\right). \tag{9.31}$$

This can be simplified by the following considerations. From eqn. (9.21) we obtain

$$\frac{d\mathbf{A}}{dt} = \frac{\partial \mathbf{A}}{\partial t} + (\mathbf{u} \,.\, \nabla)\,\mathbf{A}. \tag{9.32}$$

The term $\mathbf{u} \times \operatorname{curl} \mathbf{A}$ can be transformed by the vector identity

$$\operatorname{grad}'(\mathbf{A} \,.\, \mathbf{u}) = (\mathbf{u} \,.\, \nabla)\,\mathbf{A} + \mathbf{u} \times \operatorname{curl} \mathbf{A}, \tag{9.33}$$

where grad′ is to be taken as the gradient at constant $\mathbf{u}$. Hence

$$\mathbf{F} = Q\left(-\operatorname{grad} V + \operatorname{grad}'(\mathbf{A} \,.\, \mathbf{u}) - \frac{d\mathbf{A}}{dt}\right). \tag{9.34}$$

The x-component of this force is

$$F_x = Q\left(-\frac{\partial V}{\partial x} + \frac{\partial'(\mathbf{A} \,.\, \mathbf{u})}{\partial x} - \frac{d}{dt}\frac{\partial(\mathbf{A} \,.\, \mathbf{u})}{\partial u_x}\right). \tag{9.35}$$

In Lagrangian mechanics† we have

$$F_i = \frac{\partial L}{\partial q_i} - \frac{d}{dt}\left(\frac{\partial L}{\partial \dot{q}_i}\right), \tag{9.36}$$

where q_i and $\dot{q}_i$ are the general coordinates and $L = T - U$ = kinetic energy − potential energy. Comparing eqns. (9.36) and (9.35) and noting that V is independent of u, we find that

$$U = QV \tag{9.37}$$

† See, for instance, J. W. Leech, *Classical Mechanics*, Methuen, 1963.

and

$$T = Q\,(\mathbf{A}\,.\,\mathbf{u}). \tag{9.38}$$

The potential energy QV needs no further explanation. The kinetic energy, on the other hand, may be unfamiliar to the reader. It is the mutual kinetic energy between the current sources and the charge Q. It should be noted that the force acts in a direction to decrease the potential energy but to increase the kinetic energy, as shown by the first two terms of eqn. (9.35).

In terms of charge density we can write

$$U = \iiint \varrho V\,dv \tag{9.39}$$

and

$$T = \iiint \varrho\,(\mathbf{A}\,.\,\mathbf{u})\,dv, \tag{9.40}$$

and, in terms of current density,

$$T = \iiint \mathbf{J}\,.\,\mathbf{A}\,dv. \tag{9.41}$$

For magnetostatic fields we can use curl $\mathbf{H} = \mathbf{J}$ to convert the last expression to

$$T = \iiint \text{curl}\,\mathbf{H}\,.\,\mathbf{A}\,dv = \iiint \mathbf{H}\,.\,\text{curl}\,\mathbf{A}\,dv - \oint\!\!\oint (\mathbf{A}\times\mathbf{H})\,.\,d\mathbf{s}, \tag{9.42}$$

and, if the volume is sufficiently large for the surface integral to be negligible, we have

$$T = \iiint \mathbf{H}\,.\,\mathbf{B}\,dv. \tag{9.43}$$

This is the quantity we previously, in eqn. (7.49), designated by the symbol U^*. This reinforces the view that the magnetic energy is kinetic in origin and that electromagnetic induction is an inertia effect associated with the momentum of moving charges.

The Lagrangian formulation is particularly helpful in the discussion of the general behaviour of complicated electromagnetic devices. Other types of potential and kinetic energy can be added to U and T. U, for instance, may have terms due to gravitation and energy stored in spring systems, while T may include the kinetic energy of moving mechanical parts.

It is sometimes useful to transfer from a moving to a stationary frame of reference. By the principle of relativity only relative rectilinear velocities have measurable effects. Hence the force on a charge moving with velocity **u** across a magnetic field **B** of a stationary source must be equal to the force on a *stationary* charge when the *source* moves with velocity **u**.

In the moving frame of reference we have, from eqn. (9.34),

$$\mathbf{E}_{\text{total}} = -\text{grad}\, V + \text{grad}\,(\mathbf{A}\,.\,\mathbf{u}) - \frac{d\mathbf{A}}{dt}. \tag{9.44}$$

In the stationary frame

$$\mathbf{E}_{\text{total}} = -\text{grad}\, V' - \frac{\partial \mathbf{A}'}{\partial t}. \tag{9.45}$$

If **u** is small compared with the velocity of light† $d\mathbf{A}/dt = \partial\mathbf{A}'/\partial t$, so that

$$V' = V - \mathbf{A}\,.\,\mathbf{u}. \tag{9.46}$$

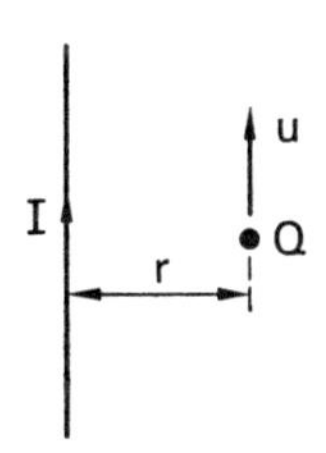

FIG. 9.9. Charge moving parallel to a long current.

A simple example illustrates this effect. In Fig. 9.9 a charge Q travels with velocity **u** parallel to a long constant current I and at a distance r from it. The only force acting on the charge is due to its motion in the magnetic field, since there is no electrostatic field and no change of **A** with time.

$$\mathbf{F} = Q\,(\mathbf{u}\times\mathbf{B}) = -\frac{QuI}{2\pi r}\,\hat{\mathbf{r}}. \tag{9.47}$$

In Fig. 9.10 the charge is stationary, but the current travels with velocity $-\mathbf{u}$. The force on either is due to the gradient of the mutual kinetic energy.

$$\mathbf{F} = Q\,\text{grad}\,(\mathbf{A}\,.\,\mathbf{u}) = Q\,\text{grad}\left(-\frac{uI}{2\pi}\ln r\right) = -\frac{QuI}{2\pi r}\,\hat{\mathbf{r}}. \tag{9.48}$$

† See Chapter 12 for a fuller discussion.

The same result could have been obtained by regarding the current and the charge as a system in which the potential energy U is zero, so that the mutual forces can be derived by inserting the kinetic energy $Q\,(\mathbf{A}\,.\,\mathbf{u})$ in

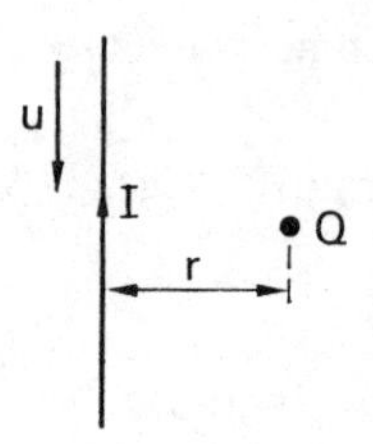

FIG. 9.10. Stationary charge and moving current.

Lagrange's equation (9.36). With this approach there is no need to involve the relativity principle. Indeed, the principle becomes a consequence of Faraday's law which leads to the idea of mutual kinetic energy.

9.7. Maxwell's Hypothesis

So far in this chapter we have been chiefly concerned with the law of electromagnetic induction and with electromotive force. Let us now turn our attention to the magnetomotive force. We start with the relationship between magnetic field strength and current, which Ampère deduced from his experiments,

$$\oint \mathbf{H}\,.\,d\mathbf{l} = I. \tag{3.26}$$

This law was subjected, about 40 years later, to a very careful analysis by Maxwell, who suggested that it was correct for systems in which there is no variation with time, but that in general the complete law should be written,

$$\oint \mathbf{H}\,.\,d\mathbf{l} = I + \frac{d\Psi}{dt}, \tag{2.48}$$

where Ψ the electric flux is given by

$$\Psi = \iint \mathbf{D}\,.\,d\mathbf{s}. \tag{9.49}$$

Equation (2.48) is known as Maxwell's hypothesis, because at the time at which he put it forward it was not possible to subject it to experiment.

The trouble is twofold: first, it is in any case difficult to measure m.m.f. because there is no magnetic current whereas e.m.f. can be measured by the flow of electric current, and, secondly, $d\Psi/dt$ is generally small and is overshadowed by the conduction current I. A full experimental test of the hypothesis depends on the availability of high-frequency generators, which can make $d\Psi/dt$ large enough to be observed.

Maxwell put forward his hypothesis because he found various unsatisfactory features in the electromagnetic relationships based on Ampère's law. It is very helpful to follow his reasoning.

The first difficulty concerns the continuity of current and charge. This continuity relationship was given in § 8.1 by the relationship

$$\operatorname{div} \mathbf{J} = -\frac{\partial \varrho}{\partial t}. \tag{8.3}$$

Readers may like to refer to Fig. 8.1 which illustrates this equation. Now Ampère's m.m.f. law can be written as

$$\operatorname{curl} \mathbf{H} = \mathbf{J}, \tag{3.32}$$

and since the curl of a vector has zero divergence [eqn. (3.48)], Ampère's law implies that

$$\operatorname{div} \mathbf{J} = 0, \tag{9.50}$$

which conflicts with the continuity equation unless there is only direct constant current.

This difficulty is especially noticeable when there is a break in a circuit. Consider Fig. 9.11a which shows a capacitor connected to an a.c. generator. If the m.m.f. integral is taken around the wire, eqn. (3.26) can be used, but if the integral is taken around the space between the plates of the capacitor there appears to be no m.m.f. because no current is linked. This is a very doubtful conclusion because eqn. (3.26) only specifies the line around which $\mathbf{H}$ is measured and not the surface bounded by that line. For instance, in Fig. 9.11b different results would apparently result from the choice of the surfaces A and B, although they are both bounded by the same perimeter. This is inconsistent even with Ampère's law.

A further difficulty faced by Maxwell concerns the propagation of electromagnetic effects. By calling $\mathbf{A}$ the electrokinetic momentum, Maxwell was drawing attention to the inertia of moving charges and magnetic fields. It seemed to him very unlikely that force could be transmitted instantaneously. There was bound to be some time delay and under alternating conditions something akin to wave motion. This meant that both

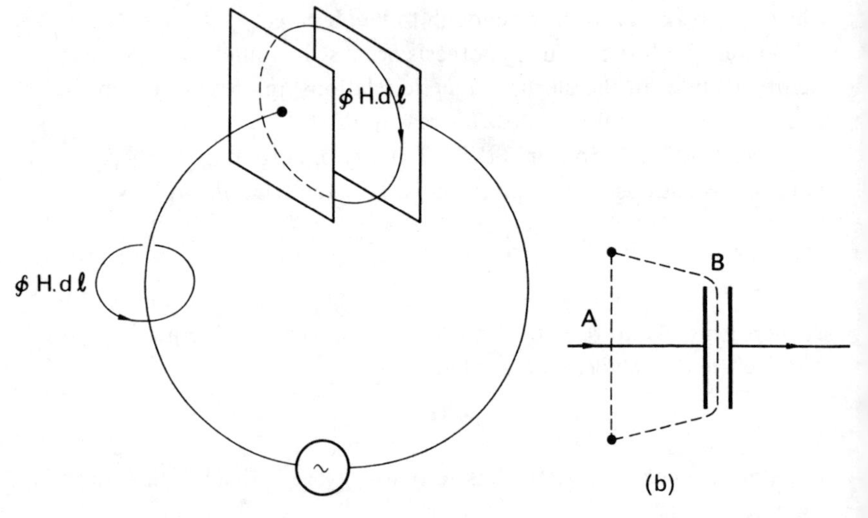

FIG. 9.11. Difficulty of the m.m.f. law applied to a capacitor.

Faraday's law in its integral form and Ampère's law would have to be modified if the length of the coil was large. For instance, in Fig. 9.12 the e.m.f. induced in the secondary coil would vary with the length of the secondary. Also the phase between m.m.f. and current, and between e.m.f. and magnetic flux, would depend on the size of the apparatus and on the frequency.

In seeking to remove these difficulties Maxwell took his clue from the mathematical symmetry of the electrical and magnetic equations. This led him to postulate that a changing electric flux should induce an m.m.f., just as a changing magnetic flux induces an e.m.f. His hypothesis is contained in eqn. (2.48) which we quoted at the beginning of this section. By one stroke of genius Maxwell removed the inconsistencies in the formulation of the subject and at the same time he laid the foundations of radio propagation.

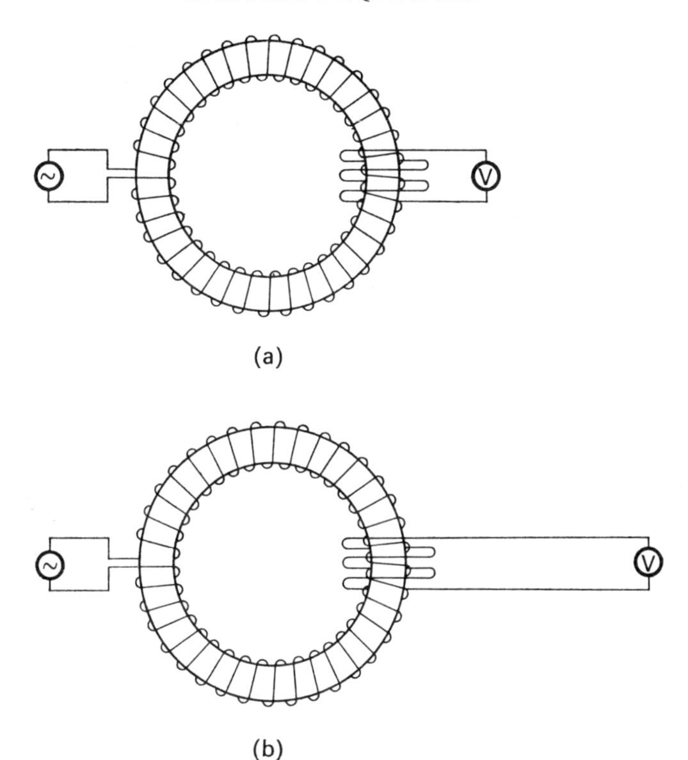

FIG. 9.12. Difficulty of the m.m.f. law applied to a toroid with secondary winding.

In order to avoid difficulties caused by the size of the loop, let us rewrite eqn. (2.48) in differential form. By Stokes's theorem,

$$\oint \mathbf{H} \,.\, d\mathbf{l} = \iint \text{curl}\, \mathbf{H} \,.\, d\mathbf{s}, \tag{3.34}$$

whence

$$\iint \text{curl}\, \mathbf{H} \,.\, d\mathbf{s} = \iint \mathbf{J} \,.\, d\mathbf{s} + \frac{d}{dt} \iint \mathbf{D} \,.\, d\mathbf{s}. \tag{9.51}$$

For a stationary circuit we can therefore write

$$\text{curl}\, \mathbf{H} = \mathbf{J} + \frac{\partial \mathbf{D}}{\partial t}. \tag{9.52}$$

Let us consider the first difficulty in the light of eqn. (9.52):

$$\text{div curl } \mathbf{H} = 0 = \text{div } \mathbf{J} + \frac{\partial \text{ div } \mathbf{D}}{\partial t} = \text{div } \mathbf{J} + \frac{\partial \varrho}{\partial t}. \tag{9.53}$$

Thus the continuity equation is correctly stated. Moreover, reference to Figs. 9.11 and 9.12 shows that around the wire

$$\oint \mathbf{H} \,.\, d\mathbf{l} = I \tag{9.54}$$

and around the capacitor

$$\oint \mathbf{H} \,.\, d\mathbf{l} = \frac{d\Psi}{dt} = \frac{dQ}{dt} = I, \tag{9.55}$$

since by Gauss's theorem the electric flux Ψ equals the charge Q on the plate of the capacitor. The first difficulty is therefore resolved.

In free space, eqn. (9.52) becomes

$$\text{curl } \mathbf{H} = \frac{\partial \mathbf{D}}{\partial t} = \varepsilon_0 \frac{\partial \mathbf{E}}{\partial t}. \tag{9.56}$$

Thus a changing electric field is always accompanied by a magnetic field of the same place. If we consider the example of Fig. 9.12 we notice that there must be some magnetic field *outside* the toroid, even if most of it is concentrated inside.

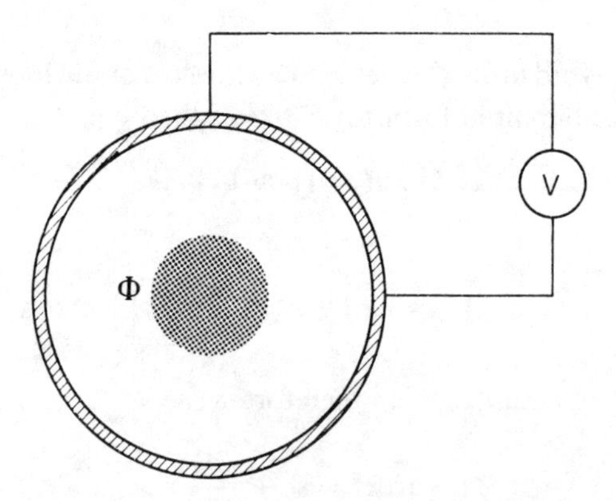

FIG. 9.13. Electric field due to changing magnetic flux.

The difficulty concerning propagation effects also is removed by Maxwell's hypothesis. Because of its great practical importance we shall discuss electromagnetic propagation more fully in the next chapter. Here it is sufficient to show that if we combine the free space m.m.f. relation of eqn. (9.56) with the e.m.f. relation

$$\text{curl}\,\mathbf{E} = -\mu_0 \frac{\partial \mathbf{H}}{\partial t}, \tag{9.57}$$

we have

$$\text{curl curl}\,\mathbf{H} = \text{grad div}\,\mathbf{H} - \nabla^2\mathbf{H} = -\mu_0\varepsilon_0 \frac{\partial^2 \mathbf{H}}{\partial t^2}. \tag{9.58}$$

Since div $\mathbf{H} = 0$, this leads to

$$\nabla^2\mathbf{H} = \mu_0\varepsilon_0 \frac{\partial^2 \mathbf{H}}{\partial t^2} \tag{9.59}$$

and, similarly,

$$\nabla^2\mathbf{E} = \mu_0\varepsilon_0 \frac{\partial^2 \mathbf{E}}{\partial t^2}. \tag{9.60}$$

It was shown in § 2.3 that these are the equations of waves propagating with the speed

$$c = 1/\sqrt{(\mu_0\varepsilon_0)}. \tag{9.61}$$

Numerically,

$$c \approx 2{\cdot}998 \times 10^8 \text{ m/s}. \tag{9.62}$$

But this is the value for the speed of light which had been calculated by astronomers, of whom the first appears to have been the Danish astronomer Roemer, who published his investigation in 1675. This caused Maxwell to make the inference that light is an electromagnetic disturbance.

9.8. Maxwell's Equations

Equation (9.52) gives the m.m.f. relationship for stationary circuits and materials. If the material is polarizable we can make use of the polarization vector $\mathbf{P}$ given by

$$\mathbf{D} = \varepsilon_0\mathbf{E} + \mathbf{P} \tag{4.32}$$

to obtain

$$\text{curl}\,\mathbf{H} = \mathbf{J} + \frac{\partial \mathbf{P}}{\partial t} + \varepsilon_0 \frac{\partial \mathbf{E}}{\partial t}. \tag{9.63}$$

P_n is the surface density of "bound" charge [see eqn. (4.23)], and we note that the bound charges contribute to the m.m.f. just as much as the "free" charges which constitute the conduction current **J**. The last term of eqn. (9.63) seems, however, to describe an entirely different effect. It is often called the "displacement current", and many writers suggest that it is necessary to add the displacement current to the effect of the moving charges. It is stated that a changing electric field behaves like a moving charge.

This is very misleading. All electric field is due to electric charge, and it makes nonsense to add the field effects to the charge effects. Thus the $d\Psi/dt$ term in eqn. (2.48) represents the local effect of electric charges. Similarly, in eqn. (9.63), which describes the relationship at a point, the first term is the local current density, the second the local polarization (or bound charge) effect, and the so-called displacement current represents the local effect of all other charges away from the point under consideration.

Let us apply these ideas to the problem of finding the m.m.f. in a material moving with a velocity **u** which is small compared with the velocity of light.

The total m.m.f. will be due to:

(1) The conduction current density **J**.
(2) Convective currents $\mathbf{u}\,(\varrho_{\text{free}} + \varrho_{\text{bound}}) = \mathbf{u}\,(\varrho - \nabla \cdot \mathbf{P})$.
(3) Currents caused by the rate of change of polarization $d\mathbf{P}/dt$.
(4) The "displacement" current density $\varepsilon_0\, \partial \mathbf{E}/\partial t$.

The only difficult term is (3) because we need to know the change of **P** due to the change of the area across which it passes as well as due to its own change in time. We had to face a similar problem in discussing Faraday's law and found that the local flux change could be written by eqn. (9.2) as $(\partial \mathbf{B}/\partial t) - \text{curl}\,(\mathbf{u} \times \mathbf{B})$.

Moreover, when we considered the change of magnetic flux we knew that $\nabla \cdot \mathbf{B} = 0$ and that there was no change of flux due to the motion of magnetic poles. Hence in our consideration of **P**, which has divergence sources, we must add the term $\mathbf{u}\,(\nabla \cdot \mathbf{P})$.

Thus, finally

$$\text{curl}\,\mathbf{H} = \mathbf{J} + \mathbf{u}\varrho_{\text{free}} - \mathbf{u}\,(\nabla \cdot \mathbf{P}) + \frac{\partial \mathbf{P}}{\partial t} - \nabla \times (\mathbf{u} \times \mathbf{P}) + \mathbf{u}(\nabla \cdot \mathbf{P}) + \varepsilon_0 \frac{\partial \mathbf{E}}{\partial t}$$

$$= \mathbf{J} + \mathbf{u}\varrho_{\text{free}} + \frac{\partial \mathbf{D}}{\partial t} + \nabla \times (\mathbf{P} \times \mathbf{u}).\dagger \tag{9.64}$$

The group of six equations formed by eqn. (9.64) combined with the local e.m.f. equation

$$\text{curl}\,\mathbf{E} = -\frac{\partial \mathbf{B}}{\partial t}, \tag{9.5}$$

the two divergence equations

$$\nabla \cdot \mathbf{B} = 0 \tag{6.35}$$

$$\nabla \cdot \mathbf{D} = \varrho_{\text{free}} \tag{4.33}$$

and the "constitutive" relations

$$\mathbf{B} = \mu_0 \mathbf{H} + \mathbf{P}^*, \tag{6.36}$$

$$\mathbf{D} = \varepsilon_0 \mathbf{E} + \mathbf{P}, \tag{4.32}$$

are called *Maxwell's equations* for a moving material. Maxwell's equations for a stationary material can be found by replacing eqn. (9.64) by eqn. (9.52). The equations for free space are

$$\left.\begin{aligned} \text{curl}\,\mathbf{H} &= \frac{\partial \mathbf{D}}{\partial t}, \\ \text{curl}\,\mathbf{E} &= -\frac{\partial \mathbf{B}}{\partial t}, \\ \text{div}\,\mathbf{B} &= 0, \\ \text{div}\,\mathbf{D} &= 0, \\ \mathbf{B} &= \mu_0 \mathbf{H}, \\ \mathbf{D} &= \mu_0 \mathbf{E}. \end{aligned}\right\} \tag{9.65}$$

† Note that **u** is the velocity of the charges and not of the m.m.f. loop. In eqn. (9.2) **u** is the velocity of the e.m.f. loop and therefore again the velocity of the charges.

Summary

The examination of Faraday's law of electromagnetic induction has led us in this chapter to the discussion of electromotive forces. These have been expressed in terms of the vector potential or electrokinetic momentum. We have shown that electromotive forces arise from changes in the mutual kinetic energy associated with moving charges.

The examination of Ampère's law of magnetomotive force has shown that this law is incomplete and applies only to direct current. We have discussed Maxwell's modification of the law and shown that this leads to the idea of electromagnetic waves. Thus changing magnetic fields are always accompanied by electric fields and changing electric fields by magnetic fields.

Exercises

9.1. In some of the early generators the rotating armature consisted of an unslotted iron cylinder to which the armature winding was fastened with string. The winding was therefore exposed to the full air-gap flux. Later the windings were placed in slots, and this meant that they lay in a very much reduced magnetic flux density. Explain in terms of flux threading and flux cutting why this change of design did not greatly affect the generated e.m.f.

9.2. A conducting ring of uniform cross-section as shown in Fig. 9.13 is linked by a magnetic flux Φ which is changing at the rate $d\Phi/dt$. The flux density is independent of the angular coordinate. The voltmeter leads are in the plane of the ring. What is the reading of the meter. Give reasons for your answer. [Ans.: $(1/4)\,(d\Phi/dt)$.]

9.3. Determine the amplitude of the induced electric field strength outside and inside a long cylindrical solenoid of mean radius a and of m.m.f. NI ampere-turns per metre. The current I alternates at a frequency f and the radius a is very much smaller than the electromagnetic wavelength. [Ans.: $\pi f \mu_0 N I a^2/r$; $\pi f \mu_0 N I r$.]

9.4. A conducting cylinder rotates about its axis at angular velocity ω in a uniform axial flux density B. Show that the induced electric field inside the cylinder has zero curl and can be derived from a scalar potential $-\frac{1}{2}\omega B r^2$. Also show that this field will give rise to a uniform charge distribution throughout the volume of density $\varrho = -2\varepsilon_0 \omega B$.

9.5. A long, straight conductor carries a constant current I. On the conductor there is threaded a thick iron tube of relative permeability μ_r. This tube moves along the conductor with velocity u. Stationary contacts touch the outer and inner surfaces of the tube and are connected by stationary wires to a voltmeter. Show that the voltmeter reading depends on I and u, but is independent of μ_r.

9.6. Show that the change of magnetic flux for a loop of travelling particles is given by

$$\oint \left(\frac{\partial \mathbf{A}}{\partial t} - \mathbf{u}\times\mathbf{B}\right) . d\mathbf{l}$$

and that this is zero in a perfectly conducting fluid.

9.7. Show that the magnetic Reynolds number can be written as $R_m = L^2/\delta^2$. Explain the physical significance of L in an electrical machine and show that a high R_m does not mean that the conductors have to be thick. The *goodness factor* of an induction machine has been defined as $G = X_m/R_2$, where X_m is the magnetizing reactance and R_2 the secondary resistance. Show that for a simple linear motor $G \propto R_m d/g$, where d is the thickness of the secondary conductor and g is the air gap.

9.8. A long, straight conductor of circular cross-section carries a current which returns symmetrically through a fluid surrounding the conductor. Show that the fluid flow is stable under small perturbations which force the fluid away from the central conductor.

9.9. By using the relationship $\oint \mathbf{H} . d\mathbf{l} = \partial \Psi/\partial t$. Show that the magnetic field of a slowly moving charge is given by

$$\mathbf{H} = \frac{Q\mathbf{u}\times\hat{\mathbf{r}}}{4\pi r^2},$$

vhere **u** is the velocity of the charge. Why is this expression likely to be incorrect for a ıapidly moving charge?

9.10. Explain the term "displacement current". At what frequency will the displacement-current density in sea water be equal to the conduction-current density, if the resistivity is 0·36 Ω m^2? At what frequency will the polarization-current density be equal to the conduction-current density, if the relative permittivity is 80? [5 $\times$ 10^{10}Hz, 6·3 $\times$ 10^8 Hz.]

9.11. When Lagrangian methods are applied to continuous systems the "equations of motion" are given by

$$\frac{\partial L}{\partial \eta} - \frac{d}{dt}\frac{\partial L}{\partial \dot{\eta}} - \frac{d}{dx}\left[\frac{\partial L}{\partial\,(\partial\eta/\partial x)}\right] - \frac{d}{dy}\left[\frac{\partial L}{\partial\,(\partial\eta/\partial y)}\right] - \frac{d}{dz}\left[\frac{\partial L}{\partial\,(\partial\eta/\partial z)}\right] = 0,$$

where η is the general coordinate and L is the Lagrangian density. For the electromagnetic field in free space

$$L = \frac{1}{2}\mu_0 H^2 - \frac{1}{2}\varepsilon_0 E^2 = \frac{1}{2}\frac{(\mathrm{curl}\,\mathbf{A})^2}{\mu_0} - \frac{1}{2}\varepsilon_0\left(\frac{\partial \mathbf{A}}{\partial t} + \nabla V\right)^2.$$

Show that by taking first V and then $\mathbf{A}$ as the general coordinate it is possible to derive the equations div $\varepsilon_0\mathbf{E} = 0$ and curl $\mathbf{H} = \varepsilon_0\,(\partial\mathbf{E}/\partial t)$.

CHAPTER 10

ELECTROMAGNETIC RADIATION I

10.1. The Electromagnetic Field in Terms of its Sources

Maxwell's equations, which we derived in the last chapter, relate the variation of the local field to the local charge and current sources. In terms of mathematics they are differential equations which have to be solved by integration. Suitable constants of integration have to be found before the solution is complete. In physical terms this means that we need to know the boundary conditions for the field as well as the local variation given in Maxwell's equations. Such a process of solution is very similar to that used in § 4.3.1 where we integrated Laplace's equation in order to find the electrostatic potential inside a rectangular duct. In that example the potential on the boundaries was known and this enabled us to find the constants of integration. We shall find the same method useful when we discuss wave guides in the next chapter.

But useful as the method undoubtedly is, it may lead us to think that to find the electromagnetic field of a particular device we need to know not only the sources of that field as given in Maxwell's equations, but also the boundary conditions. It is important to remind ourselves that in principle the whole subject of electromagnetism is the study of the interactions of stationary and moving charges. The field has no existence apart from the sources. Maxwell's equations, which refer merely to the local charge, are necessarily incomplete because they do not include the other charges at a distance. The somewhat mysterious boundary conditions are brought in to account for charges outside the region under consideration.

Previously, when we have tried to find the local field in terms of *all* the sources, we have made use of the scalar and vector potentials. We found for stationary currents and charges in free space that

$$\mathbf{E} = -\frac{\partial \mathbf{A}}{\partial t} - \operatorname{grad} V \tag{9.28}$$

and

$$\mathbf{B} = \text{curl}\,\mathbf{A}, \tag{6.25}$$

where

$$\mathbf{A} = \frac{\mu_0}{4\pi}\iiint \frac{\mathbf{J}}{r}\,dv' \tag{3.44}$$

and

$$V = \frac{1}{4\pi\varepsilon_0}\iiint \frac{\varrho}{r}\,dv'. \tag{3.22}$$

These, however, were the expressions derived on the basis of the incomplete m.m.f. law of Ampère. They imply instantaneous interaction. For instance, if ϱ varies as $\sin \omega t$, V and $\mathbf{E}$ will vary as $\sin \omega t$, and if $\mathbf{J}$ varies as $\sin \omega t$, $\mathbf{A}$ and therefore $\mathbf{B}$ will vary as $\sin \omega t$. There is no phase angle to account for a time delay as the action is propagated from the source. This was one of the difficulties faced by Maxwell, and we must inquire how the potentials will be affected by his introduction of "displacement current". We have

$$\text{curl}\,\mathbf{H} = \mathbf{J} + \frac{\partial \mathbf{D}}{\partial t}, \tag{9.52}$$

so that in free space

$$\text{curl}\,\mathbf{B} = \mu_0 \mathbf{J} + \mu_0 \varepsilon_0 \frac{\partial \mathbf{E}}{\partial t}. \tag{10.1}$$

Let us put, as before,

$$\mathbf{B} = \text{curl}\,\mathbf{A}, \tag{6.25}$$

whence

$$\text{curl curl}\,\mathbf{A} = \mu_0 \mathbf{J} + \mu_0 \varepsilon_0 \frac{\partial \mathbf{E}}{\partial t}, \tag{10.2}$$

and by eqn. (3.41) the operator curl curl can be transformed so that

$$\text{grad div}\,\mathbf{A} - \nabla^2 \mathbf{A} = \mu_0 \mathbf{J} + \mu_0 \varepsilon_0 \frac{\partial \mathbf{E}}{\partial t}. \tag{10.3}$$

Substituting the expression for $\mathbf{E}$ in eqn. (9.28),

$$\text{grad div}\,\mathbf{A} - \nabla^2 \mathbf{A} = \mu_0 \mathbf{J} - \mu_0 \varepsilon_0 \frac{\partial^2 \mathbf{A}}{\partial t^2} - \mu_0 \varepsilon_0 \,\text{grad}\, \frac{\partial V}{\partial t}. \tag{10.4}$$

It would be desirable to eliminate V from this equation, so that **A** is related to **J** only. We can do this because the divergence of **A** has not yet been defined. Let us write

$$\operatorname{div} \mathbf{A} + \mu_0 \varepsilon_0 \frac{\partial V}{\partial t} = 0, \tag{10.5}$$

which is consistent with the previous steady-state definition of div $\mathbf{A} = 0$. Then eqn. (10.4) is simplified to

$$\nabla^2 \mathbf{A} - \mu_0 \varepsilon_0 \frac{\partial^2 \mathbf{A}}{\partial t^2} = -\mu_0 \mathbf{J}, \tag{10.6}$$

which should be compared with the previous steady-state equation

$$\nabla^2 \mathbf{A} = -\mu_0 \mathbf{J}. \tag{3.42}$$

In order to find V we make use of the charge density ϱ,

$$\operatorname{div} \mathbf{D} = \varrho, \tag{3.18}$$

whence

$$\operatorname{div} \varepsilon_0 \mathbf{E} = -\varepsilon_0 \frac{\partial}{\partial t} (\operatorname{div} \mathbf{A}) - \varepsilon_0 \nabla^2 V = \varrho, \tag{10.7}$$

and, by eqn. (10.5),

$$\nabla^2 V - \mu_0 \varepsilon_0 \frac{\partial^2 V}{\partial t^2} = -\frac{\varrho}{\varepsilon_0}, \tag{10.8}$$

which should be compared with Poisson's equation

$$\nabla^2 V = -\frac{\varrho}{\varepsilon_0}. \tag{3.21}$$

If the right-hand sides of eqns. (10.6) and (10.8) were zero, they would be typical wave equations and would have solutions of the form $f(r \pm ct)$, where f expresses any functional relationship. We can, therefore, infer that the equations will be satisfied by

$$\mathbf{A} = \frac{\mu_0}{4\pi} \iiint \frac{\mathbf{J}\,[x', y', z', t \pm (r/c)]}{r}\, dx'\, dy'\, dz', \tag{10.9}$$

$$V = \frac{1}{4\pi\varepsilon_0} \iiint \frac{\varrho\,[x', y', z', t \pm (r/c)]}{r}\, dx'\, dy'\, dz', \tag{10.10}$$

where x', y', z' are the source coordinates of $\mathbf{J}$ and ϱ, and the square bracket denotes any functional relationship.

To verify eqns. (10.9) and (10.10) let us substitute (10.10) into eqn. (10.8). Consider first $\nabla^2 V$. Since this will be difficult to evaluate at $r = 0$, let us divide the volume of integration in eqn. (10.10) into two regions. The first of these is a small sphere of radius r_0 around the point at which V is to be determined. The second region is the rest of space. Hence,

$$V = V_1 + V_2 . \tag{10.11}$$

Since r/c is negligible in region (1), we can put $t' = t$

$$\begin{aligned} \nabla^2 V_1 &= \frac{1}{4\pi\varepsilon_0} \nabla^2 \iiint \frac{\varrho\,[x', y', z', t]}{r}\, dx'\, dy'\, dz' \\ &= \frac{\varrho}{4\pi\varepsilon_0} \iiint \nabla^2 \left(\frac{1}{r}\right) dv' \\ &= \frac{\varrho}{4\pi\varepsilon_0} \oiint \operatorname{grad} \left(\frac{1}{r}\right) . d\mathbf{s}' \\ &= \frac{\varrho}{4\pi\varepsilon_0} \left[- \frac{4\pi r_0^2}{r_0^2} \right] \\ &= - \frac{\varrho}{\varepsilon_0} . \end{aligned} \tag{10.12}$$

This is independent of r_0, and therefore the sphere can be made arbitrarily small and may shrink to a point.

To obtain $\nabla^2 V_2$ we note that ∇^2 operates only on r. In spherical coordinates we can therefore write

$$\nabla^2 V_2 = \operatorname{div} \operatorname{grad} V_2 = \frac{1}{r^2} \frac{\partial}{\partial r} \left(r^2 \frac{\partial V_2}{\partial r} \right) = \frac{1}{r} \frac{\partial^2}{\partial r^2} (r V_2), \tag{10.13}$$

$$\nabla^2 V_2 = \frac{1}{4\pi\varepsilon_0} \iiint \frac{1}{r} \frac{\partial^2}{\partial r^2} \varrho \left[x', y', z', t \pm \frac{r}{c} \right] dx'\, dy'\, dz' . \tag{10.14}$$

Now for any function of $t \pm (r/c)$,

$$\frac{\partial^2}{\partial r^2} = \frac{1}{c^2} \frac{\partial^2}{\partial t^2} . \tag{10.15}$$

Hence

$$\nabla^2 V_2 = \frac{1}{4\pi\varepsilon_0} \iiint \frac{1}{rc^2} \frac{\partial^2}{\partial t^2} \varrho \left[x', y', z', t \pm \frac{r}{c} \right] dx'\, dy'\, dz'$$

$$= \frac{1}{c^2} \frac{\partial^2 V}{\partial t^2}. \tag{10.16}$$

Hence

$$\nabla^2 V^2 = \nabla^2 V_1^2 + \nabla^2 V_2^2 = -\frac{\varrho}{\varepsilon_0} + \frac{1}{c^2} \frac{\partial^2 V}{\partial t^2}$$

and

$$\nabla^2 V - \frac{1}{c^2} \frac{\partial^2 V}{\partial t^2} = -\frac{\varrho}{\varepsilon_0}. \tag{10.17}$$

Since eqn. (10.17) is identical with eqn. (10.8) we have verified that eqn. (10.10) is the unique solution of eqn. (10.8) as long as all the sources are included in the integral. It follows by inspection that eqn. (10.9) satisfies eqn. (10.6).

The wave equation is equally satisfied by functions of $t - (r/c)$ and $t + (r/c)$, because the waves may be travelling forwards or backwards. But since we are working in terms of the sources, we shall discard solutions which imply movement towards the sources. Thus we shall use the *retarded potentials*

$$\mathbf{A} = \frac{\mu_0}{4\pi} \iiint \frac{[\mathbf{J}]}{r}\, dv', \tag{10.18}$$

$$V = \frac{1}{4\pi\varepsilon_0} \iiint \frac{[\varrho]}{r}\, dv' \tag{10.19}$$

where the square brackets mean that each source shall be considered at the time $t' = t - (r/c)$. This is, of course, a familiar concept in astronomy, where it is explicit that the light of a star observed on the earth at the present time gives information about the star at the much earlier time when the light was radiated from that star. The potentials $\mathbf{A}$ and V are to be calculated at any instant by using the earlier values of $\mathbf{J}$ and ϱ. The potentials are retarded by the time which it takes to travel a distance r with the velocity of light c.

We have shown that the time delay in the potentials is a direct consequence of Maxwell's hypothesis. Alternatively, Maxwell's hypothesis can be *derived* by assuming the delayed potentials. It is of historical interest that the Danish physicist L. Lorenz postulated the retarded potentials independently of Maxwell a few years after the latter had published his hypothesis.

An important point to notice is that it is incorrect to introduce a time delay into the electric or magnetic fields. These have to be calculated from the potentials. For instance, the magnetic field **B** is found from curl **A**. The operator curl implies a differentiation with respect to r, and r occurs both in the numerator and the denominator of the expression for **A**. Thus two terms result from the differentiation. If, on the other hand, we take the instantaneous potential and differentiate it, and afterwards introduce the time delay, we shall obtain only one term because r appears only in the denominator of the instantaneous potential. It is therefore not possible to speak of retarded *fields*. Nor, except in special cases, can we assert that the fields travel from the sources with the speed of light. It is the potentials, representing the electrokinetic momentum and the potential energy, which propagate at this speed.

Before we leave the general discussion of the retarded potentials it is useful to look at the relationship between the potentials. This relationship is known as the Lorentz condition.

$$\operatorname{div} \mathbf{A} + \mu_0 \varepsilon_0 \frac{\partial V}{\partial t} = 0. \tag{10.5}$$

The reader may have felt that this relation is an arbitrary one, but this is not so. The Lorentz condition ensures that the equations for **A** and V are exactly of the same form, and that **A** depends only on **J**, whereas V depends only on ϱ. Now **J** and ϱ are related by the equation of continuity

$$\operatorname{div} \mathbf{J} + \frac{\partial \varrho}{\partial t} = 0. \tag{8.3}$$

Comparison of eqn. (10.5) with (8.3) shows that the Lorentz condition is the continuity equation in terms of the potentials.

By inspection of eqns. (6.25) and (9.28) it can be seen that the magnetic and electric fields are unchanged if the potentials are transformed by the relationships

$$\mathbf{A}' = \mathbf{A} - \nabla\phi, \tag{10.20}$$

$$V' = V + \frac{\partial \phi}{\partial t}, \tag{10.21}$$

where ϕ is any function of the space and time coordinates. These transformations are known as gauge transformations. The property of gauge invariance can be used as a test whether a physical law is consistent with the known behaviour of electric charges and currents. In order to maintain the Lorentz condition the function ϕ must satisfy the wave equation

$$\nabla^2\phi - \mu_0\varepsilon_0 \frac{\partial^2 \phi}{\partial t^2} = 0. \tag{10.22}$$

10.2. Radiation Resistance

The consequences of using the delayed potentials rather than the instantaneous potentials of electrostatics and magnetostatics are startling. Instead of having to make small adjustments to the results derived from static considerations, we find that these results have to be supplemented by new and radically different phenomena.

Let us approach the subject by taking the very simple example of the field of an element carrying alternating current. We seek the field at a point P distant r from the element. The vector $\mathbf{r}$ is at right angles to the element as shown in Fig. 10.1. It may rightly be objected that a current

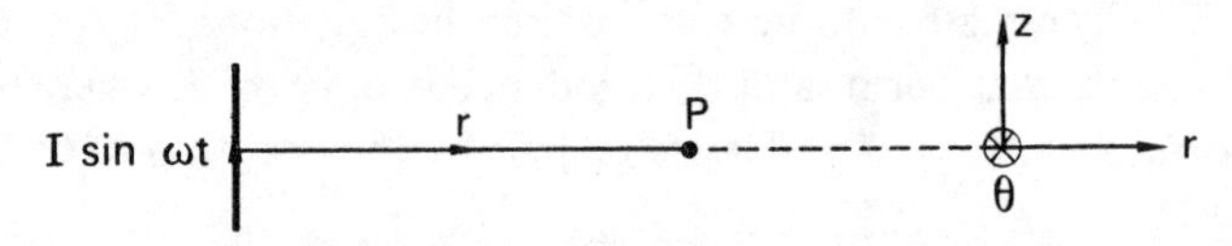

FIG. 10.1. The field of a current element.

element is a physical impossibility. The continuity of electric charge flow demands that the current be allowed to flow somewhere. We discussed this difficulty in the direct current case § 6.2.7. There we used Heaviside's stratagem and immersed the element in a conducting fluid to allow the current to flow. With alternating current we do not need Heaviside's fluid. Instead, the current can terminate in charges as illustrated in Fig. 10.2. These charges will contribute to the electric field at P (through the scalar potential) but not to the magnetic field, because they do not contribute to the vector potential. The field of the current by itself will, therefore, be a partial field. We shall discuss the field of the complete "dipole" of Fig. 10.2 in the next chapter.

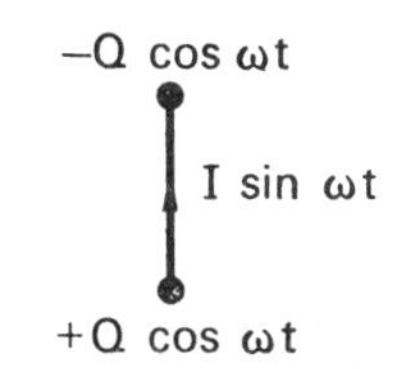

FIG. 10.2. An electric dipole.

Since there is no charge in the simplified current element of Fig. 10.1, the scalar potential is zero [eqn. (10.19)]. The vector potential can be derived from eqn. (10.18). It is parallel to the current element. We have

$$A = \frac{\mu_0}{4\pi} \frac{[I]\,\delta l}{r} = \frac{\mu_0 I\,\delta l}{4\pi r} \sin \omega\,[t - (r/c)]. \tag{10.23}$$

If we use the symbol β for the "propagation coefficient" ω/c,

$$A = \frac{\mu_0 I\,\delta l}{4\pi r} \sin(\omega t - \beta r), \tag{10.24}$$

where $\beta = \omega/c$, whence

$$A = \frac{\mu_0 I\,\delta l}{4\pi r} (\sin \omega t \cos \beta r - \cos \omega t \sin \beta r), \tag{10.25}$$

or, in the phasor notation of complex numbers,

$$\bar{A} = \frac{\mu_0 \bar{I}\,\delta l}{4\pi r} (\cos \beta r - j \sin \beta r). \tag{10.26}$$

The magnetic field can be obtained by

$$\mathbf{B} = \operatorname{curl} \mathbf{A}. \tag{6.25}$$

Using cylindrical polar coordinates we obtain

$$\bar{H}_\theta = -\frac{1}{\mu_0}\frac{\partial \bar{A}}{\partial r} = -\frac{\bar{I}\,\delta l}{4\pi}\frac{\partial}{\partial r}\left(\frac{\cos\beta r}{r} - j\frac{\sin\beta r}{r}\right)$$

$$= \frac{\bar{I}\,\delta l}{4\pi}\left[\frac{(\cos\beta r - j\sin\beta r)}{r^2} + \frac{\beta(\sin\beta r + j\cos\beta r)}{r}\right]. \qquad (10.27)$$

To check on the magnetostatic field we put $\beta = 0$ and find as in § 6.2.7

$$H_\theta = \frac{I\,\delta l}{4\pi r^2}. \qquad (10.28)$$

Equation (10.27) reveals some surprising consequences of the retardation of the potentials. The alternating magnetic field decays much less rapidly with distance than does the static field, because it varies as $1/r$ and not as $1/r^2$. Secondly, the alternating magnetic field is no longer in phase with the current but contains a component lagging on the current by 90°.

Consider now the electric field. This is parallel to **A** and hence to the current element.

$$\bar{E} = -\frac{\partial \bar{A}}{\partial t} = \frac{\omega\mu_0 \bar{I}\,\delta l}{4\pi r}(-\sin\beta r - j\cos\beta r). \qquad (10.29)$$

Close to the current element $r \to 0$, $\sin\beta r \to \beta r$ and $\cos\beta r \to 1$. Thus the electric field becomes

$$\bar{E} = \frac{\omega\mu_0 \bar{I}\,\delta l}{4\pi r}(-\beta r - j) = -\frac{\beta\omega\mu_0 \bar{I}\,\delta l}{4\pi} - \frac{j\omega\mu_0 \bar{I}\,\delta l}{4\pi r}. \qquad (10.30)$$

The second term is in phase-quadrature with the current and becomes very large as r becomes small. This is the normal "inductance" effect which would have been obtained from the instantaneous potential. The induced voltage lags 90° on the current and becomes very large when the current is compressed into a thin filament. It becomes infinite as $r \to 0$ because an infinite current density would require an infinite applied voltage.

The first term in eqn. (10.30) is, however, something quite new. The electric field is in anti-phase with the current (180°). This implies an energy input. The electric field behaves like the "back e.m.f." of an electric motor; it opposes the current flow. The instantaneous power required to keep the current flowing is

$$P_{\text{inst}} = \text{volts} \times \text{amps} = E\,\delta l\, I = \frac{\beta\omega\mu_0 I^2\,\delta l^2}{4\pi}\sin^2\omega t. \tag{10.31}$$

Hence the average power is

$$P = \frac{\beta\omega\mu_0 I^2\,\delta l^2}{8\pi}. \tag{10.32}$$

This is equivalent to an ohmic resistance

$$R = \frac{\beta\omega\mu_0\,\delta l^2}{4\pi} = \frac{\mu_0\omega^2\,\delta l^2}{4\pi c}. \tag{10.33}$$

It is called the *radiation resistance.*

In order to obtain an idea of the magnitude of this quantity we shall find it useful to introduce the wavelength λ of the electromagnetic radiation. Reference to eqn. (10.24) shows that this wavelength is given by

$$\beta\lambda = 2\pi \tag{10.34}$$

or

$$\lambda = \frac{2\pi c}{\omega} = \frac{c}{f}, \tag{10.35}$$

where f is the frequency.

Thus the radiation resistance is given by

$$R = \pi\mu_0 c\left(\frac{\delta l}{\lambda}\right)^2. \tag{10.36}$$

Hence if the element is one-tenth of a wavelength long,

$$R \doteqdot 12\,\Omega. \tag{10.37}$$

At 50 Hz the wavelength is approximately 6000 km, so our current "element" would have to be 600 km long, but at 5×10^9 Hz λ is 6 cm and an element of 6 mm would have the radiation resistance we have calculated.

It should be noticed that the radiation resistance is independent of the radius of the filament. The power depends on the square of the current and of the length. Engineers often refer to the *ampere-metres* of a radio aerial as the criterion of its power output.

Where does the power go? It does not go into heating the wire in which the current flows. We have said nothing about ohmic resistance. If the current flows in a material of finite resistivity, the ohmic resistance will have to be added to the radiation resistance. The back e.m.f. which we have calculated has arisen purely from the wave motion described by the delayed potentials. Thus we conclude that the radiation resistance represents the power which is radiated by the alternating current.

We have noted that the electric field strength [eqn. (10.30)] has two components. One is the "inductance" component and the other has been described in terms of radiation resistance. The relative magnitude is given by the ratio

$$\frac{1}{r\beta} = \frac{c}{r\omega}. \tag{10.38}$$

Hence at low frequencies the quadrature field is dominant. But it should be noted that at any frequency there is always a radiation term, however small. The "pure inductance" of circuit theory is an approximation only.

10.3. Electromagnetic Energy Transfer

10.3.1. The Poynting Vector

We have calculated the rate at which energy is radiated from a current element by considering the reaction of the field on the current. Work has to be done in pumping the current up and down. Some of this work is recovered during the cycle and we speak of reactive power. But some of the energy leaves the current source and is radiated into space. It should therefore be possible to calculate the energy by considering a surface which encloses the source and determine the rate at which the energy crosses this surface.

The total input power can be calculated from the power required to keep the current flowing,

$$P = \iiint -\mathbf{E}\,.\,\mathbf{J}\,dv. \tag{10.39}$$

Let us convert this volume integral into an expression involving only the field quantities. We have

$$\mathbf{J} = \text{curl}\,\mathbf{H} - \frac{\partial \mathbf{D}}{\partial t} \tag{9.52}$$

so that

$$P = \iiint \left(\mathbf{E} \,.\, \frac{\partial \mathbf{D}}{\partial t} - \mathbf{E} \,.\, \text{curl}\,\mathbf{H} \right) dv. \tag{10.40}$$

In order to obtain a surface term we make use of the vector identity

$$\text{div}\,(\mathbf{E} \times \mathbf{H}) = \mathbf{H} \,.\, \text{curl}\,\mathbf{E} - \mathbf{E} \,.\, \text{curl}\,\mathbf{H}. \tag{10.41}$$

Hence

$$\begin{aligned} P &= \iiint \left[\mathbf{E} \,.\, \frac{\partial \mathbf{D}}{\partial t} - \mathbf{H} \,.\, \text{curl}\,\mathbf{E} + \text{div}\,(\mathbf{E} \times \mathbf{H}) \right] dv \\ &= \iiint \left[\mathbf{E} \,.\, \frac{\partial \mathbf{D}}{\partial t} - \mathbf{H} \,.\, \text{curl}\,\mathbf{E} \right] dv + \oiint (\mathbf{E} \times \mathbf{H}) \,.\, d\mathbf{s}. \end{aligned} \tag{10.42}$$

Finally, by the use of Faraday's law,

$$\text{curl}\,\mathbf{E} = -\frac{\partial \mathbf{B}}{\partial t}, \tag{9.5}$$

$$P = \iiint \left[\mathbf{E} \,.\, \frac{\partial \mathbf{D}}{\partial t} + \mathbf{H} \,.\, \frac{\partial \mathbf{B}}{\partial t} \right] dv + \oiint (\mathbf{E} \times \mathbf{H}) \,.\, d\mathbf{s}. \tag{10.43}$$

Let us assume that the permittivity ε and the permeability μ are constant, then we can write the volume integral as

$$\iiint \left[\mathbf{E} \,.\, \frac{\partial \mathbf{D}}{\partial t} + \mathbf{H} \,.\, \frac{\partial \mathbf{B}}{\partial t} \right] dv = \frac{d}{dt} \iiint \frac{1}{2} [\mathbf{E} \,.\, \mathbf{D} + \mathbf{H} \,.\, \mathbf{B}]\, dv. \tag{10.44}$$

This expression can be regarded as the rate of change of the stored potential and kinetic energy of the field. We have to be a little cautious about this interpretation because the expressions for the field energy [eqns. (5.19) and (7.49)] were derived on the assumption that the integration was to be carried out over all space occupied by the field. In the present investigation there is field outside the volume and we may not be justified in treating $\frac{1}{2}(\mathbf{E} \,.\, \mathbf{D} + \mathbf{H} \,.\, \mathbf{B})$ as the energy density. This is an unresolved problem in the theory, but if we are dealing with average effects over a complete cycle, the integral in eqn. (10.44) is zero. The reason for this is that, if E varies as sin ωt and D varies as sin ωt, the rate of change of ED will vary as sin $2\omega t$, which has no average value. The same is true for HB. The meaning of the volume integral in terms of energy is therefore unimportant. The power equation for average effects becomes

$$\iiint -\mathbf{E}\,.\,\mathbf{J}\,dv = \oiint (\mathbf{E}\times\mathbf{H})\,.\,d\mathbf{s}. \qquad (10.45)$$

Thus the power input is equal to the outflow of the vector $\mathbf{S} = \mathbf{E}\times\mathbf{H}$ across the enclosing surface. $\mathbf{S}$ is called the Poynting vector after its discoverer, J.H.Poynting. We see that the radiated power can be obtained by the surface integral of the Poynting vector over any surface enclosing the current sources.

10.3.2. Some Features of Poynting's Energy Flow Theorem

Poynting's method can equally well be used to derive the power flow into a volume containing a sink of energy. A simple example is the cylindrical conductor shown in Fig. 10.3. A steady current I flows in the

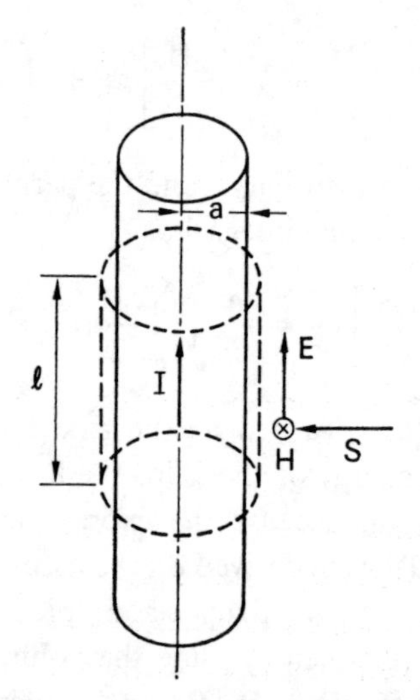

FIG. 10.3. Power flow in a cylindrical conductor.

conductor. Consider the power flow across the dotted surface. The vector **S** is perpendicular to both **E** and **H**. Hence **S** points inwards towards the axis of the wire. There is therefore no flow of **S** through the top and bottom circular areas of the dotted surface. Over the curved surface we have $E = J/\sigma$ and $H = I/2\pi a$. Also $\pi a^2 J = I$. Hence $S = I^2/2\pi^2 a^3 \sigma$. The power flow is

$$P = \oint\!\!\oint \mathbf{S} \,.\, d\mathbf{s} = \frac{I^2}{2\pi^2 a^3 \sigma} 2\pi a l = \frac{l}{\sigma \pi a^2} I^2 = RI^2. \qquad (10.46)$$

This, of course, is the ohmic loss of a length l of conductor. Thus the inflow of the Poynting vector provides a means for calculating the power to push the current through the resistive material.

The reader may now be puzzled how the Poynting vector can at one and the same time be used to calculate the outflow of radiated energy and the inflow which accounts for ohmic loss. For instance, if the current in the conductor of Fig. 10.3 alternates, it will radiate energy. Yet the calculation of eqn. (10.46) would again give the ohmic loss. Moreover, the radiated energy from a current element varies as the square of the frequency as we found in § 10.2, whereas the loss varies in quite a different manner, depending on the skin-depth.

To understand this contradiction we must refer again to § 10.2. In that section we worked out the electric field of an alternating current and we found that this field opposed the current flow. The opposing force had nothing to do with Ohm's law and we had not even discussed in what sort of conductor the current was to flow. If now we think of some material such as copper, we realize that the electric field will be governed by Ohm's law. It is also likely that there will be surface charges on the conductor to give the correct total electric field to allow the current to flow. This we discussed in § 9.4. Here is the key to the problem. The electric field of the current element [eqn. (10.30)] is the field of the current by itself. The field used in our calculation of ohmic loss is the total field of current and surface charge.

It will be remembered that in § 10.2 we referred to the electric field as a back e.m.f. like that of a motor. The motor equation can be written as

$$V - E = RI, \qquad (10.47)$$

where V is the applied potential difference and E the back e.m.f. Ohm's law demands that the combined effect of p.d. (due to surface charge) and e.m.f. shall be sufficient to overcome the resistance of the material. To obtain the output power of the motor we use EI and multiply the *partial* electric field by the current. To obtain the ohmic loss we use the *total* field $(V - E)$ multiplied by I. In the motor EI is mechanical power, whereas in a radio aerial the product of partial field and current gives the radiated power. Except for this difference the analogy is complete. Hence if in eqn. (10.45) we insert the partial field we obtain the radiated power, but if we insert the total field and integrate the Poynting vector over an area surrounding conductors we obtain the loss in these conductors.

If we seek to obtain the power by measurement of the Poynting vector we find that close to the conductor surface we can observe only the total field and hence the ohmic loss. But at a distance the effect of the surface charges becomes negligible and then total field is equal to the partial field so that we can measure the radiated power. In terms of energy the surface of integration for the distant field includes the energy source, whereas the field of the conductor surface excludes the source.

The Poynting vector is an elegant and powerful tool in calculation and it can also be used in measurement, but one further word of caution is necessary. We must not assume that $\mathbf{S}$ represents the density of power on a surface. Equation (10.45) was derived for a closed surface and we converted the volume integral of div $\mathbf{S}$ to a surface distribution of $\mathbf{S}$. We could have chosen any other vector $\mathbf{S}'$, where

$$\mathbf{S}' = \mathbf{S} + \operatorname{curl} \mathbf{Q}, \tag{10.48}$$

where $\mathbf{Q}$ is an arbitrary vector. This is so because

$$\operatorname{div} \mathbf{S}' = \operatorname{div} \mathbf{S} + \operatorname{div} \operatorname{curl} \mathbf{Q} = \operatorname{div} \mathbf{S}. \tag{10.49}$$

Many forms for the vector $\mathbf{S}'$ have been proposed. Slepian† discusses

$$\mathbf{S}' = V\left(\mathbf{J} + \frac{\partial \mathbf{D}}{\partial t}\right) + \mathbf{H} \times \frac{\partial \mathbf{A}}{\partial t}. \tag{10.50}$$

† J. Slepian, Energy flow in electric systems—the *VI* energy flow postulate. *Trans. Amer. I.E.E.*, Vol. 61, p. 835, 1942.

This has the useful property that for direct current it simplifies to $V\mathbf{J}$ and can be recognized as

$$\mathbf{S}' = \text{volts} \times \text{amps/m}^2 = \text{watts/m}^2. \qquad (10.51)$$

Also $\mathbf{S}'$ points along the current, whereas $\mathbf{S}$ is at right-angles to the current. But for radiation problems Poynting's vector $\mathbf{S} = \mathbf{E} \times \mathbf{H}$ is far simpler.†

10.3.3. The Complex Poynting Vector

Quantities which vary sinusoidally in time are called phasors and can be represented by complex numbers. This is done extensively in circuit theory and we shall find it useful to apply the method to Poynting's theorem.

We can write the electric field strength as

$$\text{Re}\, \bar{\mathbf{E}}\, e^{j\omega t} = \tfrac{1}{2}(\bar{\mathbf{E}}\, e^{j\omega t} + \bar{\mathbf{E}}^*\, e^{-j\omega t}) \qquad (10.52)$$

where Re indicates the operation of taking the real part of the expression and where the phasor $\bar{\mathbf{E}}$ is a complex vector given by

$$\bar{\mathbf{E}} = \mathbf{i}\bar{E}_x + \mathbf{j}\bar{E}_y + \mathbf{k}\bar{E}_z. \qquad (10.53)$$

$\bar{\mathbf{E}}^*$ is the complex conjugate of $\bar{\mathbf{E}}$.

Consider now the complex Poynting vector

$$\bar{\mathbf{S}} = \tfrac{1}{2}(\bar{\mathbf{E}} \times \bar{\mathbf{H}}^*). \qquad (10.54)$$

We have

$$\text{div}\, \bar{\mathbf{S}} = \text{div}\, \tfrac{1}{2}(\bar{\mathbf{E}} \times \bar{\mathbf{H}}^*) = \tfrac{1}{2}(\bar{\mathbf{H}}^* \,.\, \text{curl}\, \bar{\mathbf{E}} - \bar{\mathbf{E}} \,.\, \text{curl}\, \bar{\mathbf{H}}^*) \qquad (10.55)$$

and

$$\text{curl}\, \bar{\mathbf{E}} = -j\omega\bar{\mathbf{B}}, \qquad (10.56)$$

$$\text{curl}\, \bar{\mathbf{H}}^* = \bar{\mathbf{J}}^* - j\omega\bar{\mathbf{D}}^*. \qquad (10.57)$$

Hence

$$\text{div}\, \bar{\mathbf{S}} = -\tfrac{1}{2}\bar{\mathbf{E}} \,.\, \bar{\mathbf{J}}^* - \tfrac{1}{2}j\omega\bar{\mathbf{H}}^* \,.\, \bar{\mathbf{B}} + \tfrac{1}{2}j\omega\bar{\mathbf{E}} \,.\, \bar{\mathbf{D}}^*. \qquad (10.58)$$

† Readers may be interested to refer to a fuller discussion in a paper by the author, Electromagnetic energy transfer, *Proc. I.E.E.*, Vol. 105, Part C, Monograph 286, 1958.

Equating real and imaginary parts we obtain

$$\text{Re}\,(\text{div}\,\bar{\mathbf{S}}) = \tfrac{1}{2}\,\text{Re}\,(-\bar{\mathbf{E}}\,.\,\bar{\mathbf{J}}^*), \qquad (10.59)$$

$$\text{Im}\,(\text{div}\,\bar{\mathbf{S}}) = \tfrac{1}{2}\,\text{Im}\,[(-\bar{\mathbf{E}}\,.\,\bar{\mathbf{J}}^*)] - \tfrac{1}{2}\omega\mu\,(H)^2 + \tfrac{1}{2}\omega\varepsilon\,(E)^2. \qquad (10.60)$$

Equation (10.59) gives the average power flow and can be used to calculate the average radiated power or the ohmic loss depending on whether **E** is the partial or the total field. In circuit language it represents an equivalent resistance. Equation (10.60) is useful in problems where we consider the input of power into a material. The first term is then zero, because $\bar{\mathbf{E}}$ and $\bar{\mathbf{J}}$ are in phase. The second term represents the magnetic energy and the third the electric energy. If we denote the average magnetic energy by W_m and the average electric energy by W_e, we have

$$\text{Im}\,(\text{div}\,\bar{\mathbf{S}}) = 2\omega\,(W_m - W_e). \qquad (10.61)$$

If the displacement current is negligible, W_e is negligible and

$$\text{Im}\,(\text{div}\,\bar{\mathbf{S}}) = 2\omega W_m. \qquad (10.62)$$

Then the surface integral of $\bar{\mathbf{S}}$ can be used to calculate the inductance.†

10.4. Plane Electromagnetic Waves

In order to gain experience of the energy transfer by electromagnetic radiation let us consider the simple example of a plane wave. Figure 10.4 shows a large current sheet in which the alternating current flows in the y-direction only and is everywhere co-phased, so that we can write the line density, i.e. the current per unit length z, as $J \sin \omega t$. The practical reader may wonder how a current could be caused to flow in this manner. Such doubts are well justified, especially if the dimensions of the sheet are an appreciable fraction of the electromagnetic wavelength. Nevertheless, it is useful to discuss the field of such a current sheet: first, to acquire insight into the phenomenon of radiation and, secondly, because more complicated current and charge distributions can be built up by superposition and more complicated waves can also be analysed by means of simple components.

† See Exercise 10.11 as an application of this method.

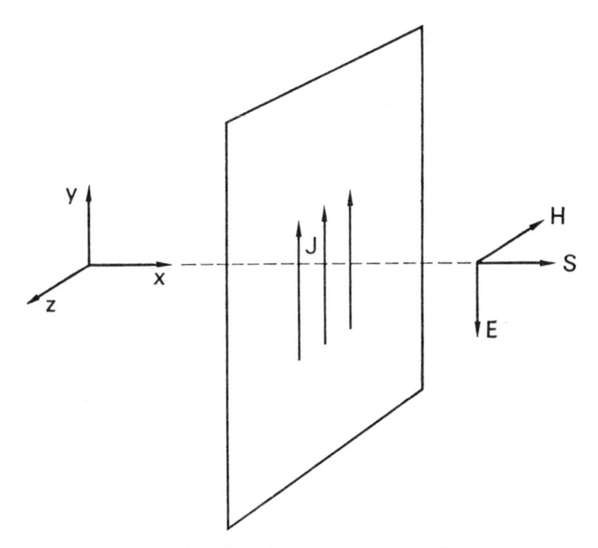

FIG. 10.4. The field of a current sheet.

Since there is no charge, there is no scalar potential. We should therefore expect to calculate the vector potential and to obtain the magnetic and electric field vectors from this potential. But in this example we have an infinite current flow and the vector potential is infinite. However, the magnetic field is not infinite and we can proceed from a knowledge of this field. At the surface of the sheet facing the positive direction of x, $H_z = -(J/2)\sin\omega t$ and on the other surface $H_z = +(J/2)\sin\omega t$, as can be shown by symmetry and the m.m.f. law. Also $H_y = H_x = 0$. In the free space outside the sheet

$$\nabla^2\mathbf{H} - \frac{1}{c^2}\frac{\partial^2\mathbf{H}}{\partial t^2} = 0. \tag{9.59}$$

This simplifies in our example to

$$\frac{\partial^2 H_z}{\partial x^2} = \frac{1}{c^2}\frac{\partial^2 H_z}{\partial t^2}, \tag{10.63}$$

which for $x > 0$ has the typical solution

$$H_z = K_1 \sin(\omega t - \beta x) + K_2 \cos(\omega t - \beta x), \tag{10.64}$$

where we have discarded the solution which fits waves travelling in the direction towards the sheet. At $x = 0$, $H_z = -(J/2) \sin \omega t$, hence $K_1 = -J/2$ and $K_2 = 0$. Thus for $x > 0$,

$$H_z = -\frac{J}{2} \sin (\omega t - \beta x), \tag{10.65}$$

also

$$\text{curl } \mathbf{H} = \varepsilon_0 \frac{\partial \mathbf{E}}{\partial t}, \tag{9.56}$$

so that

$$\frac{\partial \mathbf{E}}{\partial t} = \frac{1}{\varepsilon_0} \begin{vmatrix} \mathbf{i} & \mathbf{j} & \mathbf{k} \\ \dfrac{\partial}{\partial x} & \dfrac{\partial}{\partial y} & \dfrac{\partial}{\partial z} \\ 0 & 0 & H_z \end{vmatrix} \tag{10.66}$$

and

$$\frac{\partial E_y}{\partial t} = -\frac{1}{\varepsilon_0} \frac{\beta J}{2} \cos (\omega t - \beta x), \tag{10.67}$$

$$E_y = -\frac{\beta J}{2\varepsilon_0 \omega} \sin (\omega t - \beta x), \tag{10.68}$$

and

$$S_x = E_y H_z = \frac{\beta J^2}{4\varepsilon_0 \omega} \sin^2 (\omega t - \beta x). \tag{10.69}$$

So that the average power flow is

$$|S_x| = \frac{\beta J^2}{8\varepsilon_0 \omega} = \sqrt{\left(\frac{\mu_0}{\varepsilon_0}\right)} \frac{J^2}{8}. \tag{10.70}$$

For the left-hand side of the sheet ($x < 0$) we obtain, similarly,

$$H_z = \frac{J}{2} \sin (\omega t + \beta x), \tag{10.71}$$

$$E_y = -\frac{\beta J}{2\varepsilon_0 \omega} \sin (\omega t + \beta x), \tag{10.72}$$

$$S_x = -\sqrt{\left(\frac{\mu_0}{\varepsilon_0}\right)} \frac{J^2}{4} \sin^2 (\omega t + \beta x). \tag{10.73}$$

These results repay careful study. We note that H and E are in-phase, whereas without Maxwell's hypothesis we might have expected the e.m.f. to be in time quadrature with the magnetic field. We also notice that

$$E_y = \sqrt{\left(\frac{\mu_0}{\varepsilon_0}\right)} H_z . \tag{10.74}$$

Since E is volts/metre and H amps/metre, the ratio $\sqrt{(\mu_0/\varepsilon_0)}$ is dimensionally in ohms. We can write

$$R_0 = \sqrt{\left(\frac{\mu_0}{\varepsilon_0}\right)} = 376{\cdot}7\,\Omega . \tag{10.75}$$

R_0 is sometimes called the impedance of free space, a rather unfortunate nomenclature which ascribes to empty space something which is a property of electromagnetic waves.

We also note that power is flowing away from the current sheet to both right and left. This power is described by the Poynting vector and is "attached" to the fields. The total average power is

$$P = |S|_{x>0} + |S|_{x<0} = \sqrt{\left(\frac{\mu_0}{\varepsilon_0}\right)} \frac{J^2}{4}\ \text{W/m}^2 \tag{10.76}$$

and can be described by a radiation resistance R defined by

$$P = R\,\frac{J^2}{2} \tag{10.77}$$

so that

$$R = \frac{1}{2}\sqrt{\left(\frac{\mu_0}{\varepsilon_0}\right)} = 188{\cdot}4\,\Omega . \tag{10.78}$$

An alternative way to obtain the power is to consider the back e.m.f. on the current. At $x = 0$

$$E_y = -\frac{\beta J}{2\varepsilon_0 \omega} \sin \omega t = -\sqrt{\left(\frac{\mu_0}{\varepsilon_0}\right)} \frac{J}{2} \sin \omega t . \tag{10.79}$$

Hence the radiation resistance is given by

$$R = \frac{-E}{J} = \frac{1}{2}\sqrt{\left(\frac{\mu_0}{\varepsilon_0}\right)} = 188{\cdot}4\,\Omega \tag{10.80}$$

as before.

10.5. Radiation Pressure

In Chapters 5 and 7 we have dealt at length with the force exerted by static electric and magnetic fields. In this chapter we have found that in time-varying fields power is transferred by the field. We are, therefore, well prepared to expect force to be transmitted also by time-varying fields, and we may expect that, just as in ordinary mechanics, the force will be related to the power transferred. Let us examine this question.

Consider a plane wave impinging on a perfectly absorbing surface as illustrated in Fig. 10.5. The fields will be zero to the right of the surface and this implies that there must be electric and magnetic currents on the

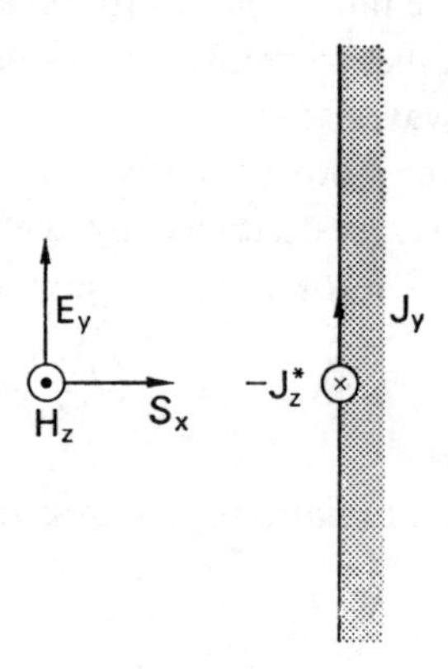

FIG. 10.5. Absorption of an electromagnetic wave.

surface. The fact that there are no magnetic currents in nature implies that in practice the wave will have to penetrate some distance into the material, but this will not affect the pressure but only its distribution. The two currents are shown in Fig. 10.5. The electric current J flows parallel to the electric field vector and the magnetic current J^* anti-parallel to the magnetic field vector.

Numerically, $J = H$ and $J^* = E$.

The pressure is given by

$$\sigma_x = \tfrac{1}{2}(\mathbf{J}\times\mathbf{B} + \mathbf{J}^*\times\mathbf{D}) \qquad (10.81)$$

$$= \tfrac{1}{2}(\mu_0 H^2 + \varepsilon_0 E^2). \qquad (10.82)$$

For sinusoidal time variation the average pressure is

$$|\sigma| = \tfrac{1}{4}(\mu_0 |H|^2 + \varepsilon_0 |E|^2). \tag{10.83}$$

This is reminiscent of Poynting's theorem. We can put

$$|\mathbf{S}| = |\mathbf{E}| \times |\mathbf{H}| \tag{10.84}$$

and use eqn. (10.74) to obtain

$$|\sigma| = \frac{|S|}{c}. \tag{10.85}$$

Thus the pressure is given by dividing the Poynting vector by the velocity of the wave. Now in mechanics we know that the force is equal to the rate of change of momentum. We are thus led to associate with the electromagnetic wave a momentum density per unit volume $\mathbf{S}/c^2$, because the momentum in a volume $c\,dt$ is then $\mathbf{S}\,dt/c$ and the pressure is given by eqn. (10.85). The Poynting vector thus gives not only the power but also the momentum of the electromagnetic wave. This is a highly significant result and shows once again the close relationship between electromagnetism and mechanics.

The radiation pressure is generally small. As an example consider the radiation due to sunlight. On a warm day this is of the order of 1 kW/m². We then have

$$\sigma = \frac{1000}{3 \times 10^8} = \frac{1}{3 \times 10^5} \quad \text{N/m}^2. \tag{10.86}$$

Our treatment of radiation pressure can readily be extended. If, for instance, the wave impinges on a perfectly reflecting surface, which reduces E to zero and doubles H, as illustrated in Fig. 10.6, we can consider the field to be made up of two travelling waves and the pressure will therefore be doubled.

More complicated waves can be built up by superposition, and waves that impinge obliquely can be resolved into their normal and tangential components. The latter will not contribute to the pressure, as may be seen by reference to Fig. 10.7. We have

$$\sigma_x = \tfrac{1}{2}(\mathbf{J} \times \mathbf{B} - \mathbf{E}q)_x - \tfrac{1}{2}(\mu_0 H^2 - \varepsilon_0 E^2) = 0. \tag{10.87}$$

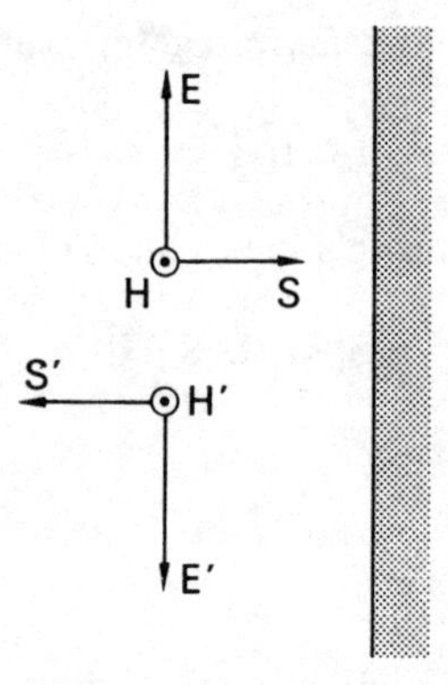

FIG. 10.6. Reflection of an electromagnetic wave.

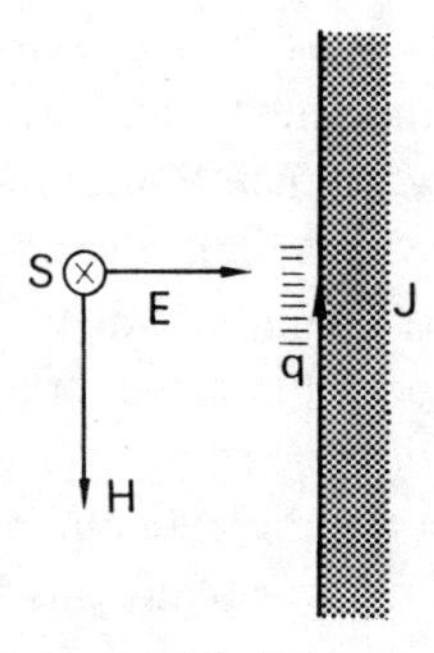

FIG. 10.7. A tangential electromagnetic wave.

Thus only the Poynting vector perpendicular to the surface contributes to the pressure. This is as expected because no momentum is transferred across the surface if the vector is parallel to the surface.

10.6. Reflection and Refraction at a Plane Boundary

In the last section we have briefly discussed the absorption and total reflection of a plane wave. Let us now consider the more general case which arises when a plane wave meets a plane boundary between two materials having permeabilities ε_1 and ε_2 (Fig. 10.8).

The incident field can be written

$$\overline{\mathbf{E}} = E_0\, e^{j(\omega t - \mathbf{k}\cdot\mathbf{r})}, \tag{10.88}$$

where

$$|\mathbf{k}| = \omega\sqrt{(\mu_1\varepsilon_1)} \tag{10.89}$$

and **k** defines the direction of propagation of the wave.

By the use of eqn. (10.74), which relates H and E for a plane wave,

$$\mathbf{H} = \frac{\mathbf{k}\times\mathbf{E}}{\omega\mu_1}. \tag{10.90}$$

Similarly, for the refracted wave,

$$\overline{\mathbf{E}}' = \mathbf{E}_0' \, e^{j(\omega t - \mathbf{k}'\cdot\mathbf{r})}, \quad \overline{\mathbf{H}}' = \frac{\mathbf{k}'\times\overline{\mathbf{E}}'}{\omega\mu_2}, \tag{10.91}$$

and for the reflected wave,

$$\overline{\mathbf{E}}'' = \mathbf{E}_0'' \, e^{j(\omega t - \mathbf{k}''\cdot\mathbf{r})}, \quad \overline{\mathbf{H}}'' = \frac{\mathbf{k}''\times\overline{\mathbf{E}}''}{\omega\mu_1}. \tag{10.92}$$

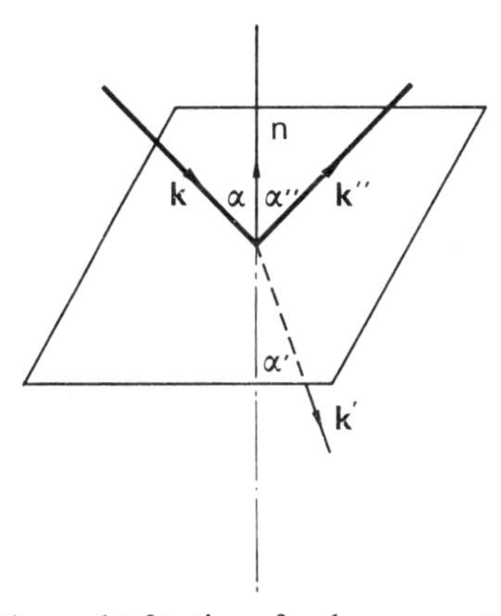

FIG. 10.8. Reflection and refraction of a plane wave at a plane boundary.

At the surface the tangential electric field strength must be continuous because there is no surface magnetic current. This implies that the frequency is unchanged, which we have already assumed in writing the expressions for the field. Secondly, the entire exponentials must be the same, because the continuity condition holds at every instant. Hence

$$\mathbf{k}\,.\,\mathbf{r} = \mathbf{k}'\,.\,\mathbf{r} - \mathbf{k}''\,.\,\mathbf{r}. \tag{10.93}$$

This implies that the vectors **k**, **k**′, and **k**″ are co-planar.

Now let **r** be a vector in the plane of the boundary, so that

$$\hat{\mathbf{n}} \cdot \mathbf{r} = 0. \tag{10.94}$$

Equation (10.93) can be rewritten

$$(\mathbf{k} - \mathbf{k}') \cdot \mathbf{r} = 0, \quad (\mathbf{k} - \mathbf{k}'') \cdot \mathbf{r} = 0. \tag{10.95}$$

This implies that **k**, **k**′, and **k**″ are co-planar with $\hat{\mathbf{n}}$. Thus the vectors of the reflected and refracted waves lie in the plane of incidence.

Also from eqn. (10.93) it follows that

$$\alpha = \alpha'' \tag{10.96}$$

and

$$\frac{\sin\alpha}{\sin\alpha'} = \frac{|\mathbf{k}'|}{|\mathbf{k}|} = \sqrt{\left(\frac{\mu_2\varepsilon_2}{\mu_1\varepsilon_1}\right)} = \frac{n_2}{n_1}. \tag{10.97}$$

Equation (10.97) is known as Snell's law. The *refractive indices* of regions 1 and 2 are defined as $n_1 = \sqrt{(\mu_1\varepsilon_1)}$ and $n_2 = \sqrt{(\mu_2\varepsilon_2)}$. The relationships between the amplitudes of the field quantities can be obtained from the boundary conditions

$$\hat{\mathbf{n}} \times (\overline{\mathbf{E}} + \overline{\mathbf{E}}'') = \hat{\mathbf{n}} \times \overline{\mathbf{E}}' \tag{10.98}$$

and

$$\hat{\mathbf{n}} \times (\overline{\mathbf{H}} + \overline{\mathbf{H}}'') = \hat{\mathbf{n}} \times \overline{\mathbf{H}}', \tag{10.99}$$

where we have assumed that there is no surface electric current. Using the relation between the electric and magnetic fields of plane waves [eqn. (10.90)] we can transform eqn. (10.99) to read

$$\hat{\mathbf{n}} \times (\mathbf{k} \times \overline{\mathbf{E}} + \mathbf{k}'' \times \overline{\mathbf{E}}'')/\mu_1 = \hat{\mathbf{n}} \times (\mathbf{k}' \times \overline{\mathbf{E}}')/\mu_2. \tag{10.100}$$

Any arbitrary plane wave can be built up from components in which:

(a) $\overline{\mathbf{E}}$ is at right angles to the plane of incidence; or

(b) $\overline{\mathbf{E}}$ is in the plane of incidence.

(a) $\bar{\mathbf{E}}$ *at right angles to the plane of incidence* (Fig. 10.9)

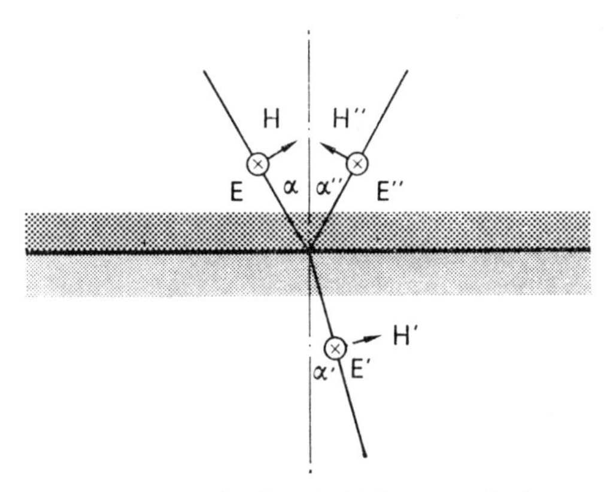

FIG. 10.9. Plane wave with electric field perpendicular to plane of incidence.

From eqns. (10.98) and (10.100)

$$\bar{E} + \bar{E}'' = \bar{E}' \tag{10.101}$$

and

$$\frac{k}{\mu_1}(\bar{E} + \bar{E}'')\cos\alpha = \frac{k'}{\mu_2}\bar{E}'\cos\alpha', \tag{10.102}$$

whence

$$\bar{E}' = \frac{2\bar{E}\cos\alpha}{\cos\alpha + \cos\alpha'\sqrt{[(\mu_1\varepsilon_2)/(\mu_2\varepsilon_1)]}} \tag{10.103}$$

$$\bar{E}'' = \bar{E}\frac{\cos\alpha - \cos\alpha'\sqrt{[(\mu_1\varepsilon_2)/(\mu_2\varepsilon_1)]}}{\cos\alpha + \cos\alpha'\sqrt{[(\mu_1\varepsilon_2)/(\mu_2\varepsilon_1)]}}. \tag{10.104}$$

If we eliminate the angle α' by the use of eqn. (10.97),

$$\bar{E}' = \frac{2\bar{E}\cos\alpha}{\cos\alpha + \sqrt{[(\mu_1\varepsilon_2/\mu_2\varepsilon_1) - (\mu_1/\mu_2)^2\sin^2\alpha]}}, \tag{10.105}$$

$$\bar{E}'' = \bar{E}\frac{\cos\alpha - \sqrt{[(\mu_1\varepsilon_2/\mu_2\varepsilon_1) - (\mu_1/\mu_2)^2\sin^2\alpha]}}{\cos\alpha + \sqrt{[(\mu_1\varepsilon_2/\mu_2\varepsilon_1) - (\mu_1/\mu_2)^2\sin^2\alpha]}}. \tag{10.106}$$

(b) $\bar{\mathbf{E}}$ *in the plane of incidence* (Fig. 10.10)

Equations (10.98) and (10.100) now become

$$(\bar{E} - \bar{E}'')\cos\alpha = \bar{E}'\cos\alpha', \tag{10.107}$$

$$(\bar{E} + \bar{E}'')\sqrt{(\varepsilon_1/\mu_1)} = \bar{E}'\sqrt{(\varepsilon_2/\mu_2)}, \tag{10.108}$$

whence

$$\bar{E}' = \frac{2\bar{E}\cos\alpha\sqrt{(\mu_2\varepsilon_1/\mu_1\varepsilon_2)}}{\cos\alpha + \sqrt{[\mu_2\varepsilon_1/\mu_1\varepsilon_2 - (\varepsilon_1/\varepsilon_2)^2\sin^2\alpha]}} \tag{10.109}$$

and

$$\bar{E}'' = \bar{E}\frac{\cos\alpha - \sqrt{[\mu_2\varepsilon_1/\mu_1\varepsilon_2 - (\varepsilon_1/\varepsilon_2)^2\sin^2\alpha]}}{\cos\alpha + \sqrt{[\mu_2\varepsilon_1/\mu_1\varepsilon_2 - (\varepsilon_1/\varepsilon_2)^2\sin^2\alpha]}}. \tag{10.110}$$

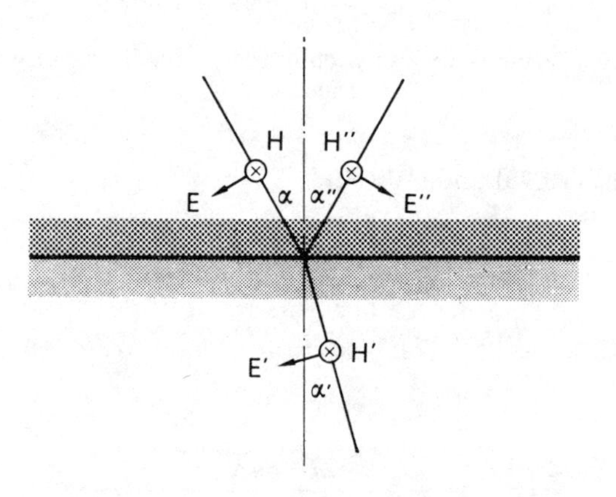

FIG. 10.10. Plane wave with electric field in plane of incidence.

Equations (10.105), (10.106), (10.109), and (10.110) are known as Fresnel's formulae. They have several interesting properties. Thus for dielectric materials where $\mu_1 = \mu_2 = \mu_0$ there is always reflection in case (a) unless the two materials have also the same permittivity.

However, in case (b) there is zero reflection if

$$\tan\alpha = \sqrt{\left(\frac{\varepsilon_2}{\varepsilon_1}\right)}. \tag{10.111}$$

This is called the Brewster angle.

Thus at this "polarizing" angle only waves of type (a) will be reflected. Use can be made of this property to produce a beam of polarized light.

There is a whole range of angles of incidence for which Snell's law [eqn. (10.97)] would give values for $\sin\alpha'$ greater than unity, if $n_1 > n_2$. Consider a plane wave of type (a). We then have for a dielectric interface

$$\bar{E}'' = \bar{E}\,\frac{\cos\alpha - j\sqrt{[\sin^2\alpha - (n_2/n_1)^2]}}{\cos\alpha + j\sqrt{[\sin^2\alpha - (n_2/n_1)^2]}}, \tag{10.112}$$

which we have written in this form, because $\sin\alpha > n_2/n_1$. It is clear from eqn. (10.113) that the amplitude of the reflected wave $\bar{E}''$ is the same as the amplitude of the incident wave $\bar{E}$, but the phase has changed by an angle ψ, where

$$\tan\left(\frac{\psi}{2}\right) = \frac{\sqrt{[\sin^2\alpha - (n_2/n_1)^2]}}{\cos\alpha}. \tag{10.113}$$

The reflection and transmission coefficients are defined as

$$\bar{R} = \left|\frac{\bar{E}''}{\bar{E}}\right|, \psi, \quad \bar{T} = \left|\frac{\bar{E}'}{\bar{E}}\right|, \phi, \tag{10.114}$$

where ϕ and ψ denote the phase angles.

For the case which we have just discussed, where $\sin\alpha > (n_2/n_1)$, it can readily be shown that

$$|\bar{R}| = 1, \quad |\bar{T}| = 0. \tag{10.115}$$

There is a field which penetrates into the material below the interface, but there is no average power flow into this material. The phenomenon is described as *total reflection*.

10.7. Waves in Conducting Material

The wave equation can be derived from Maxwell's equations and Ohm's law. In the absence of free charge we have

$$\nabla^2\mathbf{E} - \mu\sigma\,\frac{\partial\mathbf{E}}{\partial t} - \mu\varepsilon\,\frac{\partial^2\mathbf{E}}{\partial t^2} = 0. \tag{10.116}$$

For a plane wave with sinusoidal time variation this becomes

$$\frac{\partial^2 \overline{\mathbf{E}}}{\partial x^2} - j\omega\mu\sigma\overline{\mathbf{E}} + \omega^2\mu\varepsilon\overline{\mathbf{E}} = 0 \tag{10.117}$$

or

$$\frac{\partial^2 \overline{\mathbf{E}}}{\partial x^2} = \gamma^2 \overline{\mathbf{E}}, \tag{10.118}$$

where the propagation constant γ is given by

$$\gamma = \alpha + j\beta = \sqrt{(-\omega^2\mu\varepsilon + j\omega\mu\sigma)}. \tag{10.119}$$

Thus the propagation constant has a real as well as an imaginary component. This implies that there will be exponential decay as well as wave propagation.

It is now possible to apply the treatment of § 10.6 to waves of the form

$$\bar{E} = E_0\, e^{-\alpha x}\, e^{j(\omega t - \beta x)}. \tag{10.120}$$

However, for frequencies below the optical region we can with metallic conductors make the approximation $\sigma \gg \omega\varepsilon$. We then have

$$\alpha = \beta = \sqrt{\left(\frac{\mu\sigma\omega}{2}\right)}. \tag{10.121}$$

This, of course, is the quantity $1/\delta$ where δ is the skin-depth discussed in Chapter 8. This is as expected because the approximation $\sigma \gg \omega\varepsilon$ has converted the wave equation into the simple diffusion equation. Discussion of the energy dissipated when waves impinge on metallic boundaries can therefore be based on the treatment of eddy currents in Chapter 8.

If σ and $\omega\varepsilon$ are of the same order of magnitude, the expressions for the field given in § 10.6 can be modified by writing $\varepsilon - j\sigma/\omega$ instead of ε. An example of particular interest is the reflection of radio waves from the ground. We can then write $\mu_1 = \mu_2 = \mu_0$, $\varepsilon_1 = \varepsilon_0$, $\varepsilon_2 = \varepsilon_r\varepsilon_0$, $\sigma_1 = 0$ and $\sigma_2 = \sigma$. If the electric field is at right angles to the plane of incidence, i.e. for a horizontally polarized wave we find

$$\frac{\bar{E}''}{\bar{E}} = \frac{\cos\alpha - \sqrt{|(\varepsilon_r - j\sigma/\omega\varepsilon_0) - \sin^2\alpha|}}{\cos\alpha + \sqrt{|(\varepsilon_r - j\sigma/\omega\varepsilon_0) - \sin^2\alpha|}}, \tag{10.122}$$

and putting

$$\frac{1}{\omega\varepsilon_0} = \frac{\lambda}{2\pi c\varepsilon_0} = \frac{1}{2\pi}\sqrt{\left(\frac{\mu_0}{\varepsilon_0}\right)}\lambda \doteqdot 60\lambda, \qquad (10.123)$$

$$\frac{\bar{E}''}{\bar{E}} = \frac{\cos\alpha - \sqrt{|(\varepsilon_r - j\,60\lambda\sigma) - \sin^2\alpha|}}{\cos\alpha + \sqrt{|(\varepsilon_r - j\,60\lambda\sigma) - \sin^2\alpha|}}. \qquad (10.124)$$

Figure 10.11 shows the reflection coefficient for such a wave, when the frequency is 10 MHz and the ground has a conductivity of 10^{-2} siemens and relative permittivity 10, plotted against the angle of incidence α.

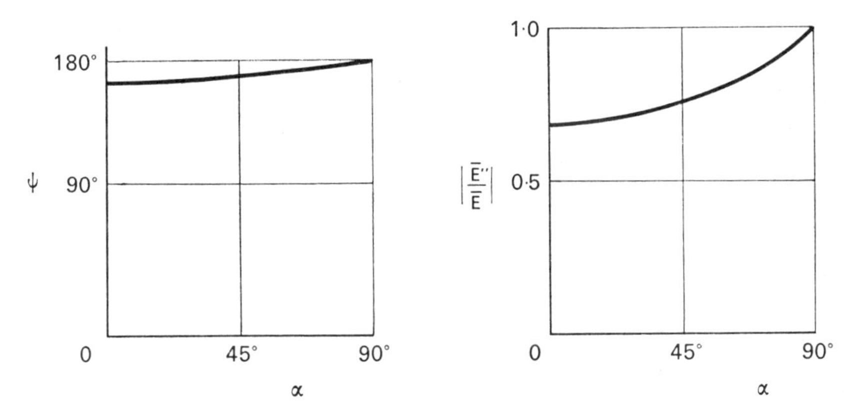

FIG. 10.11. Reflection coefficient for horizontally polarized wave.

If the electric field is in the plane of incidence the reflected field is given by the expression

$$\frac{\bar{E}''}{\bar{E}} = \frac{(\varepsilon_r - j\,60\lambda\sigma)\cos\alpha - \sqrt{|(\varepsilon_r - j\,60\lambda\sigma) - \sin^2\alpha|}}{(\varepsilon_r - j\,60\lambda\sigma)\cos\alpha + \sqrt{|(\varepsilon_r - j\,60\lambda\sigma) - \sin^2\alpha|}}. \qquad (10.125)$$

Figure 10.12 shows the reflection coefficient for the same frequency, ground conductivity and permittivity as before.

It will be noticed that the reflection coefficient for a wave in which the electric field vector lies in the plane of incidence, has a minimum value. The angle at which this occurs is called the pseudo-Brewster angle, and

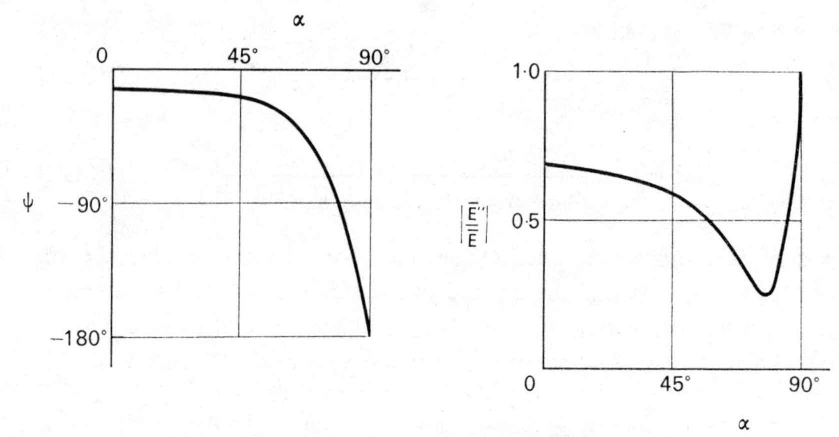

FIG. 10.12. Reflection coefficient for vertically polarized wave.

corresponds to the condition for zero reflection which occurs in a lossless medium. The horizontally polarized wave has a reflection coefficient near unity at this angle. Ionospheric communication systems working with wavelengths of a few decametres, which use a large angle of incidence, therefore require horizontally polarized waves.

Summary

Maxwell's hypothesis leads to the conclusion that electromagnetic energy is transmitted at a finite velocity. The field quantities can be calculated by using the retarded potentials. The radiated energy has to be provided by the radiating source and can be observed as a radiation resistance.

The problem of energy transfer was investigated by Poynting and led him to the description of radiated power density in terms of the Poynting vector. Associated with the radiated energy there is momentum, and pressure is exerted by electromagnetic waves as they are absorbed or reflected.

Exercises

10.1. A large flat sheet lies in the plane containing the y- and z-axes of a rectangular system of coordinates. A current of line density $I \cos \beta y \, e^{j\omega t}$ A/m flows in the sheet in the direction parallel to the z-axis. Show that the scalar potential of the electromagnetic field is zero. Show also that the vector potential has the form

$$A_z = A_1 \cos \beta y \, e^{j(\omega t - mx)}$$

or

$$A_z = A_2 \cos \beta y \, e^{-mx} e^{\omega t}$$

according as $m^2 > 1$ or $m^2 < 1$, where $m^2 = (\omega/c)^2 - \beta^2$.

10.2. Determine the constants A_1 and A_2 in the problem of Exercise 10.1. [Ans.: $-j\mu_0 I/2m$; $-\mu_0 I/2m$.]

10.3. Determine the electric and magnetic field due to the current distribution in Exercise 10.1 for $x > 0$.

$$\left[\text{Ans.:} \quad \begin{cases} E_z = -\mu_0 \dfrac{\omega I}{2m} \cos \beta y \, e^{j(\omega t - mx)}, \\[2ex] H_y = \dfrac{I}{2} \cos \beta y \, e^{j(\omega t - mx)}, \\[2ex] H_x = \dfrac{j\beta I}{2m} \sin \beta y \, e^{j(\omega t - mx)}, \end{cases}\right.$$

$$\text{or} \quad \left. \begin{cases} E_z = \dfrac{j\mu_0 \omega I}{2m} \cos \beta y \, e^{-mx} e^{j\omega t}, \\[2ex] H_y = \dfrac{I}{2} \cos \beta y \, e^{-mx} e^{j\omega t}, \\[2ex] H_x = -\dfrac{\beta I}{2m} \sin \beta y \, e^{-mx} e^{j\omega t}. \end{cases} \right]$$

10.4. Using the result of Exercise 10.3, show that the current sheet will radiate energy only if $\omega^2/c^2 > \beta^2$, i.e. if the wavelength of the spatial distribution is greater than the electromagnetic wavelength $2\pi c/\omega$. Show also that the effective wavelength of the radiation field is

$$\frac{2\pi}{\sqrt{(|\omega/c|^2 - \beta^2)}}.$$

10.5. Show that the radiation resistance of the current sheet of Exercise 10.1 is

$$R = \frac{\mu_0 \omega}{2\sqrt{(|\omega/c|^2 - \beta^2)}} \ \Omega.$$

Sketch the variation of R with frequency.

10.6. Show that the average values of the Poynting vector of the radiation from the current sheet of Exercise 10.1 is

$$|S_x| = \frac{\mu_0 \omega I^2}{8m} (\cos \beta y)^2 \quad \text{W/m}^2.$$

Compare the amount of radiated power in Exercises 10.5 and 10.6.

10.7. Discuss the electromagnetic radiation from a plane grid of parallel currents $I e^{j\omega t}$. Take the currents as flowing parallel to the z-axis on the y-axis spaced a distance g apart as shown in Fig. 10.13. [*Hint:* The grid of currents of strength I can be replaced by a Fourier series of line density i, where

$$\frac{i}{I} = \frac{1}{g} + \frac{2}{g} \sum_{n=1}^{\infty} \cos\left(\frac{2\pi n y}{g}\right), \quad \text{where } n \text{ is odd.]}$$

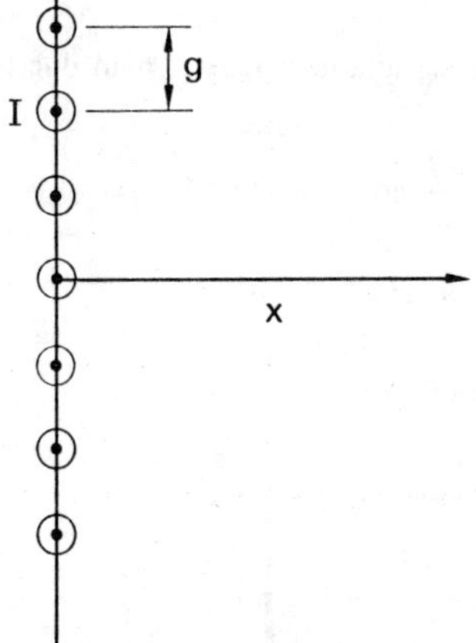

FIG. 10.13. Radiation from a grid of currents.

10.8. Explain why the delayed potentials **A** and V are dependent on each other. Show that they can be replaced by the single vector $\mathbf{\Pi}$, called the Hertz vector, where

$$\mathbf{A} = \frac{1}{c^2}\frac{\partial \mathbf{\Pi}}{\partial t},$$

$$V = -\operatorname{div} \mathbf{\Pi}.$$

Show also that if

$$\mathbf{C} = \operatorname{curl} \mathbf{\Pi}$$

the magnetic and electric fields are given by

$$\mathbf{B} = \frac{1}{c^2}\frac{\partial \mathbf{C}}{\partial t},$$

$$\mathbf{E} = \operatorname{curl} \mathbf{C}.$$

What is the source of $\mathbf{\Pi}$? [Ans.: The so-called polarization vector **p** defined by $\varrho_{\text{free}} = -\operatorname{div} \mathbf{p}$ and $\mathbf{J} = \partial \mathbf{p}/\partial t$.]

10.9. A coaxial cable carries a constant current I. The resistance of the cable conductor is negligible and there is a constant potential difference V between the outer and inner conductor. The inner conductor has radius a and the outer conductor has an internal radius b. Determine the Poynting vector and the total power flow. Discuss the power flow if the resistance of the conductors is taken into account.

[Ans.: $S = \dfrac{VI}{2\pi r^2 \ln (b/a)}$ along the cable. Power $= \int_a^b S\, 2\pi r\, dr = VI$. There will be a radial component of S to give the ohmic losses—see § 10.3.2.]

10.10. Define the term *complex Poynting vector*. Show that if displacement current is neglected the equivalent impedance of a conductor is given by

$$Z = R + jX = -\frac{1}{I^2} \oiint \bar{\mathbf{S}} \,.\, d\mathbf{s}.$$

10.11. Use the complex Poynting vector to show that the impedance of a flat sheet of thickness $2b$ carrying current I per unit width is given by

$$Z = \frac{1}{2\sigma\delta} \left| \frac{\sinh 2b/\delta + \sin 2b/\delta}{\cosh 2b/\delta - \cos 2b/\delta} + j \frac{\sinh 2b/\delta - \sin 2b/\delta}{\cosh 2b/\delta - \cos 2b/\delta} \right|,$$

where σ is the conductivity and δ the skin-depth. [*Hint:* Refer to § 8.4. Find $\bar{\mathbf{S}}$ on the surface from $\mathbf{H}$ and $\mathbf{E} = \mathbf{J}/\sigma$ at the surface.)

10.12. A plane electromagnetic wave impinges normally on a flat surface. Analyse the conditions when the radiation is (a) totally reflected, or (b) totally absorbed. If the breakdown strength in air is 3×10^6 V/m, what is the maximum pressure which can be exerted on a sheet of perfectly conducting metal by a radio wave? [Ans.: 20 N/m^2.]

CHAPTER 11

ELECTROMAGNETIC RADIATION II

11.1. Radiation from an Electric Dipole

In § 10.2 we examined the field of a current element. But we realized that such a current element cannot exist without electric charges at its ends to make the current flow possible. Let us now examine a complete electric dipole consisting of a current element and two charges as shown in Fig. 11.1. Let the magnitude of the current be I, the charges $\pm Q$, the length l, and the dipole moment $Ql = P$. Let the time variation be harmonic and let us write it as $e^{j\omega t}$. By continuity if the current is $Ie^{j\omega t}$, the

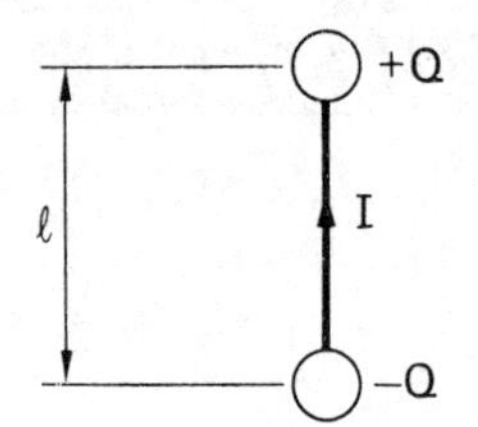

FIG. 11.1. An electric dipole.

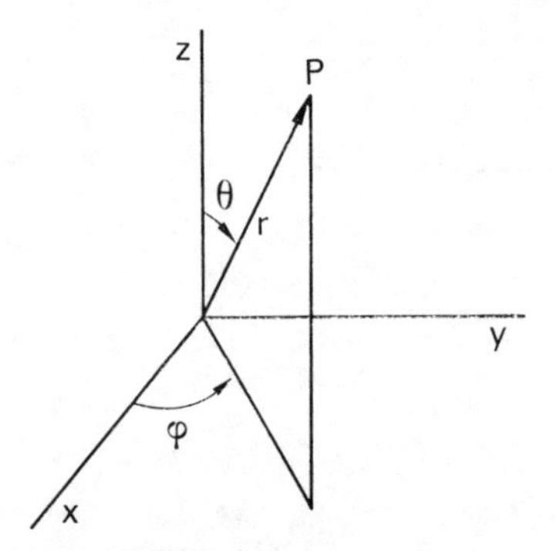

FIG. 11.2. Spherical coordinates.

charge will be

$$Q\,e^{j\omega t} = \int I e^{j\omega t}\,dt = -\frac{jI}{\omega}\,e^{j\omega t}. \tag{11.1}$$

The delayed vector potential is given by

$$A_z = \frac{\mu_0 Il}{4\pi r}\,e^{j(\omega t-\beta r)} = \left(\frac{j\omega\mu_0 P}{4\pi r}\right) e^{j(\omega t-\beta r)}. \tag{11.2}$$

In spherical coordinates (Fig. 11.2) this is

$$\left.\begin{aligned} A_r &= \frac{j\omega\mu_0 P\cos\theta}{4\pi r}\,e^{j(\omega t-\beta r)},\\ A_\theta &= -\frac{j\omega\mu_0 P\sin\theta}{4\pi r}\,e^{j(\omega t-\beta r)},\\ A_\phi &= 0. \end{aligned}\right\} \tag{11.3}$$

The only component of magnetic field strength is

$$\begin{aligned} H_\phi &= \frac{1}{\mu_0}(\text{curl } A)_\phi = \frac{1}{\mu_0 r}\left[\frac{\partial(rA_\theta)}{\partial r} - \frac{\partial A_r}{\partial\theta}\right]\\ &= -\frac{\omega P\sin\theta}{4\pi}\left(\frac{\beta}{r} - \frac{j}{r^2}\right) e^{j(\omega t-\beta r)}. \end{aligned} \tag{11.4}$$

The delayed scalar potential can be derived by the use of § 4.1.2,

$$\begin{aligned} V &= -\frac{P\cos\theta}{4\pi\varepsilon_0}\frac{\partial}{\partial r}\left[\frac{e^{j(\omega t-\beta r)}}{r}\right]\\ &= \frac{P\cos\theta}{4\pi\varepsilon_0}\left(\frac{1}{r^2} + \frac{j\beta}{r}\right] e^{j(\omega t-\beta r)}. \end{aligned} \tag{11.5}$$

The electric field strength is given by

$$\begin{aligned} E_r &= -\frac{\partial A_r}{\partial t} - \frac{\partial V}{\partial r}\\ &= \frac{P\cos\theta}{4\pi}\left(+\frac{\omega^2\mu_0}{r} - \frac{\beta^2}{\varepsilon_0 r} + \frac{2j\beta}{\varepsilon_0 r^2} + \frac{2}{\varepsilon_0 r^3}\right) e^{j(\omega t-\beta r)}\\ &= \frac{P\cos\theta}{2\pi\varepsilon_0}\left(\frac{j\beta}{r^2} + \frac{1}{r^3}\right) e^{j(\omega t-\beta r)}, \end{aligned} \tag{11.6}$$

Since $\beta^2/\varepsilon_0 = \omega^2/\varepsilon_0 c^2 = \omega^2\mu_0$,

$$E_\theta = -\frac{\partial A_\theta}{\partial t} - \frac{1}{r}\frac{\partial V}{\partial \theta}$$

$$= \frac{P\sin\theta}{4\pi}\left(-\frac{\omega^2\mu_0}{r} + \frac{j\beta}{\varepsilon_0 r^2} + \frac{1}{\varepsilon_0 r^3}\right)e^{j(\omega t - \beta r)}, \qquad (11.7)$$

and

$$E_\phi = -\frac{\partial A_\phi}{\partial t} - \frac{1}{r\sin\theta}\frac{\partial V}{\partial \phi} = 0. \qquad (11.8)$$

As r becomes large we can neglect all terms except those in $1/r$. The radiation field then becomes

$$E_\theta = \sqrt{\left(\frac{\mu_0}{\varepsilon_0}\right)}H_\phi = -\frac{\omega^2\mu_0 P\sin\theta}{4\pi r}e^{j(\omega t - \beta r)}, \qquad (11.9)$$

$$E_r = E_\phi = H_r = H_\theta = 0. \qquad (11.10)$$

This is a spherical field distribution with the electric field as circles of longitude and the magnetic field as circles of latitude. The field is a maximum on the equator and is zero at the poles. The Poynting vector is radially outwards (Fig. 11.3).

$$S_r = E_\theta H_\phi. \qquad (11.11)$$

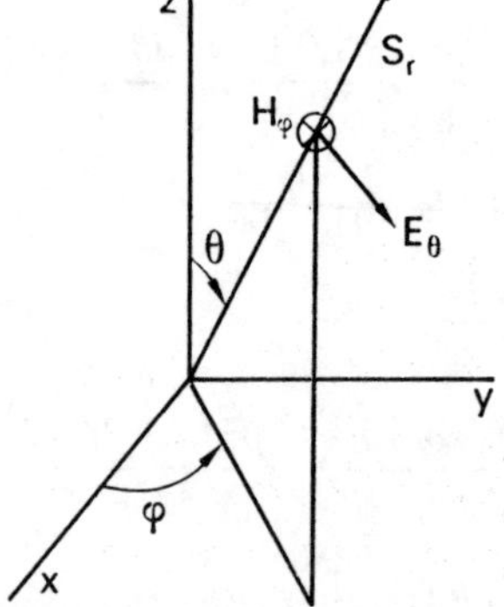

FIG. 11.3. The Poynting vector in spherical coordinates.

The average power flow per unit area is

$$|S| = \frac{\omega^4 \mu_0 P^2 \sin^2 \theta}{32\pi^2 c r^2}. \tag{11.12}$$

The total power flow

$$\int_0^\pi \int_0^{2\pi} |S|\, r^2 \sin\theta \, d\theta \, d\phi = \frac{\omega^4 \mu_0 P^2}{16\pi c} \int_0^\pi \sin^3 \theta \, d\theta = \frac{\omega^4 \mu_0 P^2}{12\pi c}. \tag{11.13}$$

Now $P = Ql$, so that in magnitude $P = Il/\omega$. We can write the power in terms of the current and a radiation resistance R

$$R \frac{I^2}{2} = \frac{\omega^4 \mu_0 P^2}{12\pi c} = \frac{\omega^2 \mu_0 l^2}{12\pi c} I^2,$$

whence

$$R = \frac{\omega^2 \mu_0 l^2}{6\pi c}\ \Omega. \tag{11.14}$$

In terms of the wavelength $\lambda = 2\pi/\beta = 2\pi c/\omega$,

$$R = \sqrt{\left(\frac{\mu_0}{\varepsilon_0}\right)} \frac{2\pi}{3} \left(\frac{l}{\lambda}\right)^2 \Omega = 790 \left(\frac{l}{\lambda}\right)^2 \Omega. \tag{11.15}$$

This can also be obtained from the field close to the dipole. Let us consider E_r for small values of r:

$$\begin{aligned} E_r &= \frac{P \cos\theta}{2\pi\varepsilon_0} \left(j \frac{\beta}{r^2} + \frac{1}{r^3} \right) e^{-j\beta r} e^{j\omega t} \\ &= -j \frac{Il \cos\theta}{2\pi\varepsilon_0 \omega} \left(j \frac{\beta}{r^2} + \frac{1}{r^3} \right) \left(1 - j\beta r - \frac{\beta^2 r^2}{2} + j \frac{\beta^3 r^3}{6} \cdots \right) e^{j\omega t}. \end{aligned} \tag{11.16}$$

The dominant real term is

$$E_r = -\frac{Il \cos\theta}{2\pi\varepsilon_0 \omega} \frac{\beta^3}{3}. \tag{11.17}$$

Along the dipole $\cos\theta = 1$ and the dominant real term in E is independent of r. This component of E is in anti-phase with I. Thus the radiation resistance is

$$R = -\frac{El}{I} = \frac{l^2\beta^3}{6\pi\varepsilon_0\omega} = \sqrt{\left(\frac{\mu_0}{\varepsilon_0}\right)}\frac{2\pi}{3}\left(\frac{l}{\lambda}\right)^2 \tag{11.18}$$

as was obtained previously by the use of the Poynting vector.

Before leaving the discussion of the electric dipole it is interesting to look at expressions for the field at low frequencies.

As $\beta \to 0$,

$$H_\phi \to \frac{j\omega P \sin\theta}{4\pi r^2}, \tag{11.19}$$

which in terms of I gives

$$H_\phi = \frac{Il \sin\theta}{4\pi r^2}, \tag{11.20}$$

which is the well-known expression for the field of a "Heaviside" current element [eqn. (2.36)]. It is useful to know that the magnetic field of a current element is the limiting case of the magnetic field of a dipole as the frequency becomes small.

Similarly, the terms in $1/r^3$ in eqns. (11.6) and (11.7) give the electric field of a static electric dipole.

11.1. Radiation from a Small Current Loop

By a small loop we mean that the radius a of the loop is very much smaller than the wavelength λ. This enables us to assume a constant current around the loop at any instance. Secondly, we mean that we shall investigate the field only at large distances, so that $r \gg a$.

Because the current is constant there will be no charge associated with it. Hence the scalar potential is zero. To obtain the vector potential we note that by symmetry A will be disposed in circles parallel to the loop. Thus the only component of A will be A_ϕ. Let us calculate A_ϕ at the point P, which without loss of generality can be taken to lie in the y-, z-plane (Fig. 11.4). The loop is taken to lie in the x-, y-plane.

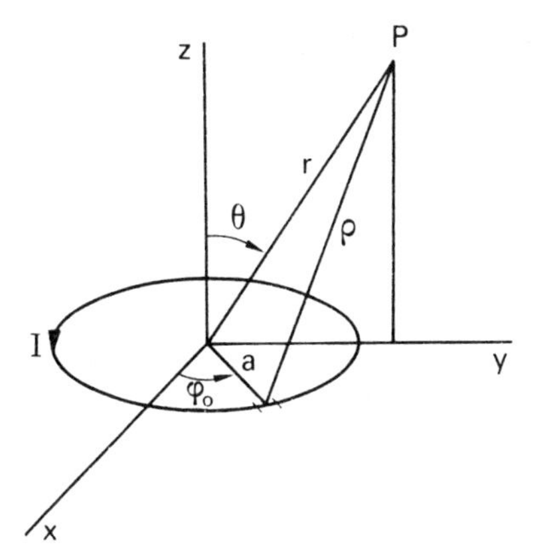

FIG. 11.4. A small current loop.

Consider the contribution to A_ϕ made by a typical current element at an angle ϕ_0 and at a distance ϱ from P. We can write

$$A_\phi = \frac{\mu_0 I}{4\pi}\int_0^{2\pi}\frac{a\sin\phi_0}{\varrho}\,e^{j(\omega t-\beta\varrho)}\,d\phi_0. \tag{11.21}$$

Since r and ϱ are very much greater than a, we can resolve along r and write

$$\varrho = r - a\sin\phi_0\sin\theta. \tag{11.22}$$

We can also substitute r for ϱ in the denominator of eqn. (11.21) because we are neglecting terms in $(a/r)^2$.

Hence

$$A_\phi = \frac{\mu_0 Ia}{4\pi r}\,e^{j(\omega t-\beta r)}\int_0^{2\pi} e^{(j\beta a\sin\phi_0\sin\theta)}\sin\phi_0\,d\phi_0. \tag{11.23}$$

Since $\beta a = 2\pi a/\lambda$ is very much smaller than unity,

$$A_\phi = \frac{\mu_0 Ia}{4\pi r}\,e^{j(\omega t-\beta r)}\int_0^{2\pi}|\sin\phi_0 + j\beta a\sin\theta\,(\sin\phi_0)^2|\,d\phi_0, \tag{11.24}$$

whence

$$A_\phi = \frac{j\mu_0 Ia^2\beta\sin\theta}{4r}\,e^{j(\omega t-\beta r)}. \tag{11.25}$$

The electric field strength is given by

$$\left.\begin{aligned} E_\phi &= -\frac{\partial A_\phi}{\partial t} = \frac{\omega\mu_0 I a^2\beta \sin\theta}{4r}\, e^{j(\omega t-\beta r)}, \\ E_\theta &= E_r = 0. \end{aligned}\right\} \tag{11.26}$$

The magnetic field strength is given by

$$\left.\begin{aligned} H_r &= \frac{1}{\mu_0 r\sin\theta}\left[\frac{\partial(\sin\theta A_\phi)}{\partial\theta}\right], \\ H_\theta &= -\frac{1}{\mu_0 r}\,\frac{\partial(rA_\phi)}{\partial r}, \\ H_\phi &= 0. \end{aligned}\right\} \tag{11.27}$$

Thus

$$\left.\begin{aligned} H_r &= \frac{jIa^2\beta\cos\theta}{2r^2}\, e^{j(\omega t-\beta r)}, \\ H_\theta &= -\frac{Ia^2\beta^2\sin\theta}{4r}\, e^{j(\omega t-\beta r)}, \\ H_\phi &= 0. \end{aligned}\right\} \tag{11.28}$$

Since H_r decays as $1/r^2$, we can neglect it at large distances. We then have for the radiation field,

$$E_\phi = -\sqrt{\left(\frac{\mu_0}{\varepsilon_0}\right)}\, H_\theta. \tag{11.29}$$

The Poynting vector is positive in the r-direction, since

$$S_r = -E_\phi H_\theta. \tag{11.30}$$

If we compare the field of the current loop with that of the electric dipole discussed in § 11.1, we notice that the current loop behaves as a *magnetic* dipole of moment

$$P^* = \mu_0 I\pi a^2. \tag{11.31}$$

Thus Ampère's equivalence between magnetic dipoles and current loops can be used to calculate radiation as long as the dimensions of the loop are small compared with the electric wavelength. This is an extremely important result and can be put to good use in the calculation of radiation fields.

It should also be noted that this result does not depend on the exact shape of the loop. We have discussed a circular loop, but the result can be extended to a loop of arbitrary shape. Equation (11.31) then becomes

$$P^* = \mu_0 Is, \tag{11.32}$$

where s is the area of the loop.

11.3. Radiation from a Linear Antenna with a Known Current Distribution

The dipole discussed in § 11.1 can be used as a basic unit from which can be built any arbitrary distribution of current. In this way we can consider antennas whose length is appreciable compared with a wavelength.

Let us examine the field of an antenna of length $2h$ at a distant point (Fig. 11.5). The contribution from the dipole at height z is found from eqn. (11.9):

$$\begin{aligned}\delta E_\theta &= -\frac{\omega^2 \mu_0 P \sin\theta}{4\pi r} e^{j(\omega t - \beta r)} \\ &= j\frac{1}{2}\sqrt{\left(\frac{\mu_0}{\varepsilon_0}\right)}\frac{\delta z}{\lambda}\frac{I\sin\theta}{r} e^{j(\omega t - \beta r)},\end{aligned} \tag{11.33}$$

where the time variation of the current is $e^{j\omega t}$.

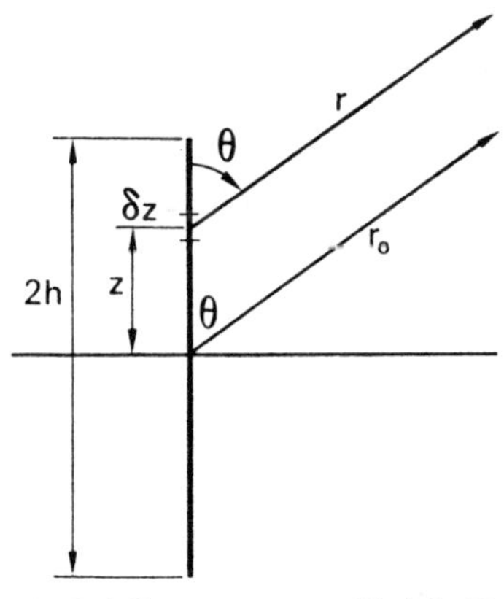

FIG. 11.5. A linear antenna of height $2h$.

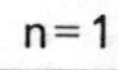

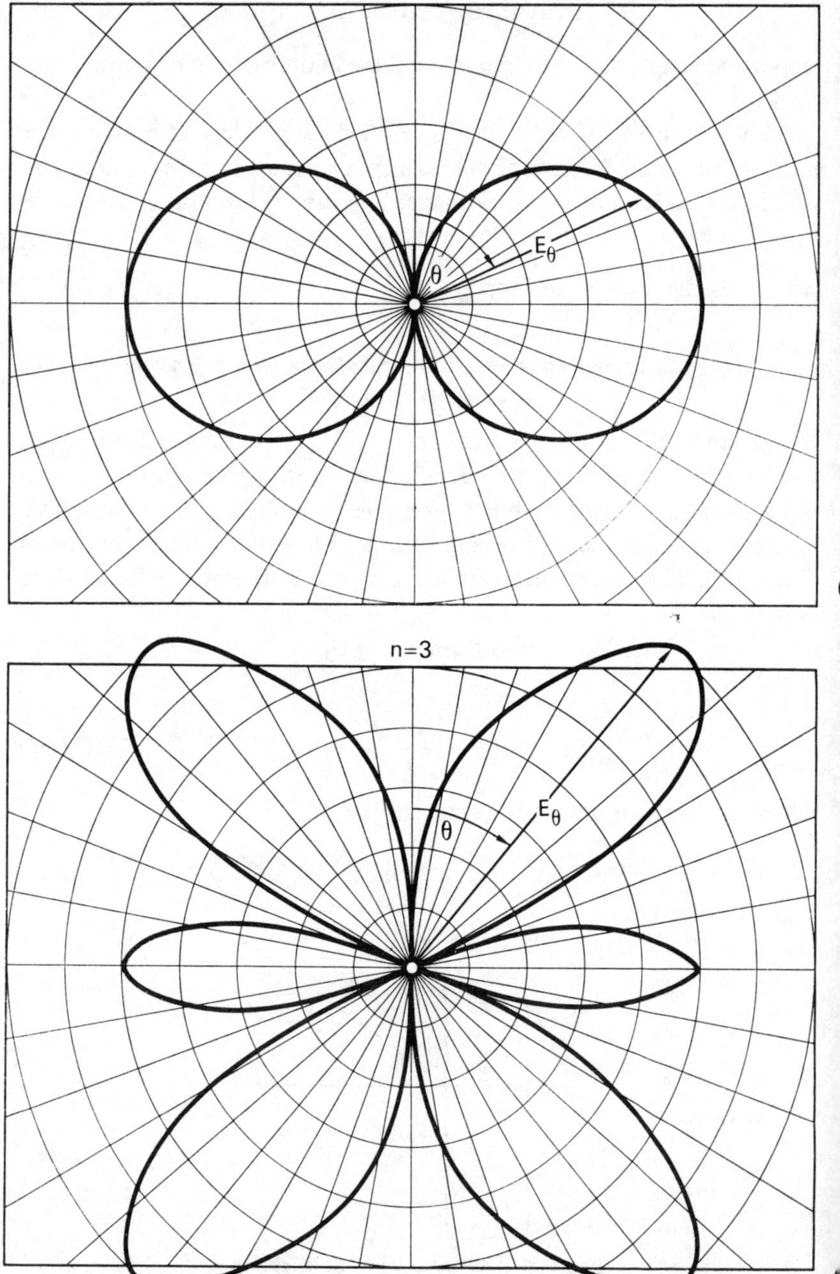

FIG. 11.6. The field of a linear antenna.

The field at a distant point must be obtained by integration

$$E_\theta = j\frac{1}{2}\sqrt{\left(\frac{\mu_0}{\varepsilon_0}\right)}\frac{\sin\theta}{\lambda}\,e^{j(\omega t-\beta r_0)}\int_{-h}^{+h}\frac{I(z)}{r_0}\,e^{j\beta z\cos\theta}\,dz. \qquad (11.34)$$

As an example suppose the current varies as $I_0 \cos \beta z$ and suppose $2h$ is an odd number of half wavelengths. Then the limits of the integral in eqn. (11.34) must be $\pm n\lambda/4$, where n is an odd number. We can make use of the relationship

$$\int e^{az}\cos bz\,dz = \frac{e^{az}}{a^2+b^2}(a\cos bz + b\sin bz), \qquad (11.35)$$

so that the integral in eqn. (11.34) becomes

$$\int_{-n(\lambda/4)}^{+n(\lambda/4)}\frac{I_0}{r_0}\cos\beta z\,e^{j\beta z\cos\theta}\,dz = \frac{I_0}{r_0}\,\frac{\cos[n(\pi/2)\cos\theta]}{\beta\sin^2\theta}$$

and the magnitude of E_θ is given by

$$|E_\theta| = \frac{1}{2\pi}\sqrt{\left(\frac{\mu_0}{\varepsilon_0}\right)}\frac{I_0}{r_0}\,\frac{\cos[n(\pi/2)\cos\theta]}{\sin\theta}. \qquad (11.36)$$

Figure 11.6 shows the variation of E_θ with θ for $n = 1$ and $n = 3$. It will be noticed that for $n = 3$ the antenna has become more directive, with its field concentrated in various beams.

11.4. The Influence of a Conducting Plane on the Radiation Field

In § 10.7 we discussed the reflection of a plane wave from the ground and noticed that in spite of the relatively poor conductivity of soil in comparison with metal, it is frequently permissible to regard the ground as a perfectly reflecting sheet. In this chapter we have been calculating the radiation from known current and charge distributions, and it is important to find the effect of the ground in such cases.

To do so we make use of the theory of images which has already been mentioned in electrostatics and magnetostatics. We replace the effect of the currents and charges induced in the ground by an equivalent image source located below the ground surface. We then add the field of the image source to the field of the antenna to obtain the resultant field of an antenna above a conducting plane. Consider this with reference to Fig. 11.7. The image of a charge in a perfect conductor is of equal strength and opposite sign as shown in Fig. 11.7a. This is so because such an arrangement gives zero electric field strength along the conducting surface and is

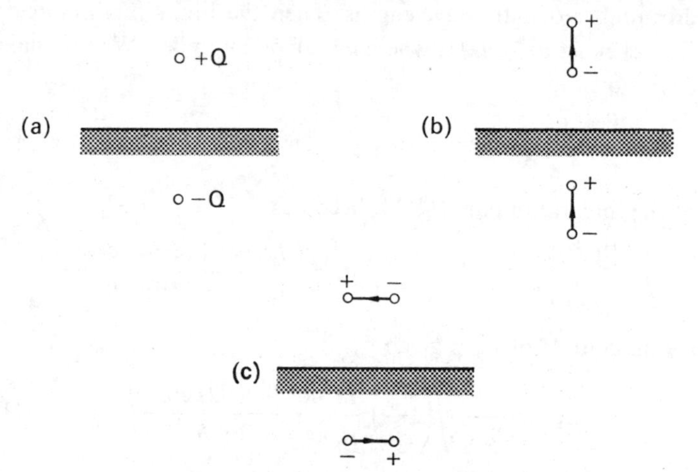

FIG. 11.7. Images of charges and dipoles.

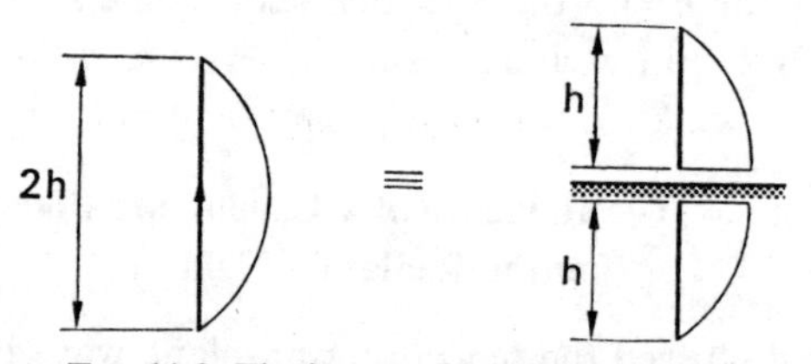

FIG. 11.8. The image of a linear antenna.

thus correct both for constant charges and charges varying in magnitude. When the charges are altering we have to take the dipole as the fundamental unit, and Fig. 11.7b shows the image of a vertical dipole. Inspection shows that this is the correct image as far as the charges are concerned. To make sure that there is also zero tangential field at the surface of the ground due to the current element, we refer to eqn. (11.9) and note that this condition is satisfied. A similar investigation shows that the image of a horiontal dipole is as shown in Fig. 11.7c.

Referring now to § 11.3 we see that the cosine distribution of current on an isolated antenna of height $2h$ could equally be achieved by feeding the antenna with current at its foot and placing it vertically on a conducting plane as illustrated in Fig. 11.8. Thus the ground doubles the effective height of such an antenna.

11.5. Radiation Pattern and Power Gain of Antenna Arrays

In § 11.3 we found the distant field of a single antenna and noticed a considerable change in the radiation pattern when the length of the antenna was changed from $\lambda/2$ to $3\lambda/2$. The pattern for the shorter antenna consists of two broad beams of radiation, whereas the longer antenna has six beams which are very much narrower. We can regard the longer antenna as being made up of an array of three half-wave antennas placed end to end, the current in the middle antenna being in the opposite direction to the current in the outer ones. It is clear that the radiation pattern is strongly affected by the interaction of the three members of the array.

This is potentially a very useful property of arrays. Often it is required to have a strong signal in a particular direction only. Power radiated in other direction is then not only wasteful but also highly undesirable because it causes interference. The construction of highly directive arrays is something of an art as well as a science, but it is worth while to investigate the general principles.

Consider two antennas with co-phased currents parallel to each other and at a distance g. Figure 11.9 shows them in plan, and we shall examine the radiation pattern in the equatorial plane of the antennas. The path difference of the radiated field due to the spacing g is $g \sin\phi$, where ϕ is measured from the normal to the array. This path difference corresponds to a phase angle $(g \sin\phi)\, 2\pi/\lambda$, and the resultant field can be obtained from the phasor diagram of Fig. 11.10 as

$$E = 2E_0 \cos(g\pi \sin\phi/\lambda). \tag{11.37}$$

Figure 11.11 shows the polar diagram of this array with $g = \lambda/2$. It is the same as the polar diagram of the antenna, Fig. 11.5, but the angle has a different significance. Clearly the polar diagram in the equa-

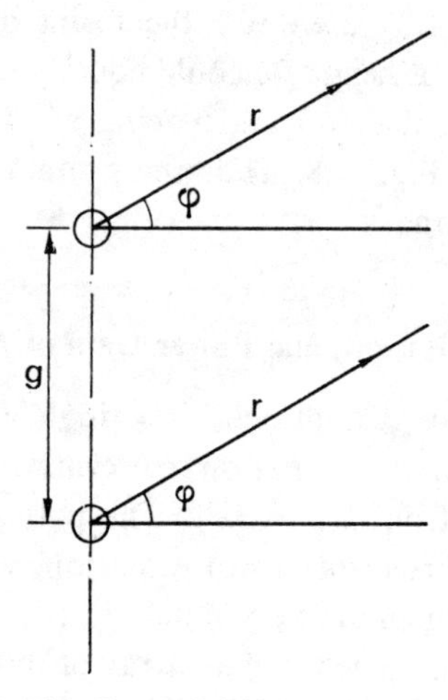

FIG. 11.9. The distant field of two parallel antennas.

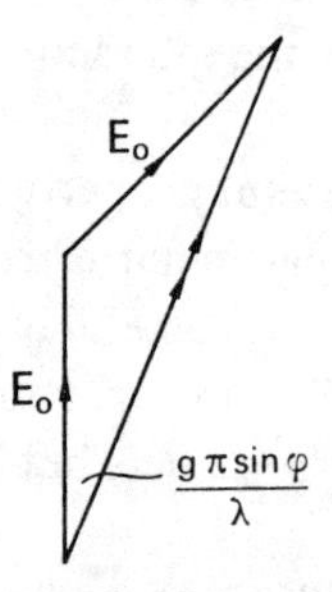

FIG. 11.10. The resultant field of two parallel antennas.

torial plane for a single antenna would be a circle, because it would radiate equally in all directions. Thus the addition of the second antenna has narrowed the beam and the array does not radiate at all in the direction $\phi = \pm\pi/2$. The half-wave spacing of the two members produces a phase angle of 180° in these directions and the field of one antenna is cancelled by the other. If the spacing between the antennas is increased, the number of "lobes" of the polar diagram increases. Figure 11.2 shows the diagram for two antennas spaced at $g = \lambda$. The polar diagram can also be changed by introducing a phase-angle between the currents in the antennas.

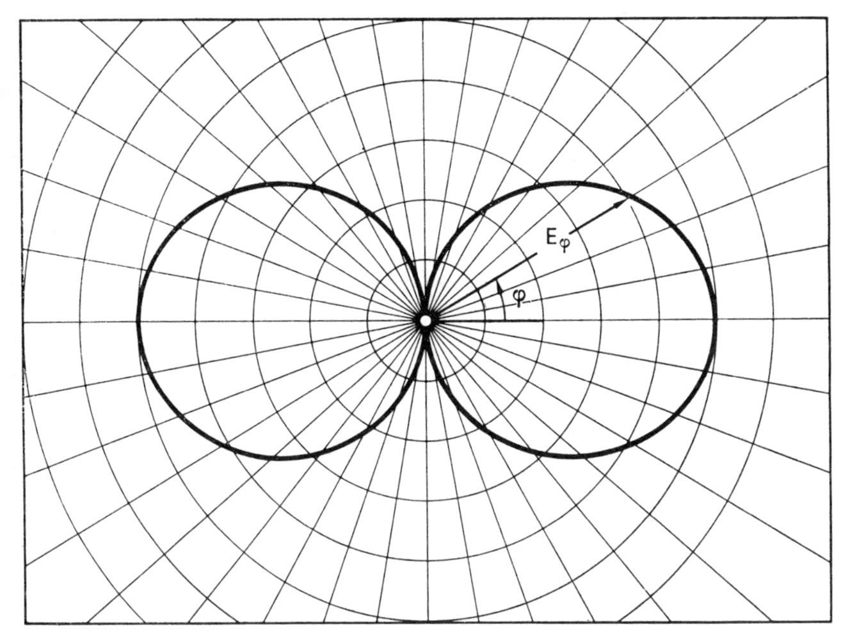

FIG. 11.11. Polar diagram of the field of two parallel antennas a half wave-length apart.

The effect of introducing more members into the array is very marked. Suppose there are N co-phased currents spaced at $g = \lambda/2$. Figure 11.13 shows the phasor diagram. By simple geometry we see that the resultant field E is given by

$$\frac{E}{NE_0} = \frac{\sin[(N\pi \sin\phi/2)]}{N\sin(\pi\sin\phi/2)} \tag{11.38}$$

The field is zero on bearings given by

$$\frac{N\pi\,\sin\phi}{2} = \pi,\ 2\pi,\ 3\pi,\ \ldots,$$

$$\sin\phi = \frac{2}{N},\ \frac{4}{N},\ \frac{6}{N},\ \ldots \tag{11.39}$$

Thus the larger N is, the more individual beams or *side lobes* there will be.

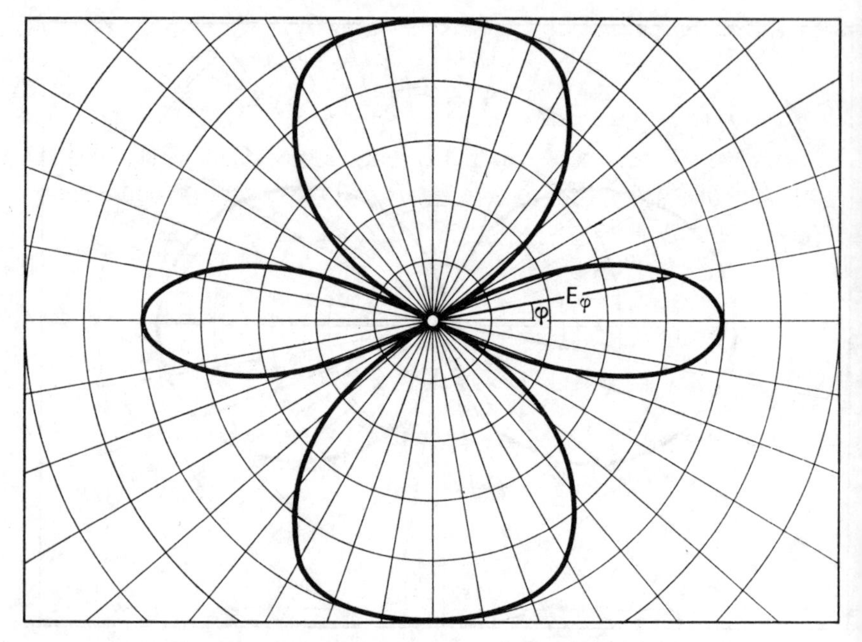

FIG. 11.12. Polar diagram of the field of two parallel antennas a wavelength apart.

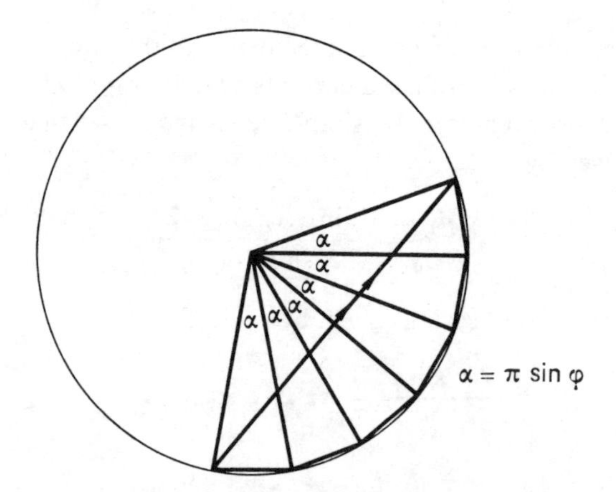

FIG. 11.13. Phasor diagram for array of N antennas.

If N is large we can take the first zero as given by

$$\sin\phi = \phi = \frac{2}{N}. \tag{11.40}$$

Thus the main beam can be concentrated into a very small angle and the array can be made highly directive. Figure 11.14 shows the radiation pattern for an array of 16 members.

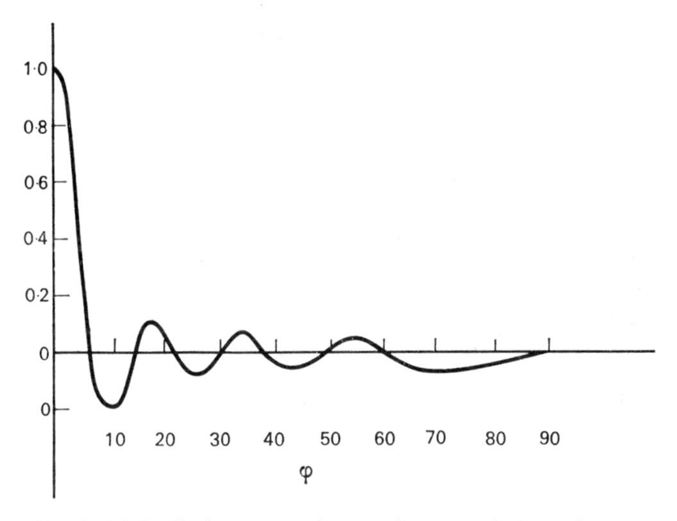

FIG. 11.14. Radiation pattern for curtain array of 16 members.

Associated with such a pattern there is a considerable saving in energy. The energy radiated from the array is largely confined to the main beam; the side lobes contribute very little, especially since the power is proportional to the square of the electric field strength.

The *power gain* of an array is defined as the ratio

$$\frac{\text{Power required by a single antenna}}{\text{Power required by the array}}$$

for the same maximum field strength. The power gain can be calculated either by the back e.m.f. method or by the Poynting vector. Since we are dealing with the distant field we shall use the latter method. As an example we choose the array of 16 members spaced at $g = \lambda/2$. So far we have been concerned with the radiation in the equatorial plane, and in order to avoid the complication of the length of the antennas and the form of the current distribution along this length, we shall take the Poynting integral over a cylindrical surface as illustrated in Fig. 11.15. The squares of the field distribution are given in Fig. 11.16 and the power gain is therefore the ratio of the areas shown in Fig. 11.16a and b. In the example the power gain is approximately 22.

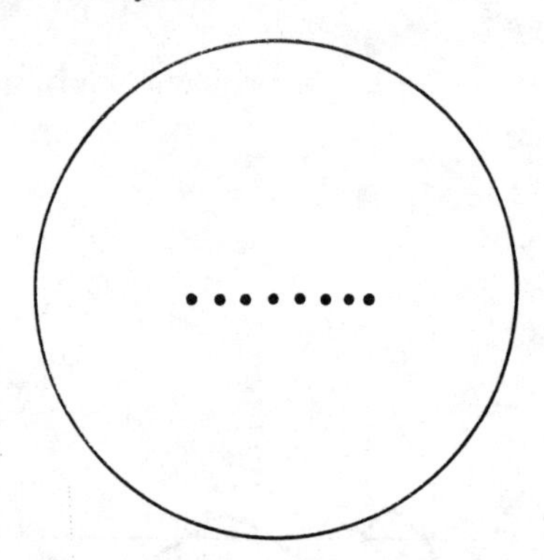

FIG. 11.15. Poynting integral for an array.

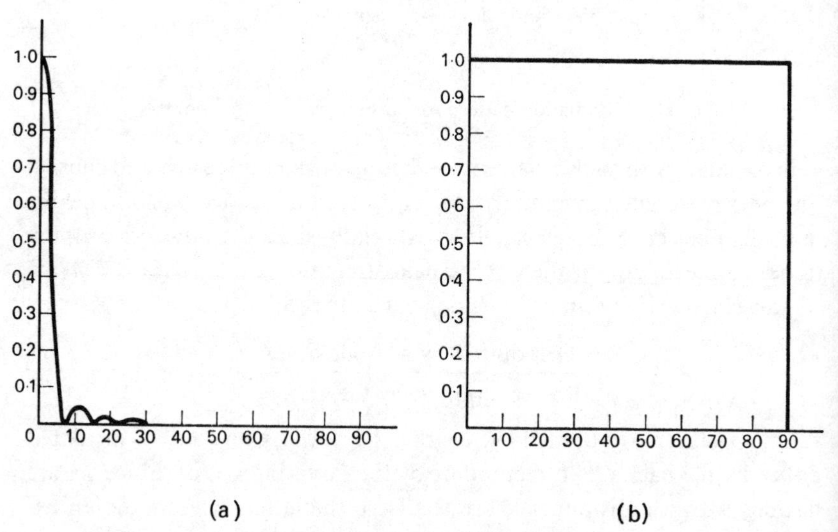

FIG. 11.16. Power gain of array.

11.6. Radiation from an Aperture

At microwave frequencies (say 10^9–10^{12} Hz) it is inconvenient to construct directional arrays by energizing a large number of separate elements. Instead it becomes possible to use reflectors and optical techniques. In such antennas the distribution of currents and charges is unknown, but it is possible to estimate the field at the surface of an aperture. We must, therefore, consider how to calculate the radiation pattern from the fields at the aperture. Let us consider the problem in general terms. Figure 11.17

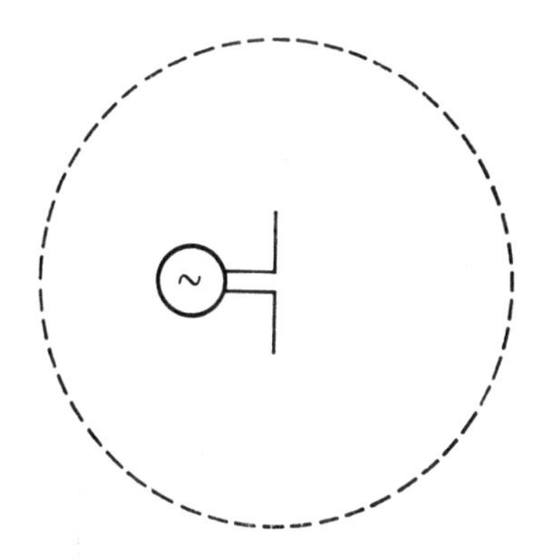

FIG. 11.17. Radiating source surrounded by a surface.

shows a radiating source surrounded by a surface s. The field outside s will be undisturbed if it is terminated by the correct current and charge distributions on s, in the same way in which we terminated the electric and magnetic field in §§ 5.4 and 7.4. As far as the region outside s is concerned we can therefore replace the source inside s by a surface distribution on s, but it must be remembered that it is the total surface distribution on s which replaces the effect of the source inside s. Unfortunately, this presents a difficulty when we come to discuss the problem of an aperture. Consider Fig. 11.18 which shows a surface s made up of s_1 and s_2. The part s_1 is impervious to radiation, but s_2 is an aperture. Can we find the radiation pattern by considering s_2 in isolation?

In principle we can do so, if we know the field distribution on the surface of s_2, because on s_1 this distribution is zero. So the total distribution on s would be correctly specified by a knowledge of s_2 only. But the problem of finding the distribution on s_2 is full of difficulties, particularly near the edges. We therefore have to assume a simplified field distribution on s_2 and to ignore the edge effects. This assumption is reasonable, if the dimensions of the aperture are considerably larger than the electromagnetic wavelength.

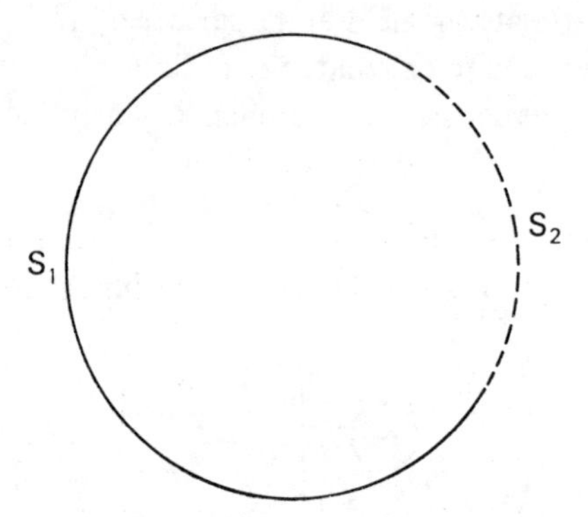

FIG. 11.18. A surface with an aperture.

To find the radiation pattern we replace the tangential H by a surface distribution of electric current and the normal H by a surface pole density. Also we replace tangential E by a surface distribution of magnetic current and normal E by a surface charge density. To ensure that the continuity equations are satisfied there would also have to be electric line charge and magnetic pole strength around the edges of the aperture, but the effect of these is often neglected. This means that the fields determined from the simple surface sources above do not in general satisfy Maxwell's equations.

The reader may question the necessity for both magnetic and electric sources, since magnetic currents and poles do not occur in nature. The reason for their introduction into our discussion is inherent in the fact that we seek to terminate the field at a surface of zero thickness. This implies discontinuities in the field, which have to be replaced by magnetic sources. In a practical case the field would penetrate through the surface. However, the idea of a source distribution on a surface is extremely convenient for the purpose of calculation.

We can now define the sources as follows. The line densities of electric and magnetic surface current are given by

$$\left.\begin{aligned} \mathbf{J} &= \hat{\mathbf{n}} \times \mathbf{H}, \\ \mathbf{J}^* &= -\hat{\mathbf{n}} \times \mathbf{E}, \end{aligned}\right\} \tag{11.41}$$

where $\hat{\mathbf{n}}$ is the outward normal to the surface. Similarly, the charge and pole densities are given by

$$\left.\begin{aligned} q &= \varepsilon\hat{\mathbf{n}} \,.\, \mathbf{E}, \\ q^* &= \mu\hat{\mathbf{n}} \,.\, \mathbf{H}. \end{aligned}\right\} \tag{11.42}$$

From the fields on s we obtain the sources, and from the sources the delayed potentials for the sources. The potentials for the electric sources are:

$$\left.\begin{aligned} \mathbf{A} &= \frac{\mu}{4\pi}\iint \frac{|\mathbf{J}|}{r} \,.\, d\mathbf{s}, \\ V &= \frac{1}{4\pi\varepsilon}\iint \frac{|\varrho|}{r}\, ds. \end{aligned}\right\} \tag{11.43}$$

The potentials for the magnetic sources are defined as:

$$\left.\begin{aligned} \mathbf{A}^* &= \frac{\varepsilon}{4\pi}\iint \frac{|\mathbf{J}^*|}{r} \,.\, d\mathbf{s}, \\ V^* &= \frac{1}{4\pi\mu}\iint \frac{|\varrho^*|}{r}\, ds. \end{aligned}\right\} \tag{11.44}$$

Remembering that $\mathbf{H} = (1/\mu)$ curl $\mathbf{A}$ in the absence of magnetic sources, we can write

$$\left.\begin{aligned} \mathbf{E} &= -\frac{\partial \mathbf{A}}{\partial t} - \text{grad}\, V - \frac{1}{\varepsilon}\,\text{curl}\, \mathbf{A}^*, \\ \mathbf{H} &= -\frac{\partial \mathbf{A}^*}{\partial t} - \text{grad}\, V^* + \frac{1}{\mu}\,\text{curl}\, \mathbf{A}, \end{aligned}\right\} \tag{11.45}$$

where the curl terms give the interaction between the electric and magnetic fields. As a simple example consider an element $\delta x\, \delta y$ of the surface of a plane wave on which there is constant density of electric and magnetic

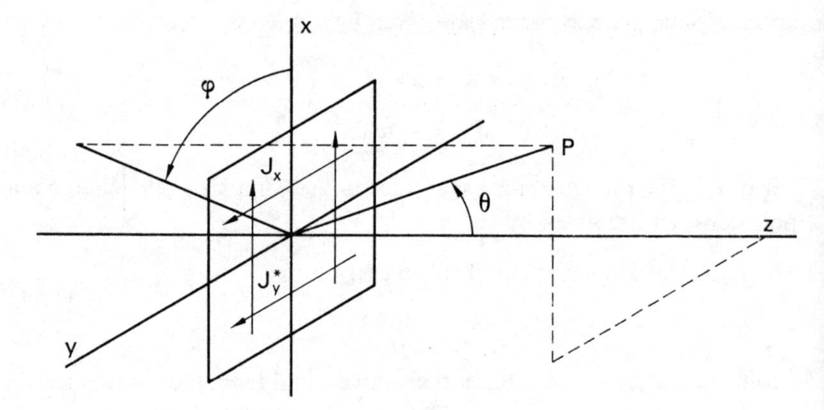

FIG. 11.19. Radiation from a Huygens' source.

current, and no charge or pole strength (Fig. 11.19). Such an element is called a Huygens' source. Consider the field at the point P, having spherical coordinates r, θ, ϕ, when the element is located at the origin.

In terms of the surface fields $E_x = E_0$ and $H_y = H_0$ we have

$$\left. \begin{aligned} J_x &= -H_0, \\ J_y^* &= E_0, \end{aligned} \right\} \tag{11.46}$$

whence

$$\left. \begin{aligned} A_x &= -\frac{\mu}{4\pi} H_0 \,\delta x\, \delta y \frac{e^{j(\omega t - \beta r)}}{r}, \\ A_y^* &= \frac{\varepsilon}{4\pi} E_0 \,\delta x\, \delta y \frac{e^{j(\omega t - \beta r)}}{r} \end{aligned} \right\} \tag{11.47}$$

also

$$\left. \begin{aligned} q = q^* &= 0, \\ V = V^* &= 0. \end{aligned} \right\} \tag{11.48}$$

Since the Huygens' source is part of a plane wave,

$$E_0 = \sqrt{\left(\frac{\mu}{\varepsilon}\right)} H_0. \tag{11.49}$$

Hence after some reduction the terms in $1/r$ are given by

$$\left.\begin{aligned} E_\theta &= j\,\frac{E_0\,\delta x\,\delta y}{2\lambda r}\,(\cos\phi\cos\theta+\cos\phi)\,e^{j(\omega t-\beta r)},\\ E_\psi &= -j\,\frac{E_0\,\delta x\,\delta y}{2\lambda r}\,(\sin\phi+\sin\phi\cos\theta)\,e^{j(\omega t-\beta r)}. \end{aligned}\right\} \tag{11.50}$$

The corresponding magnetic fields are given by

$$\left.\begin{aligned} H_\phi &= \sqrt{\left(\frac{\varepsilon}{\mu}\right)}\,E_\theta,\\ H_\theta &= \sqrt{\left(\frac{\varepsilon}{\mu}\right)}\,E_\phi. \end{aligned}\right\} \tag{11.51}$$

A useful check is to find the field behind the surface. This is given by $\theta = \pi$, which gives $E_\theta = E_\phi = 0$. Thus the surface radiates outwards only, which is as it should be.

The directive properties of an aperture can be illustrated by considering a Huygens' source of finite dimensions. Let such a source be located in the xy-plane and let its width in the direction x be $2a$ and in the y-direction $2b$. Let us consider the electric field in the plane given by $\phi = 0$. We then have

$$\begin{aligned} E_\theta &= j\,\frac{E_0\,(1+\cos\theta)}{2\lambda r}\,e^{j(\omega t-\beta r)}\int_{-a}^{+a}\int_{-b}^{+b} e^{-j\beta\sin\theta\,x}\,dx\,dy\\ &= E_0\,\frac{(1+\cos\theta)}{2\lambda r}\,e^{j(\omega t-\beta r)}\,(2b)\,\frac{2\sin(a\beta\sin\theta)}{\beta\sin\theta}\\ &= \frac{k}{r}\,e^{j(\omega t-\beta r)}\,(1+\cos\theta)\,\frac{\sin(a\beta\sin\theta)}{a\beta\sin\theta}. \end{aligned} \tag{11.52}$$

In the neighbourhood of $\theta = 0$ the term $(1 + \cos\theta)$ varies relatively slowly and the radiation pattern is dominated by the term

$$\frac{\sin(a\beta\sin\theta)}{a\beta\sin\theta} = \frac{\sin|2\pi\sin\theta\,(a/\lambda)|}{2\pi\sin\theta\,(a/\lambda)} = \frac{\sin u}{u}. \tag{11.53}$$

This function is plotted in Fig. 11.20. The first zero of the function shows half the width of the main beam to be given by the angle θ, where

$$\sin\theta = \frac{\lambda}{2a}. \qquad (11.54)$$

Thus the beam is narrowed by increasing the width of the aperture.

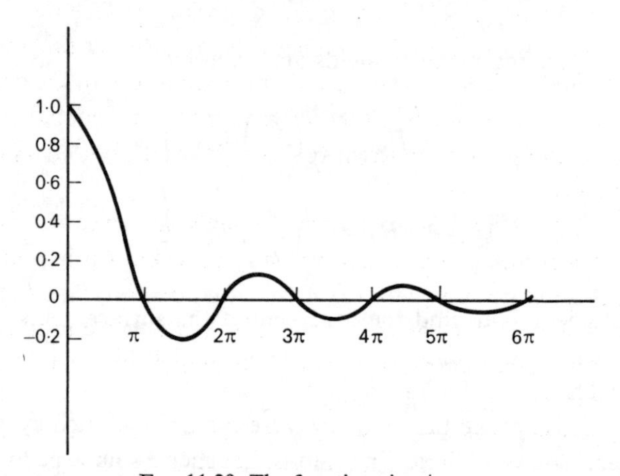

FIG. 11.20. The function sin u/u.

11.7. Waves Guided by Metallic Boundaries

So far in this chapter we have been discussing the radiation of electromagnetic energy from various types of antenna. We now need to look at the problem of conveying the energy from place to place without losing it on the way by radiation. At low frequencies ordinary circuit elements serve this purpose, but if the electromagnetic wavelength is small enough to be of the order of the size of the apparatus, radiation becomes troublesome. At microwave frequencies† (10^9–10^{12} Hz, λ = 30 cm/0·3 mm) the problem is overcome by piping the energy through *waveguides*. The radiation is confined within highly conducting boundaries, and apart from some ohmic loss in the conductors the energy is conveyed substantially without loss.

† For a full treatment see the companion volume in this series, *Microwaves* by A. J. Baden Fuller, Pergamon, 1969.

The most convenient method for the analysis of waveguides is the inverse of the method we used with antennas. Instead of starting with charge and current we discuss waveguides in terms of the field inside the guide. Of course there will be charges and currents on the conducting boundaries and these are the sources of the field. But it is easier to predict the field pattern rather than the source distributions. The reason is a mathematical one. If we assume perfectly conducting walls the electric field must be normal to the walls and the magnetic field must be tangential. With metallic boundaries this is a very good approximation for the field pattern, even if the conductivity is finite, since very little tangential electric field is needed to cause a large current to flow. Hence the boundary conditions for E and H are simple.

Such a state of affairs is ideal where the field pattern can be obtained by the solution of differential equations subject to known boundary conditions. We therefore start with the wave equations for E and H and then look for possible solutions within the waveguide. If we need the charges and currents we obtain them from the normal electric field and the tangential magnetic field at the surface.

But before launching into mathematical analysis we shall follow our usual practice and discuss the matter in general terms. The simplest wave is the one discussed in § 10.4, in which both the **E** vector and the **H** vector are perpendicular to the direction of propagation. Types of wave are called "modes" and the plane wave is the TEM (transverse electric magnetic) mode. Figure 11.21 shows such a wave confined between parallel conducting planes. It is clear that the boundary conditions at the surfaces are correct. To check, it is worth while looking at the charges and currents. On the surface of the top conductor there will be a charge distribution of

$$q = -D_y \quad \text{C/m}^2 \tag{11.55}$$

and a current distribution

$$J_z = H_x \quad \text{A/m}. \tag{11.56}$$

The continuity equation of q and J is

$$\operatorname{div} \mathbf{J} + \frac{\partial q}{\partial t} = 0. \tag{11.57}$$

Hence

$$\frac{\partial H_x}{\partial z} = \frac{\partial D_y}{\partial t}. \tag{11.58}$$

But this is the Maxwell equation curl $\mathbf{H} = \partial \mathbf{D}/\partial t$, and thus we have shown that the fields can be terminated by this charge and current. Thus the TEM mode is a possible one. The direction of propagation of the wave is

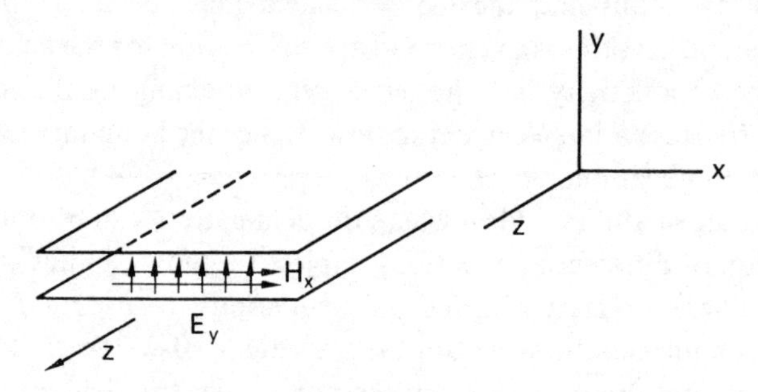

FIG. 11.21. A plane wave between conducting boundaries.

arbitrary. By choosing E_y and H_x positive we have chosen a wave for which the Poynting vector lies in the z-direction. The type of wave of Fig. 11.21 could be achieved practically by a wave between two large cylindrical conductors as illustrated in Fig. 11.22. This arrangement is similar to a con-

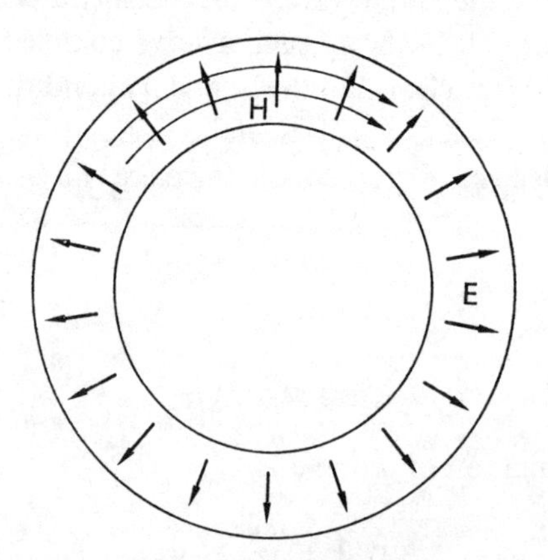

FIG. 11.22. A TEM wave between cylindrical boundaries.

centric cable. The TEM mode is often referred to as the transmission line mode.

Since E must always be perpendicular to the boundary, it is not possible to have a TEM wave in a tubular waveguide. In such guides different wave patterns are needed and these are called waveguide modes. We shall show that many such modes are possible, but before doing so let us discuss some common features of these modes. Figure 11.23 shows a

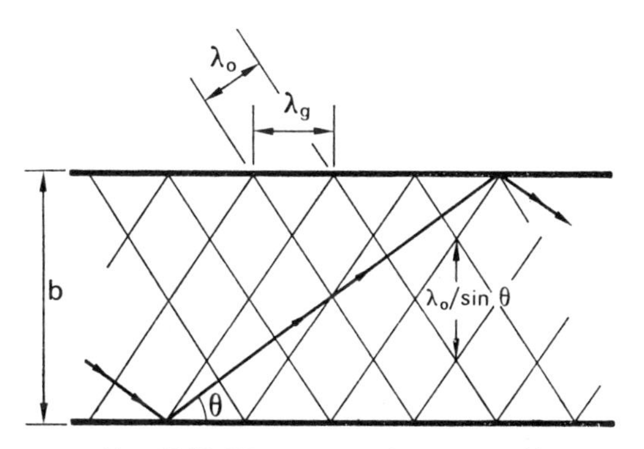

FIG. 11.23. Wave pattern in a waveguide.

wave which impinges on the walls of a waveguide. Lines of equal phase are also drawn perpendicular to the direction of the wave, and these lines are spaced at a distance of a wavelength λ_0. The angle of incidence of the wave is θ and the spacing between the boundaries is b. In order that there shall be the same boundary conditions (e.g., zero tangential electric field) at both the top and bottom boundaries, an integral number of half wavelengths must be contained between these boundaries. Thus from Fig. 11.23

$$\frac{n\lambda_0}{2 \sin \theta} = b. \tag{11.59}$$

When θ is 90° the wave will not propagate. Thus for any value of n the smallest width between the plates which makes propagation possible is

$$b = \frac{n\lambda_0}{2}. \tag{11.60}$$

For a given b this means that $\lambda_0 = 2b/n$ is the longest wavelength. It is called the cut-off wavelength and denoted by λ_c. This also determines the *cut-off frequency*

$$f_c = \frac{nc}{2b} = \frac{n}{2b\sqrt{\mu\varepsilon}}. \tag{11.61}$$

The wavelength in the direction of the waveguide is called the *waveguide wavelength* and is given by

$$\lambda_g = \frac{\lambda_0}{\cos\theta} = \frac{\lambda_0}{\sqrt{|1-(\lambda_0/\lambda_c)^2|}}. \tag{11.62}$$

Associated with λ_g is the *waveguide phase velocity*

$$v_p = \frac{c}{\cos\theta} = \frac{c}{\sqrt{|1-(\lambda_0/\lambda_c)^2|}}. \tag{11.63}$$

Thus the phase velocity is greater than c. However, this is not the velocity at which energy is transmitted, because the waves follow the zigzag path shown in Fig. 11.23. If we call this velocity the group velocity we have

$$v_g = c\cos\theta \tag{11.64}$$

and

$$v_p v_g = c^2. \tag{11.65}$$

11.8. Waves in a Rectangular Waveguide

Let us now discuss in detail the important practical case of waves in a rectangular waveguide. The wave equation for the electric field is

$$\nabla^2\mathbf{E} = \mu\varepsilon\frac{\partial^2\mathbf{E}}{\partial t^2}. \tag{11.66}$$

This vector equation consists of three scalar equations in E_x, E_y, and E_z. Let us choose z as the direction along the guide (Fig. 11.24) and examine E_z. Equation (11.66) becomes

$$\frac{\partial^2 E_z}{\partial x^2} + \frac{\partial^2 E_z}{\partial y^2} + \frac{\partial^2 E_z}{\partial z^2} = \mu\varepsilon\frac{\partial^2 E_z}{\partial t^2}. \tag{11.67}$$

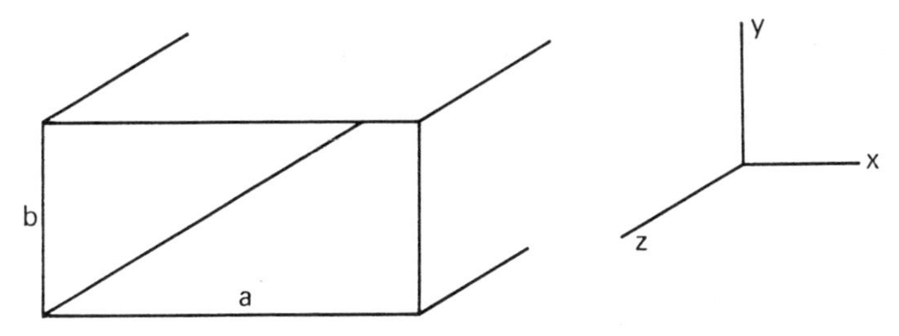

FIG. 11.24. A rectangular waveguide.

This equation can be solved by the method of separation of variables discussed in § 4.3.1. We assume that the wave is propagating in the z-direction, so that the dependence on t and z can be written

$$E_z = E_0 \, e^{j(\omega t - \beta z)} \tag{11.68}$$

so that

$$\frac{\partial^2}{\partial z^2} = -\beta^2 \quad \text{and} \quad \frac{\partial^2}{\partial t^2} = -\omega^2. \tag{11.69}$$

The x and y dependence of E_z can be chosen as

$$\left.\begin{matrix}\sin \\ \cos\end{matrix}\right\} px, \quad \left.\begin{matrix}\sin \\ \cos\end{matrix}\right\} qy,$$

so that from eqn. (11.67)

$$-p^2 - q^2 - \beta^2 = -\omega^2 \mu \varepsilon. \tag{11.70}$$

There will be propagation if β is real, but if β is imaginary the wave will attenuate and not propagate. The limiting case is $\beta = 0$. This defines the cut-off frequency

$$\omega_c = \sqrt{\left(\frac{p^2 + q^2}{\mu \varepsilon}\right)} \tag{11.71}$$

and the corresponding maximum wavelength

$$\lambda_c = \frac{2\pi}{\sqrt{(p^2 + q^2)}}. \tag{11.72}$$

The propagating wavelength λ_g is given by

$$\lambda_g = \frac{2\pi}{\beta} \tag{11.73}$$

and the wavelength of a plane wave is

$$\lambda = \frac{2\pi}{\omega\sqrt{(\mu\varepsilon)}}. \tag{11.74}$$

Hence from eqn. (11.71)

$$\frac{1}{\lambda_c^2} + \frac{1}{\lambda_g^2} = \frac{1}{\lambda^2} \tag{11.75}$$

or

$$\lambda_g = \frac{\lambda}{\sqrt{|1-(\lambda/\lambda_c)^2|}}. \tag{11.76}$$

We must now adapt the constants p and q to fit the boundary conditions. Since E_z must be zero at $x = y = 0$ and $x = a$ and $y = b$ (Fig. 11.24), the expression for E_z becomes

$$E_z = E_0 \sin\frac{m\pi x}{a} \sin\frac{n\pi y}{b} e^{j(\omega t - \beta z)}, \tag{11.77}$$

where m and n are integers. We shall have to examine whether Maxwell's equations permit expressions for E_z and E_y which will also satisfy the boundary conditions.

Before doing so let us examine the general dependence of the field components on each other. There are three components for the electric field and three for the magnetic field. We can express the dependence of E_x on the other five components as

$$E_x = f(E_y, E_z, H_x, H_y, H_z). \tag{11.78}$$

This, however, can be simplified considerably. We notice that **E** and **H** are related to each other by the curl equations. Thus E_x does not depend on H_x. Also the three components of **E** are related by the divergence relation for the electric field. Thus we can eliminate E_y. Hence

$$E_x = f'(E_z, H_y, H_z). \tag{11.79}$$

But H_y depends on E_x and E_z, so that, finally,

$$E_x = f''(E_z, H_z). \tag{11.80}$$

Similarly, E_y, H_x, and H_y can be shown to depend only on E_z and H_z. Moreover, all the equations are linear and we therefore arrive at two independent sets of waves, one in terms of E_z and the other in terms of H_z. For the first set H_z is zero and such waves are called TM modes (transverse magnetic field). Similarly, waves which have zero E_z are called TE modes (transverse electric).

The dependence on E_z and H_z is written out in full as follows:

$$\left.\begin{aligned}
E_x &= -\frac{j}{p^2+q^2}\left(\beta\frac{\partial E_z}{\partial x} + \omega\mu\frac{\partial H_z}{\partial y}\right),\\
E_y &= \frac{j}{p^2+q^2}\left(-\beta\frac{\partial E_z}{\partial y} + \omega\mu\frac{\partial H_z}{\partial x}\right),\\
H_x &= \frac{j}{p^2+q^2}\left(\omega\varepsilon\frac{\partial E_z}{\partial y} - \beta\frac{\partial H_z}{\partial x}\right),\\
H_y &= -\frac{j}{p^2+q^2}\left(\omega\varepsilon\frac{\partial E_z}{\partial x} + \beta\frac{\partial H_z}{\partial y}\right).
\end{aligned}\right\} \tag{11.81}$$

For TM modes we put $H_z = 0$. By the use of eqn. (11.77) we obtain:

$$\left.\begin{aligned}
E_x &= -\frac{j\beta p E_0}{p^2+q^2}\cos px \sin qy\, e^{j(\omega t-\beta z)},\\
E_y &= -\frac{j\beta q E_0}{p^2+q^2}\sin px \cos qy\, e^{j(\omega t-\beta z)},\\
E_z &= E_0 \sin px \sin qy\, e^{j(\omega t-\beta z)},\\
H_x &= \frac{j\omega\varepsilon q E_0}{p^2+q^2}\sin px \cos qy\, e^{j(\omega t-\beta z)},\\
H_y &= -\frac{j\omega\varepsilon p E_0}{p^2+q^2}\cos px \sin qy\, e^{j(\omega t-\beta z)},\\
H_z &= 0,
\end{aligned}\right\} \tag{11.82}$$

where $p = m\pi/a$ and $q = n\pi/b$ as before.

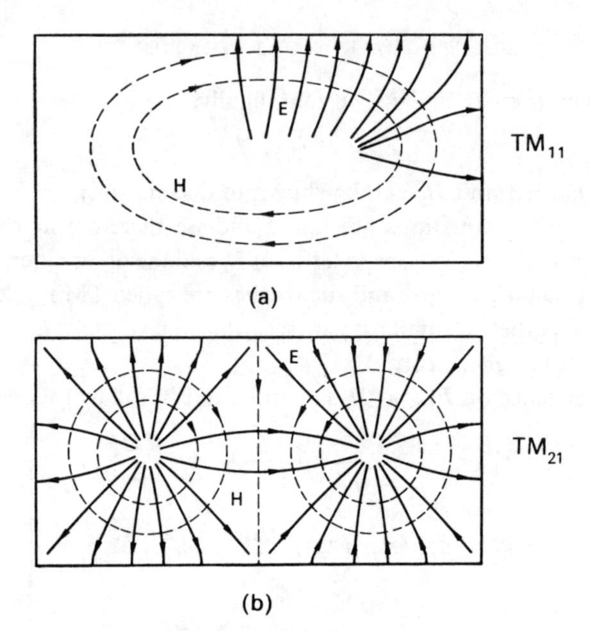

FIG. 11.25. TM modes.

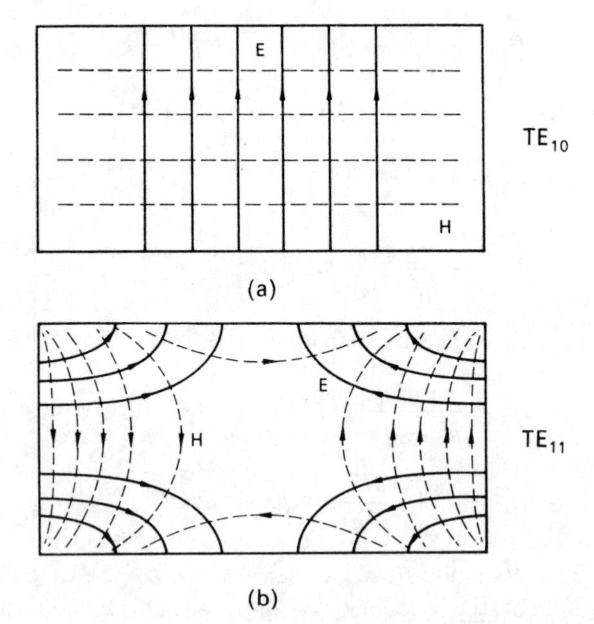

FIG. 11.26. TE modes.

We notice that tangential E and normal H are zero at the boundaries, so that the boundary conditions are satisfied. Hence eqn. (11.82) describes a possible set of wave patterns or modes. The modes are classified by the integers m and n. Thus if $m = 1$ and $n = 3$, we speak of the TM_{13} mode. The simplest mode is the TM_{11} mode, because there is no field at all if either p or q is zero. Reference to eqns. (11.71) and (11.72) shows that the lower modes correspond to lower frequencies. In any practical case there will be many modes present, depending on the way in which the wave is launched from a source. But the lowest mode is likely to be dominant and it is often called the dominant mode. Figure 11.25 illustrates the field pattern of a TM_{11} and TM_{21} mode.

For TE modes the fields are expressed in terms of H_z. We have:

$$\left.\begin{aligned}
E_x &= \frac{j\omega\mu q H_0}{p^2 + q^2} \cos px \sin qy \, e^{j(\omega t - \beta z)}, \\
E_y &= -\frac{j\omega\mu p H_0}{p^2 + q^2} \sin px \cos qy \, e^{j(\omega t - \beta z)}, \\
E_z &= 0, \\
H_x &= \frac{j\beta p H_0}{p^2 + q^2} \sin px \cos qy \, e^{j(\omega t - \beta z)}, \\
H_y &= \frac{j\beta q H_0}{p^2 + q^2} \cos px \sin qy \, e^{j(\omega t - \beta z)}, \\
H_z &= H_0 \cos px \cos qy \, e^{j(\omega t - \beta z)}.
\end{aligned}\right\} \qquad (11.83)$$

The dominant mode is TE_{10} (or TE_{01}) for which E_x and H_y are zero, so that the only components of the field are E_y, H_x, and H_z. Figure 11.26 illustrates the TE_{10} and TE_{11} modes.

Summary

In this chapter we have examined the electromagnetic radiation from some simple distributions of alternating current and charge. These investigations provide an introduction to the study of radio transmitters. We have studied the field radiated from an array of elements and have also investigated radiation from an aperture. Finally, we have discussed the propagation of electromagnetic waves in metallic waveguides.

Exercises

11.1. In § 11.1 the radiation from an electric dipole has been analysed in terms of the potentials **A** and V. As an alternative use the Hertz vector defined in Exercise 10.8 to show that

$$\Pi = \frac{Q\mathbf{l}}{4\pi\varepsilon_0 r} e^{j(\omega t-\beta r)}$$

and derive the electric and magnetic fields from the Hertz vector.

11.2. Determine the magnetic field of an oscillating electric dipole and show that it can be divided into a radiation component and an induction component. At what distance are the two components equal in magnitude? At what distance is the radiation field (a) 100 times the induction field, (b) 1/100 times the induction field? Discuss these results in relation to the radiation from apparatus at power frequencies. [Ans.: $r = c/\omega = \lambda/2\pi$; $r = 15{\cdot}9\lambda$; $r = \lambda/629$ at 50 Hz; $\lambda \doteqdot 6000$ km.]

11.3. Ampère demonstrated the equivalence between *a* loop of constant current and a magnetic dipole as far as the *distant* field is concerned. What additional condition is needed to extend this equivalence to alternating currents? Use this equivalence principle to obtain the distant electric field of a loop of alternating current from the magnetic field of an electric dipole. [Ans.: The dimensions of the loop must be small compared to the electromagnetic wavelength; see eqns. (11.4) and (11.26).]

11.4. Determine the radiation resistance of a loop of radius a carrying current of frequency f, where $a \ll c/f$. [Ans.: $320\pi^6\,(a/\lambda)^4 = 320\pi^6\,(af/c)^4$.]

11.5. Explain carefully why the radiation from a vertical antenna close to the ground is strengthened while that of a horizontal antenna is weakened.

11.6. Show that the width of the main beam of a curtain array of antennas carrying co-phased current varies inversely as the number of members of the array. Estimate the magnitude of the maximum side lobe of a large curtain array of co-phased members spaced at $\lambda/2$ between members. [Ans.: 21%.]

11.7. Show that a curtain array of co-phased currents spaced at $\lambda/2$ has no side lobes if the magnitude of the currents follows the binomial coefficients, e.g. 1, 2, 1; 1, 3, 3, 1; 1, 4, 6, 4, 1. [*Hint:* Consider the field of two equal currents spaced at $\lambda/2$ and then proceed by superposition.]

11.8. Show that at a conducting boundary the ratio of the tangential component of magnetic field to the normal component is of the order of the ratio of the wavelength of the field to the skin-depth.

11.9. Explain what is meant by the *cut-off frequency* of a waveguide. Show that the lowest cut-off frequency for a TM mode is $\sqrt{2}$ times the lowest cut-off frequency for a TE mode if the cross-section of the waveguide is a square.

11.10. Show that in a waveguide operating in the TE_{10} mode there will be charge distributions of $q = \mp\varepsilon E_y$ C/m^2 and longitudinal currents $J_z = \pm H_x$ A/m on the top and bottom of the guide and also transverse currents $J_x = \mp H_z$ A/m on the top and bottom and $J_y = H_z$ A/m on the sides. Show also that the attenuation due to loss in the conducting walls will increase as the dimension b is reduced, the power transfer being kept constant.

CHAPTER 12

ELECTROMAGNETISM AND RELATIVITY

12.1. Inertial Frames and the Principle of Relativity

At the centre of the study of mechanics there is the idea of *inertia*. Newton's laws of motion are a consequence of assuming that matter has the property of inertia and that forces have to be applied to change the motion. Nowadays there is something almost obvious about this way of looking at motion, but we are being wise after the event. For 2000 years, from Aristotle to Galileo, force was associated with uniform motion and not with acceleration. The concept of inertia had not been discovered.

Coupled with this concept, but often overlooked, there is the idea of an *inertial frame* in which observations are being made. Newton's laws are true in an inertial frame. The surface of the earth furnishes a good approximation to such a frame because the forces due to the earth's rotation are small. By definition there are no forces due to the constant linear motion of such an inertial frame. Indeed, absolute velocity cannot be observed at all; only relative velocity is measurable. Moreover, if two inertial frames are moving relative to each other at constant velocity, the laws of mechanics will be identical in both frames. Consider Fig. 12.1 which shows two inertial frames Σ and Σ'. The frame Σ' moves with a constant velocity u

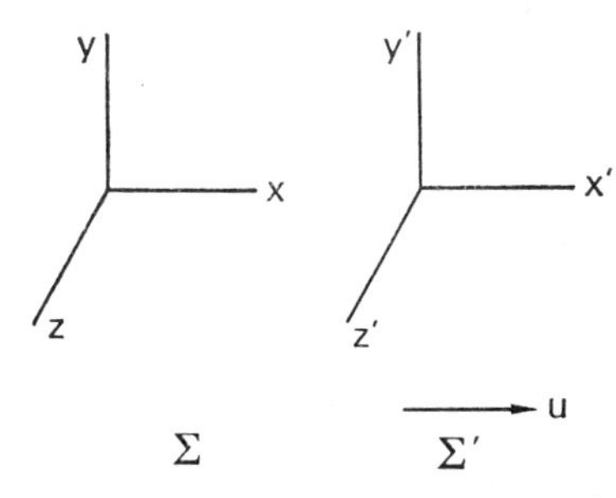

FIG. 12.1. Inertial frames in relative motion.

relative to Σ and u is parallel to the x- and x'-axes of the frames. The transformation equations for the coordinate systems are

$$\left.\begin{aligned} x' &= x - ut, \\ y' &= y, \\ z' &= z. \end{aligned}\right\} \tag{12.1}$$

If, moreover, it is assumed as obvious that $t' = t$, because time flows uniformly, we can translate the equations of motion from (x, y, z, t) coordinates to (x', y', z', t') coordinates. This is called a *Galilean* transformation in memory of Galileo. It can easily be seen that accelerations are not affected by the transformation. Thus *Newton's laws are invariant with respect to a Galilean transformation.* This statement is called the principle of Galilean relativity.

In this chapter we shall show that this principle does not apply to electrodynamics. This faces us with three possibilities.

(a) A relativity principle exists in mechanics but not in electrodynamics. There is an inertial frame in which the equations of electrodynamics have the form developed by Maxwell.
(b) A relativity principle exists in mechanics and electrodynamics, but Maxwell's formulation of electrodynamics is incorrect.
(c) A relativity principle exists in mechanics and electrodynamics but Newton's laws are incorrect.

The fourth possibility that mechanics and electrodynamics have different relativity principles has been dismissed because it would make nonsense of the formulation of electromagnetism in terms of the mechanical concepts of force and momentum. In any case we know that the so-called mechanical properties of matter are indissolubly linked with the interaction of charged particles.

The choice amongst the three possibilities (a), (b), and (c) will depend both on experimental evidence and on the manner in which the experimental evidence is obtained. The choice will hinge on the internal consistency of the proposed formulation and on its usefulness in scientific work. There is no uniqueness theorem which can be applied to a scientific conceptual scheme. Anybody is free to propose an alternative, but their labours will be arduous. Relativity theory has some very queer features, as we shall see, but in the absence of any alternatives which can stand the test of experiment, we shall have to accept it gratefully.

12.2. First-order Velocity Effects in Electromagnetism

The principle of Galilean relativity arose from an investigation into the effects of uniform motion. In order to find out if it is possible to apply the principle to electrodynamics we shall similarly start with a consideration of velocity effects. The first thing we notice is that there is a force on a charge moving through a magnetic field

$$\mathbf{F} = Q\mathbf{u}\times\mathbf{B}. \tag{12.2}$$

On the face of it this runs counter to any relativistic principle. But eqn. (12.2) does not give the total force. For this we turn to the Lorentz equation

$$\mathbf{F} = Q\,(\mathbf{E} + \mathbf{u}\times\mathbf{B}). \tag{9.30}$$

Hence, if we transform from a frame Σ, in which the fields are **E** and **B** and in which a charge Q has a velocity **u**, to a frame Σ' in which the charge is at rest, the force would be

$$\mathbf{F}' = Q\mathbf{E}'. \tag{12.3}$$

For Galilean relativity to apply, the force should not be altered and we should have

$$\mathbf{E}' = \mathbf{E} + \mathbf{u}\times\mathbf{B}. \tag{9.3}$$

This we found to be consistent with Faraday's experiments on electromagnetic induction, where

$$\text{curl}\,\mathbf{E} = -\frac{\partial\mathbf{B}}{\partial t} \tag{9.5}$$

and

$$\text{curl}\,\mathbf{E}' = -\frac{\partial\mathbf{B}}{\partial t} + \text{curl}\,(\mathbf{u}\times\mathbf{B}). \tag{9.2}$$

Thus the Lorentz force and Faraday's law are consistent with the relativity principle of mechanics, as long as the electric field **E** is not itself altered by the motion of the charge. In other words, the charge must not move so fast as to make it difficult for the effect of distant charges to catch up with it. We shall find that this assumption is justified as long as we can neglect terms in $(u/c)^2$. We conclude that to this order of accuracy electrodynamics fits into the Newtonian framework. The magnetic field, far from being in contradiction to the relativistic principle, is needed to ensure that the principle is obeyed. Motion through a magnetic field produces an additional electric field. There are books on electromagnetism which start with relativity as axiomatic and deduce the magnetic field as a necessary consequence. This is a perfectly justified procedure. We have gone the other way about because the idea of a magnetic field is easier to grasp than that of inertial frames of reference. Moreover magnetic fields are much more easily investigated by experiment.

An investigation of the m.m.f. law provides us with the transformation of the magnetic field from a frame Σ in which a pole (or dipole) is moving to a frame Σ' in which it is stationary. We have in free space

$$\mathbf{H}' = \mathbf{H} - \mathbf{u} \times \mathbf{D}. \tag{12.4}$$

Thus we conclude that a magnetic source moving through an electric field observes an additional magnetic field. As always we can think of the magnetic source either as a pole, or a dipole or as a current loop. Equation (12.4) is restricted to free space because we wish to avoid the complications due to conduction, convection, and polarization currents.

The transformation equations (9.3) and (12.4) for the electric and magnetic field are not independent. This faces us with a difficulty. We have

$$\mathbf{E}' = \mathbf{E} + \mathbf{u} \times \mathbf{B}, \tag{9.3}$$

so that by the relativity principle we should expect

$$\mathbf{E} = \mathbf{E}' - \mathbf{u} \times \mathbf{B}'. \tag{12.5}$$

But eqns. (9.3) and (12.5) can only both be correct if $\mathbf{B} = \mathbf{B}'$, whereas we know from eqn. (12.4) that

$$\mathbf{B}' = \mathbf{B} - \frac{1}{c^2}(\mathbf{u} \times \mathbf{E}). \tag{12.6}$$

Substituting eqn. (12.6) in eqn. (12.5), we obtain

$$\mathbf{E} = \mathbf{E}' - \mathbf{u}\times\mathbf{B} + \frac{1}{c^2}\mathbf{u}\times(\mathbf{u}\times\mathbf{E}). \tag{12.7}$$

There is, therefore, an inconsistency in the equations and eqns. (9.3) and (12.4) are not relativistically correct. But the unwanted term is of the order $(u/c)^2$. As long as we neglect second-order velocity terms, we can retain eqns. (9.3) and (12.4). To this order of accuracy they are consistent and fit into the framework of Newtonian mechanics and Galilean relativity.

Let us now test our findings by some experimental situations. Figure 12.2 shows a point charge Q in the (empty) space between the plates of a large

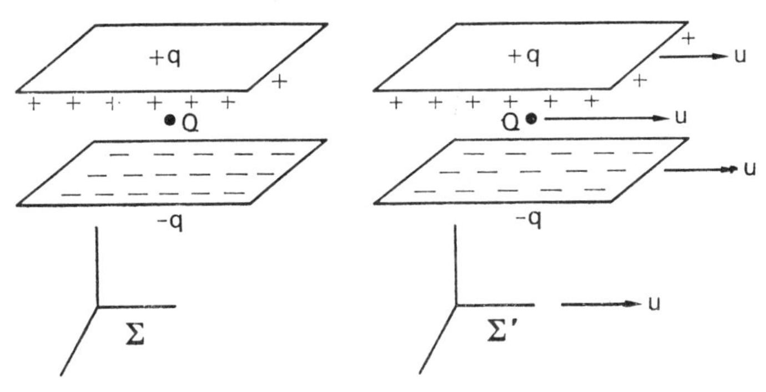

FIG. 12.2. A point charge between the plates of a charged capacitor in two inertial frames.

charged capacitor. In the frame Σ the point charge and the capacitor are at rest. There is an electric field $\mathbf{E}$ and zero magnetic field. The force on the point charge Q is $\mathbf{F} = Q\mathbf{E}$. Consider next the frame Σ'. In this frame the capacitor and the point charge are again stationary, but the whole assembly including Σ' is moving with $\mathbf{u}$ relative to Σ. Remembering that the velocity in eqn. (9.3) is relative to the frame in which it is measured we deduce $\mathbf{E}' = \mathbf{E}$ and by the same argument $\mathbf{H}' = \mathbf{H} = 0$. Thus the motion produces no measurable effect, in accordance with the relativistic principle.

Consider the same situation from the viewpoint of an observer in a frame Σ'' (Fig. 12.3). The capacitor and point charge are now moving with $\mathbf{u}$

relative to this frame. There is a component magnetic field in Σ'' since the moving charges constitute a current. But the total magnetic field has to be obtained from eqn. (12.4). We have, since q is charge density per unit area and $q\mathbf{u}$ the effective current density per unit length, $H'' = H' - uD = qu - qu = 0$, so that the magnetic field is unchanged. The electric field in Σ'' is modified by the $\mathbf{u} \times \mathbf{B}$ term, but since $\mathbf{B}$ is itself proportional to the velocity $\mathbf{u}$ of the charge on the plates of the capacitor, the term $\mathbf{u} \times \mathbf{B}$ is a second-order velocity term and we shall neglect it. Thus the electric field is $\mathbf{E}'' = \mathbf{E}$ as before.

The generation of magnetic fields by moving electric charges is so familiar as to be readily accepted. The converse process in which an elec-

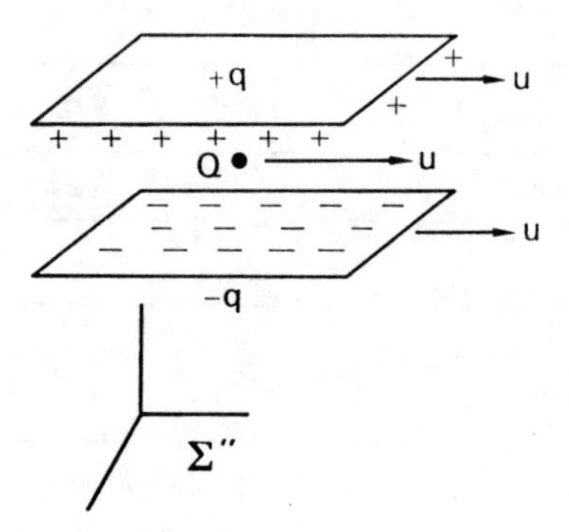

FIG. 12.3. Another view of the problem of Fig. 12.2.

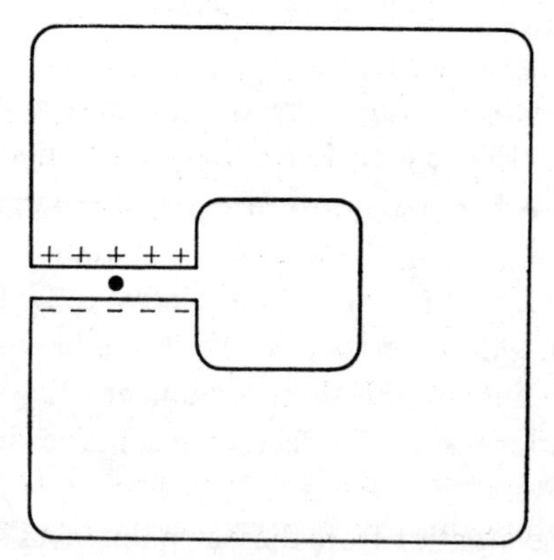

FIG. 12.4. A large magnet with a parallel-sided gap.

tric field appears near a moving magnetic source is not so easily understood. Such a situation would arise if the capacitor were replaced by a magnet as shown in Fig. 12.4. The trouble stems from the fact that we have no magnetic conductors with which to experiment. Faraday's e.m.f. law is easily tested, but the m.m.f. law cannot be tested in the same simple way. It is therefore worth while to look in detail at a particular example of a moving magnetic source. Consider Fig. 12.5 which shows two flat con-

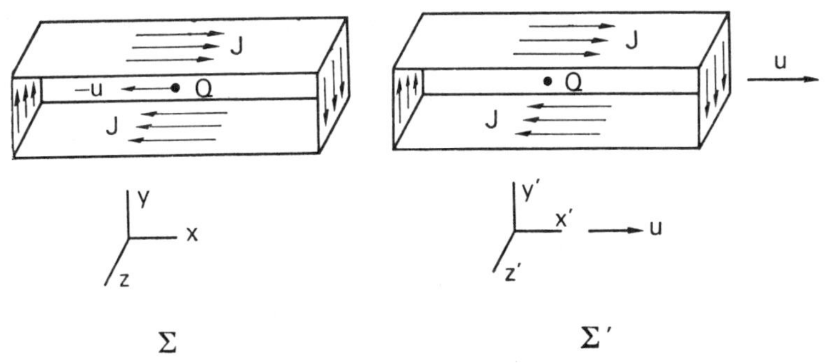

FIG. 12.5. A point charge between two current sheets.

ductors of large area forming the "go" and "return" path of a circuit carrying constant current. The current density per unit width is J. Between the two conductors there is a point charge Q. In the frame Σ the current is stationary and the charge has a velocity $-u_x$. The magnetic field in Σ is $H_z = -J$ and there is zero electric field. The force on the charge is $F_y = -Q\mu_0 u_x J$. In the frame Σ' the magnetic field is unchanged, neglecting a second-order velocity effect. If we postulate an electric field $E_y = -\mu_0 u_x J$ the relativistic principle is satisfied. Some writers seek to explain this electric field as being due to a set of "apparent" charges caused by the motion of the current. In our simple example the apparent charge would have a density per unit area

$$q = \varepsilon_0 E_y = -\varepsilon_0 \mu_0 u_x J = -uJ/c^2. \tag{12.8}$$

The top plate would have a positive charge and the bottom a negative charge.

The concept of apparent charge associated with moving magnetic sources is of doubtful value in the treatment which neglects second-order velocity terms, which we are adopting in this section. It is better to accept the fact that just as moving charges are accompanied by magnetic fields, so moving currents and poles are accompanied by electric fields. The physical reason in terms of energy for this electric field has already been touched on in § 9.6. In the system of Fig. 12.5 there exists mutual kinetic energy between the current and the charge. This energy is due to the coupling of the electrokinetic momentum **A** with the velocity **u** and is given by

$$T = Q\,(\mathbf{A}\,.\,\mathbf{u}). \tag{9.38}$$

Making use of eqn. (9.36) we find that this energy gives rise to the forces

$$\mathbf{F} = Q \operatorname{grad}(\mathbf{A}\,.\,\mathbf{u}) - Q\frac{d\mathbf{A}}{dt}. \tag{12.9}$$

We thus have the electric field

$$\mathbf{E} = \operatorname{grad}(\mathbf{A}\,.\,\mathbf{u}) - \frac{d\mathbf{A}}{dt}. \tag{12.10}$$

Since **u** is constant we can write this

$$\mathbf{E} = (\mathbf{u}\,.\,\nabla)\,\mathbf{A} + \mathbf{u}\times \operatorname{curl}\mathbf{A} - \frac{d\mathbf{A}}{dt}. \tag{12.11}$$

But

$$\frac{d\mathbf{A}}{dt} = \frac{\partial\mathbf{A}}{\partial t} + (\mathbf{u}\,.\,\nabla)\,\mathbf{A}, \tag{9.32}$$

so that the electric field due to the kinetic energy is

$$\mathbf{E} = -\frac{\partial\mathbf{A}}{\partial t} + \mathbf{u}\times\mathbf{B} \tag{12.12}$$

which, of course, is known from Faraday's law. In the example of Fig. 12.5 **A** does not change with time, because the current is constant, so that we are left with

$$\mathbf{E} = \mathbf{u}\times\mathbf{B}. \tag{12.13}$$

We have derived this equation without reference to either Σ or Σ'. All we have done is to consider the system formed by the current and the charge. From consideration of the kinetic energy of the system it follows that a moving current is accompanied by the electric field of eqn. (12.13) or its simpler special case of eqn. (12.14).

Our discussion has shown that there is no conflict between Galilean relativity and electrodynamics as far as first-order velocity effects are concerned. The only difficulty is that the division between electric and magnetic effects depends on the inertial frame in which the fields are measured. The total force is independent of the frame. Most engineering applications deal with velocities which are much smaller than the speed of light. In these applications the principle of Galilean relativity can be used and Einstein's principle need not be invoked.

12.3. The Lorentz Transformation

In spite of the comforting conclusion of the last section there is a fundamental conflict between electrodynamics and Galilean relativity. We have already noticed that there is an inconsistency in the field equations and several times we have had to neglect second-order velocity effects.

At the heart of the conflict there is the curious feature which we have noticed in Maxwell's equations: the velocity c is a characteristic of the theory. If electromagnetic waves are propagated with this velocity in an inertial frame Σ, it would seem obvious that in another frame Σ' moving relative to Σ the velocity should be different. Thus in sound waves in air the velocity of propagation depends on the speed of the air. If electromagnetic waves occur in some medium, which has often been called the aether, then their speed in a particular frame must depend on the velocity of the aether relative to this frame. The inertial frame which moved with the aether would then have a special significance and there would be no relativity principle for electrodynamics. This question was investigated theoretically and experimentally in the last quarter of the nineteenth century and the first few years of the twentieth century.† Two crucial experiments were those of Michelson and Morley who compared the speed of light in two directions at right angles, and of Trouton and Noble who attempted to measure the torque on a parallel plate capacitor. They reasoned that the motion through the aether would make the moving charged plates into current elements and there should be a torque as illustrated in Fig. 12.6.

All the attempts to find effects due to aether velocity failed. This led to

† Readers who are interested in the history of science are referred to the masterly two-volume study by E. T. Whittaker, *History of the Theories of Aether and Electricity*, Nelson, 1951 and 1953.

the conviction that such effects do not exist. Poincaré in 1904 enunciated the hypothesis that the relativistic principle applies to electrodynamics. Einstein generalized the hypothesis and proposed that the velocity of light *in vacuo* should be accepted as a fundamental physical constant.

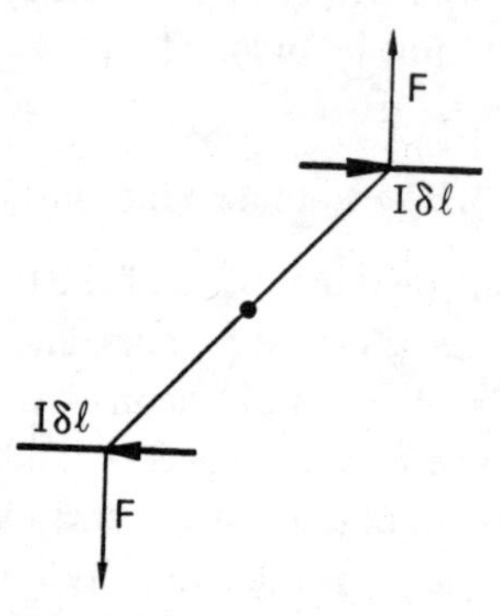

FIG. 12.6. Torque between two current elements.

On the face of it the Poincaré–Einstein hypothesis is in accordance with common sense, but its consequences are by no means obvious. Let us find a transformation equation between two inertial frames Σ and Σ'. As before we choose the relative velocity u between Σ and Σ' to be parallel to the x- and x'-axes. Let us next choose the time t and t' so that at $t = 0$ and $t' = 0$ the x- and x'-axes coincide. This implies that if $y = 0$, $z = 0$ it always follows that $y' = 0$, $z' = 0$. Hence the transformation equations for y' and z' must be of the form

$$\begin{aligned} y' &= \alpha y + \beta z, \\ z' &= \gamma y + \delta z. \end{aligned} \tag{12.14}$$

The y- and z-directions have equal status, so that also

$$\begin{aligned} y' &= \varepsilon y, \\ z' &= \varepsilon z. \end{aligned} \tag{12.15}$$

Moreover, we must put $\varepsilon = 1$, because otherwise at $t = t' = 0$ we could detect a difference between Σ and Σ' and this would violate the principle of relativity. Thus

$$\begin{aligned} y' &= y, \\ z' &= z. \end{aligned} \tag{12.16}$$

To find the transformation equations of x, x' and t, t' we note that they must ensure symmetry in the velocity between Σ and Σ', i.e. Σ' moves with $+u$ relative to Σ and Σ moves with $-u$ relative to Σ'. Thus

$$\begin{aligned} x' &= \gamma(x - ut), \\ x &= \gamma(x' + ut'). \end{aligned} \tag{12.17}$$

To find γ we make use of the postulate that the velocity of light is c in Σ and Σ'. Let a flash of light be emitted at the origin of coordinates when $t = t' = 0$. The light travels as a spherical wave in *both* systems. Thus

$$\begin{aligned} r^2 &= x^2 + y^2 + z^2 = c^2t^2, \\ r'^2 &= x'^2 + y'^2 + z'^2 = c^2t'^2, \end{aligned} \tag{12.18}$$

and by the use of eqn. (12.16)

$$x^2 - c^2t^2 = x'^2 - c^2t'^2. \tag{12.19}$$

From eqns. (12.17) and (12.19) we find

$$\gamma = \frac{1}{\sqrt{[1 - (u/c)^2]}} \tag{12.20}$$

and

$$\left.\begin{aligned} x' &= \frac{x - ut}{\sqrt{[1 - (u^2/c^2)]}}, \\ t' &= \frac{t - (u/c^2)\,x}{\sqrt{[1 - (u^2/c^2)]}}. \end{aligned}\right\} \tag{12.21}$$

Eqns. (12.16) and (12.21) enable us to transform from the Σ to the Σ' frame. Collecting the results we have

$$\left.\begin{aligned} x' &= \frac{x - ut}{\sqrt{[1 - (u^2/c^2)]}}, \\ y' &= y, \\ z' &= z, \\ t' &= \frac{t - (u/c^2)x}{\sqrt{[1 - (u^2/c^2)]}}. \end{aligned}\right\} \tag{12.22}$$

Equation (12.22) is called the *Lorentz transformation*. The inverse transformation can be obtained by changing the sign of u. Thus

$$\left.\begin{aligned} x &= \frac{x' + ut'}{\sqrt{[1 - (u^2/c^2)]}}, \\ y &= y', \\ z &= z', \\ t &= \frac{t' + (u/c^2)\,x'}{\sqrt{[1 - (u^2/c^2)]}}. \end{aligned}\right\} \tag{12.23}$$

In accepting the Lorentz transformation we follow possibility (c) of § 12.1. Maxwell's theory is accepted as it stands, but Newtonian mechanics has to be modified.

12.4. Consequences of the Lorentz Transformation

The consequences of the Lorentz transformation are more easily seen if we find the transformation of x to x' and t to t'. Remembering that at $t = t' = 0$ the axes of x and x' coincide we have

$$x_2' - x_1' = \frac{(x_2 - x_1)}{\sqrt{[1 - (u^2/c^2)]}}. \tag{12.24}$$

Thus a length parallel to the direction of motion in Σ seems to have shrunk in length when observed from Σ'. This is also true if the length in Σ' is observed from Σ, which, of course, is required by the relativity principle. Otherwise one could detect which is Σ and which is Σ'. This transformation is at variance with the Galilean transformation in which all lengths remain invariant.

More surprising still is the transformation of the time t to t'. Common sense suggests, and it was taken as axiomatic by Newton, that time flows uniformly in all inertial frames. But this conflicts with the constancy of the velocity of light. We now have to accept that the measurement of time is dependent on the choice of inertial frame. Space and time can no longer be treated separately. We cannot assert that an event is taking place at a particular time in all frames of reference. The place and the time have to be specified together in one system of space and time coordinates. Equally we cannot talk about a particular place without referring to the time; the event has to be specified by "here and now", not by "here" only or "now" only. Reflection shows that this is not unreasonable. We see moving objects by the light waves sent out by them, hence our knowledge even of spatial coordinates depends on the time taken by the light to reach us. We have already used this fact in defining delayed potentials to calculate electromagnetic effects.

To compare the time in Σ and Σ' consider two clocks in Σ synchronized by light signals and one clock in Σ' which is coincident with the first clock in Σ at the beginning of the time interval to be measured and coincident with the second clock in Σ at the end of this interval. At the beginning a light source at the common point in Σ and Σ' at which the clocks are

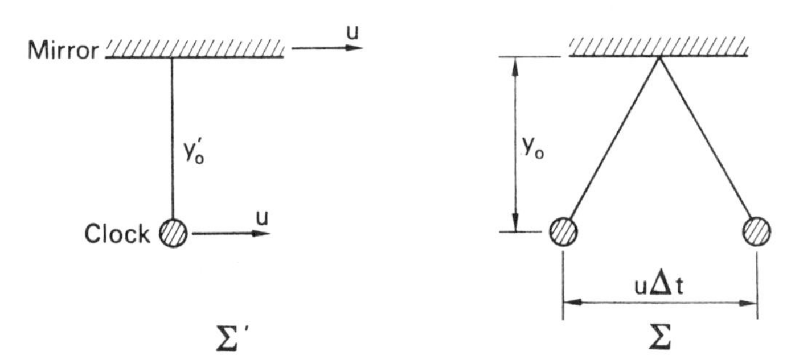

FIG. 12.7. The dilatation of time.

located emits a flash of light. There is a mirror which reflects the light (Fig. 12.7). The light reaches the second clock in Σ just as it is coincident with the clock in Σ'. Referring to Fig. 12.7 we see that

$$\left.\begin{aligned} c\,\Delta t' &= 2y_0', \\ c\,\Delta t &= 2\sqrt{|y_0^2 + (u\,\Delta t/2)^2|}. \end{aligned}\right\} \qquad (12.25)$$

Since $y_0' = y_0$, we obtain

$$\Delta t = \frac{\Delta t'}{\sqrt{[1 - (u^2/c^2)]}}. \qquad (12.26)$$

The time interval $\Delta t'$ measured in the frame in which the clock is moving with the frame is called the *proper time*. Equation (12.26) shows that the time interval observed in Σ is longer than the proper time. Considerable support for the time dilatation of eqn. (12.26) comes from experiments with high-velocity particles, called mesons, which seem to decay very much less rapidly in flight through the atmosphere than when they are at rest. It should be noticed that in our discussion we have assumed that there are such things as clocks which behave in a uniform manner. Similarly, we have assumed that there are reproducible standards of length. For the first physicists turn to radioactive decay periods and for the second to the wavelength of the radiation from certain atoms.

Comparing the Lorentz transformation and the Galilean transformation we note that if $u^2/c^2 \ll 1$ the Lorentz transformation becomes the same as the Galilean one. Thus the differences between the relativity principles of Newton and Einstein lie in terms of the order u^2/c^2. Except in special subjects like nuclear engineering, the Galilean transformation can be used as a good approximation.

12.5. The Covariant Formulation of Electromagnetism

We have deduced the Lorentz transformation from Einstein's relativity postulate. Conversely we can use this transformation to test whether a statement is correctly formulated so as to conform with relativity. As an example we can write out the electric and magnetic field quantities in the frame $\Sigma\,(x, y, z, t)$ and transform the equations to the frame $\Sigma'\,(x', y', z', t')$. When this is done we see that the equations are of the same form as long as

$$\left.\begin{aligned} E'_x &= E_x, \\ E'_y &= \frac{E_y - uB_z}{\sqrt{[1-(u^2/c^2)]}}, \\ E'_z &= \frac{E_z + uB_y}{\sqrt{[1-(u^2/c^2)]}}, \\ B'_x &= B_x \\ B'_y &= \frac{B_y + (u/c^2)\,E_z}{\sqrt{[1-(u^2/c^2)]}}, \\ B'_z &= \frac{B_z - (u/c^2)\,E_y}{\sqrt{[1-(u^2/c^2)]}}. \end{aligned}\right\} \tag{12.27}$$

This is in accordance with the approximate results of § 12.2 if we can neglect u^2/c^2 compared with unity.

The procedure of transforming the coordinates from Σ to Σ' is cumbersome. There is, however, a very much more powerful method due to Minkowski. The details of the method are beyond the scope of this book,† but we can give a brief outline. It will be remembered that we introduced vectors into the formulation of electromagnetism because they enabled us to write one equation instead of three. But vector language does more than to reduce the labour. Vector equations are independent of the choice of axes, so that the axes can be changed without changing the form of the vector equations. This invariance to the choice of axes is a property of vectors. A quantity that fits into a vector equation has this invariance and by writing the equation in vector form we are saved the trouble of having to test which quantities are vectors and which are not. This does not matter greatly in three-dimensional space, but it becomes enormously important in relativistic electromagnetism.

We have seen that the Lorentz transformation treats space and time as a single set of coordinates. Let us recognize this formally by writing

$$\left.\begin{aligned} x &= x_1, \\ y &= x_2, \\ z &= x_3, \\ jct &= x_4. \end{aligned}\right\} \tag{12.28}$$

Then, since $x^2 + y^2 + z^2 - c^2t^2 = 0$ describes a light wave travelling outwards from the origin, we know that

$$x_1^2 + x_2^2 + x_3^2 + x_4^2 = r^2 \tag{12.29}$$

must be invariant under a Lorentz transformation. We therefore regard r as a length in a four-dimensional vector space. Thus we look for *four-vectors* in order to find a vector language for four dimensions. Any relationship which is correctly formulated in four-vectors will be invariant under a Lorentz transformation and is said to be *Lorentz covariant*.

† For a full treatment see W.K.H.Panofsky and M.Phillips, *Classical Electricity and Magnetism*, Addison Wesley, 1955, or R.Becker and F.Sauter, *Electromagnetic Fields and Interactions*, Vol. 1, Blackie, 1964.

As an example consider the current density and charge density. We can write the four-vectors J as:

$$\left.\begin{aligned} J_1 &= \varrho v_x/c, \\ J_2 &= \varrho v_y/c, \\ J_3 &= \varrho v_z/c, \\ J_4 &= j\varrho. \end{aligned}\right\} \tag{12.30}$$

If the four-vector divergence is given by

$$\frac{\partial J_1}{\partial x_1} + \frac{\partial J_2}{\partial x_2} + \frac{\partial J_3}{\partial x_3} + \frac{\partial J_4}{\partial x_4},$$

then the equation of continuity of charge

$$\operatorname{div} \mathbf{J} + \frac{\partial \varrho}{\partial t} = 0 \tag{8.3}$$

is the same as putting the four-vector divergence equal to zero. The three-dimensional J is the spatial part of the four-vector, the *proper*† charge $j\varrho$ is the time part. Since we know how the space and time coordinates transform, we deduce at once that

$$\left.\begin{aligned} J' &= \frac{\varrho v}{c\sqrt{[1-(u^2/c^2)]}}, \\ \varrho' &= \frac{\varrho}{\sqrt{[1-(u^2/c^2)]}}. \end{aligned}\right\} \tag{12.31}$$

We can deduce from this that total charge is invariant under a Lorentz transformation. Since a volume Δv undergoes a Lorentz contraction in one dimension only we have

$$\Delta v' = \Delta v \sqrt{[1-(u^2/c^2)]}. \tag{12.32}$$

Combining this with the second part of eqn. (12.32) we have

$$\varrho' \, \Delta v' = \varrho \, \Delta v. \tag{12.33}$$

Another four-vector is the potential

$$A = A_x, A_y, A_z, jcV, \tag{12.34}$$

† The *proper* charge density is measured in a frame in which it is at rest.

and the two wave equations for the scalar and vector potentials are combined into a single four-vector equation.

Force and momentum are not components of a four-vector, but we can define a four-vector momentum by

$$p = \left[\frac{m\mathbf{v}c}{\sqrt{[1-(u^2/c^2)]}}, \frac{jmc^2}{\sqrt{[1-(u^2/c^2)]}}\right]. \tag{12.35}$$

This shows that mass is not Lorentz invariant, but transforms as

$$m' = \frac{m}{\sqrt{[1-(u^2/c^2)]}}. \tag{12.36}$$

The time part of the four-vector momentum is the energy

$$E = m'c^2. \tag{12.37}$$

Thus energy and momentum are shown to be interdependent. The division between energy and momentum depends on the frame of the observer. The laws of conservation of energy and momentum are not independent.

The full power of the covariant method is best shown in tensor formulation. Many of the relationships which we have derived in this book can be unified in tensor language. Nevertheless, for their application one still has to resort to a particular system of coordinates.

Summary

In Newtonian mechanics forces are caused by acceleration and not by uniform rectilinear velocity. The laws of mechanics are therefore independent of such velocity and have the same form for observers travelling at relative uniform velocity to each other.

In electromagnetism the velocity of light enters into the equations and such a velocity effect is at variance with Newtonian mechanics. Einstein proposed the relativistic principle that the velocity of light should be taken as a universal constant independent of the velocity of the observer. This leads to a transformation of the coordinates of space and time, which is known as the Lorentz transformation.

The Lorentz transformation embodies some unfamiliar ideas about space and time, but these only become significant if the velocity of the observer of some experiment is an appreciable fraction of the velocity of light. Thus in most engineering applications Newtonian mechanics is applicable.

Exercises

12.1. Show that the wave equation

$$\frac{\partial^2 \phi}{\partial x^2} = \frac{1}{c^2}\frac{\partial^2 \phi}{\partial t^2}$$

is invariant under a Lorentz transformation but not under a Galilean transformation.

12.2. Show that the Galilean and Lorentz transformations are identical if the velocity of light is taken as being infinite. Show that the principle of causality would be violated if a signal travelled faster than the speed of light.

12.3. Show that Faraday's law is invariant under a Lorentz transformation if the field components are transformed in accordance with eqn. (12.27). Hence show that if $u^2/c^2 \ll 1$ the correct transformations for relative motion $\mathbf{u}$ are

$$\begin{cases} \mathbf{E}' = \mathbf{E} + \mathbf{u}\times\mathbf{B}, \\ \mathbf{B}' = \mathbf{B} - \dfrac{\mathbf{u}\times\mathbf{E}}{c^2}. \end{cases}$$

12.4. Show that if $\mathbf{E}$ and $\mathbf{B}$ are perpendicular in one inertial frame, they are perpendicular in all inertial frames.

12.5. If $\mathbf{E}$ is perpendicular to $\mathbf{B}$, show that there is in general an inertial frame in which the field is *either* purely electric *or* purely magnetic unless $\mathbf{E} = c\mathbf{B}$.

12.6. Show that the current and charge densities described by eqn. (12.31) constitute a four-vector, i.e. that they are invariant under a Lorentz transformation.

12.7. Show that the delayed potentials $\mathbf{A}$ and V constitute a four-vector (cA_x, cA_y, cA_z, jV).

12.8. Show by considering $\mathbf{u} = d\mathbf{r}/dt$ that the velocity $\mathbf{u}$ is not the spatial part of a four-vector, but that $\mathbf{u}/\sqrt{[1-(u^2/c^2)]}$ is. What is the time component of the four-vector? [Ans.: $jc/\sqrt{[1-(u^2/c^2)]}$.]

12.9. In Exercise 9.4 we discussed the rotation of a conducting cylinder in a uniform axial magnetic field. This rotation gives rise to a charge distribution in the cylinder of density $\varrho = -2\varepsilon_0 \omega B$. Show that if the cylinder is a magnet, no free charge is displaced, but the external field is unchanged.†

† This is discussed in detail in E. G. Cullwick, *Electromagnetism and Relativity*, Longmans, 1957. Reference should be made to this book for a full treatment of relativistic effects in electromagnetic induction.

CHAPTER 13

EXPERIMENTAL AND NUMERICAL METHODS FOR THE SOLUTION OF ELECTROMAGNETIC PROBLEMS

13.1. Physical and Geometrical Factors in Electromagnetism

There are two reasons for the study of electromagnetism. The first of these, which has been dominant in most of this book, is that the study of this subject enables us to understand the principles underlying all electrical devices. Without a grasp of these principles each device represents an isolated body of knowledge. It is as if a doctor had to treat each case as completely unconnected with others because he had never learned about the general subject of infections. But, of course, the patient will demand more than such general knowledge. He wants to be cured of his particular disease. In the same way the second reason for the study of electromagnetism is that it can be usefully applied to the analysis and design of pieces of apparatus or of systems. The reader of this book may feel that we have paid too little attention to the application of the subject. This chapter is written to explain the reasons for this apparent lack of balance and also to point out a way in which the balance may be restored.

There are several reasons why we have dealt more with the structure than with the application of electromagnetism. Until our minds are clear on the interaction of static and moving electric charges and the accompanying electric and magnetic fields, we cannot make any significant progress with the study of complicated devices. The general principles of electromagnetism are best brought out by choosing very simple configurations. Thus in electrostatics we start with a point charge because the law of force between point charges is simple. This is in spite of the fact that engineers do not generally deal with point charges but with charged bodies of varying shape. The purpose in choosing a point charge is to isolate the physical interaction from the geometrical factors. By saying that the law between point charges is simple we mean not only that eqn. (2.1) is straightforward, but that the geometrical factors can be dealt with by using this law as a basic building-brick. The law enables us to make a physical statement before having to make a geometrical one. Thus we can proceed from the inverse square law at once to the general statement

$$V = \frac{1}{4\pi\varepsilon_0} \iiint \frac{\varrho}{r}\, dv'. \tag{4.43}$$

This statement is important because of its generality, but when we come to apply it to a particular problem it is little more than an expression of hope that somebody will carry out the integration.

In § 4.1 we endeavoured to meet the difficulty by providing ourselves with a collection of useful charge distributions, and in § 6.2 of useful magnetic pole and current distributions. By superposition we can then build up more complicated structures. This method is not to be despised and the engineer who has in his possession such a collection finds himself in a strong position. Nevertheless, it is irksome to be restricted to simple geometrical shapes, when the restriction is seen to lie in the mathematical computation and not in the physical process.

The position is slightly better when we turn to the differential method. Instead of the integral of eqn. (4.43) we can start with Poisson's equation

$$\nabla^2 V = -\varrho/\varepsilon_0 \tag{3.21}$$

or Laplace's equation

$$\nabla^2 V = 0. \tag{3.24}$$

These equations can be solved by the method of separation of the variables which was discussed in § 4.3.1. There is a wide choice of coordinate systems in which the variables can be separated.† The method can also be applied to the diffusion equation (see Chapter 8) and the wave equation (see Chapter 11). Nevertheless, there are severe limitations. The solution may be in terms of complicated or unfamiliar functions, and these may not be tabulated. More seriously, the method fails to deal with irregular

† See P. Moon and D. E. Spencer, *Field Theory for Engineers*, Van Nostrand, 1961, for a very helpful and extensive treatment of this method.

shapes. Thus it is possible to solve the equations for reactangular boundaries and circular boundaries, but not for rectangles with rounded corners. Intuitively one knows that the rounding of the corners will not affect the physical solution greatly, but it is difficult to assess the magnitude of the effect. The trouble is that the mathematical equations are dominated by geometrical and not by physical factors. This is generally true also in other analytical methods such as the method of conformal transformation,† although this is a powerful method.

13.2. Experimental Methods for Solving Electromagnetic Problems

When presented with difficulties in analytical work an engineer turns to experiment. There are a number of useful experimental techniques which can be applied to electromagnetic problems. A full discussion is beyond the scope of this book and we can do little more than list some of the methods.‡

13.2.1. The Method of Curvilinear Squares

The only experimental equipment required for this method is paper and pencil and a rubber. The method is restricted to two-dimensional problems and to problems described by Poisson's and Laplace's equa-

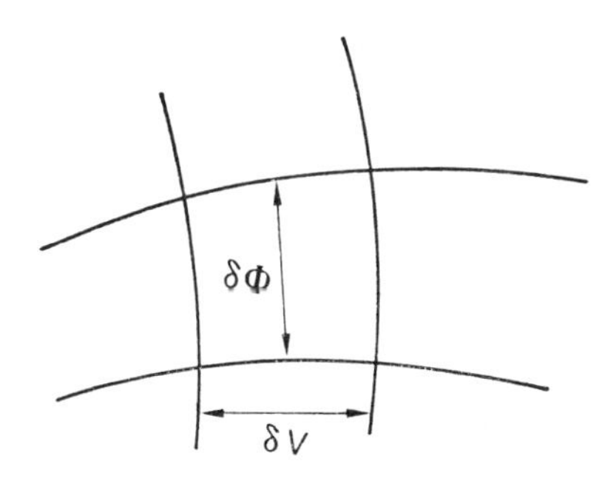

FIG. 13.1. Curvilinear squares.

† See K.J.Binns and P.J.Lawrenson, *Analysis and Computation of Electric and Magnetic Field Problems*, Pergamon, 1963.

‡ See D.Vitkovitch (editor), *Field Analysis: Experimental and Computational Methods*, Van Nostrand, 1966, for further information.

tions. As already discussed in § 4.1.8 we can draw equipotentials and flux lines at right angles to each other. Consider Fig. 13.1 which shows some lines in an electrostatic field. By definition

$$-\delta V = E\,\delta x, \tag{13.1}$$

$$\delta\Phi = D\,\delta y = \varepsilon E\,\delta y, \tag{13.2}$$

where the flux Φ is taken per unit depth into the paper. If we arbitrarily choose $\delta x = \delta y$ we have

$$\left|\frac{\delta\Phi}{\delta V}\right| = \varepsilon. \tag{13.3}$$

If, therefore, we map the field in curvilinear squares, the total flux and the total potential difference can be found by counting the number of either the flux lines or the equipotentials. The method starts with drawing equipotentials and then fitting flux lines at right angles. Where the equipotentials and flux lines do not form curvilinear squares, the equipotentials have to be redrawn. In parts of the field where V and Φ are changing rapidly the accuracy is improved by drawing smaller squares. In experienced hands this is a quick and powerful method. It can be applied to two-dimensional problems in electrostatics, magnetostatics and steady current flow. Other examples occur in mechanical and civil engineering.

13.2.2. Conducting Sheet Analogues

If the magnetic field can be neglected, the equation of current flow in a conductor is given by

$$-\text{grad}\, V = \frac{\mathbf{J}}{\sigma}. \tag{13.4}$$

Since $\text{div}\,\mathbf{J} = 0$ we obtain Laplace's equation

$$\nabla^2 V = 0. \tag{3.24}$$

Thus fields obeying Laplace's equation can be simulated by suitably scaled models involving the flow of current. This principle is used in conducting sheet analogues. A common form of the analogue uses graphitized paper. Equipotentials of arbitrary shape can be painted on the paper with silver paint. A potential difference is applied between equipotentials at opposite sides of the paper and equipotentials are then drawn on the paper by moving a probe across the paper. Flux lines can be drawn in afterwards. Since the equipotentials are correctly drawn, there is no difficulty in drawing the flux lines. The method can be used also to find the magnetic field of a known current distribution. We use the vector potential

$$\nabla^2 \mathbf{A} = -\mu \mathbf{J} \tag{3.42}$$

and away from the current sources

$$\nabla^2 \mathbf{A} = 0. \tag{13.5}$$

If the only component of A is A_z, the magnetic field is given by

$$\left.\begin{aligned} B_x &= \frac{\partial A}{\partial y}, \\ B_y &= -\frac{\partial A}{\partial x}. \end{aligned}\right\} \tag{13.6}$$

The boundary conditions of the analogue are achieved by putting equipotentials at boundaries which are impenetrable to magnetic flux, for example if there is no B_x at a boundary then A will be constant and independent of y on this boundary. Details of the method are given in the book by Vitkovitch mentioned earlier.

13.2.3. The Electrolytic Tank

The conducting sheet is more accurate than the method of curvilinear squares, but the equipment is more complicated. The electrolytic tank represents a further step in the same direction. Current flow is again used as an analogue of other types of field. Instead of conducting paper, a liquid conductor is used. Tap water is usually satisfactory. One great advantage of the tank is that it has depth and this makes it possible to investigate some three-dimensional fields. Fields of infinite extent can be simulated by a technique which makes use of a double tank†. The measurement

† A. R. Boothroyd, E. C. Cherry, and R. Makar, An electrolytic tank for the measurement of transient response and allied properties of networks, *Proc. I.E.E.*, Vol. 96. Part 1, 1949.

methods used in electrolytic tanks have been extensively studied and refined. Accuracies of $\frac{1}{4}\%$ can be achieved in the measurement of potentials. Some tanks have been connected to small digital computers and it is possible to arrange for automatic plotting of the field.

13.2.4. Network Analogues

An alternative to the electrolytic tank is the resistance network consisting of a mesh of resistances as shown in Fig. 13.2. Equipotential boundaries are simulated by connecting some of the nodes of the network

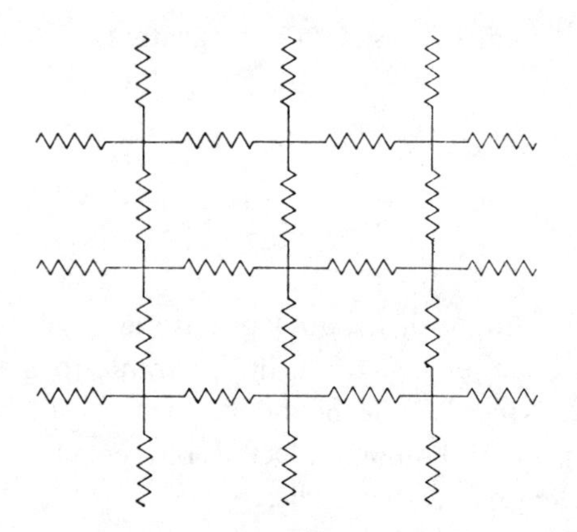

FIG. 13.2. A resistance network.

to each other. In magnetic field problems currents can be fed into the network at various nodes. The use of a network avoids many of the difficulties of measurement in electrolytic tanks, particularly those due to chemical effects. On the other hand, errors are introduced because the meshes are of finite size. Thus the field is accurately† represented only at the nodes and not throughout the entire space. Also the network is likely to be more expensive than the tank. By itself the resistance network has little to commend it.

However, the idea of a circuit analogue enables us to introduce other circuit components and by this means the network can be used to solve

† The accuracy depends on the mesh length.

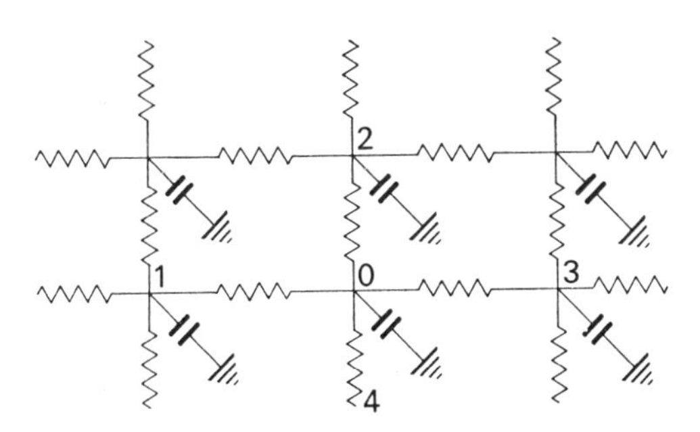

FIG. 13.3. An RC network for the diffusion equation.

time-varying problems. Consider the resistance-capacitance network of Fig. 13.3. The total current flowing into the node 0 is

$$\frac{V_1 - V_0}{R} + \frac{V_2 - V_0}{R} + \frac{V_3 - V_0}{R} + \frac{V_4 - V_0}{R} - C\frac{dV_0}{dt} = 0,$$

$$V_1 + V_2 + V_3 + V_4 - 4V_0 = RC\frac{dV_0}{dt}. \tag{13.7}$$

Consider next the diffusion equation

$$\nabla^2 \mathbf{J} = \sigma\mu_0\mu_r \frac{\partial \mathbf{J}}{\partial t}. \tag{8.11}$$

If **J** has only the z-component and we are dealing with a two-dimensional problem, eqn. (8.11) becomes

$$\frac{\partial^2 J}{\partial x^2} + \frac{\partial^2 J}{\partial y^2} = \sigma\mu_0\mu_r \frac{\partial J}{\partial t}. \tag{13.8}$$

By using Taylor's series we can write

$$J = J_0 + h\left.\frac{\partial J}{\partial x}\right|_0 + \left.\frac{h^2}{2}\frac{\partial^2 J}{\partial x^2}\right|_0 \cdots$$

Applying this to a five-point star as in eqn. (13.7) we have

$$J_1 + J_3 - 2J_0 = h^2\left.\frac{\partial^2 J}{\partial x^2}\right|_0 \tag{13.9}$$

because the terms in h cancel out, h being positive for J_1 and negative for J_2. Similarly,

$$J_2 + J_4 - 2J_0 = h^2 \left.\frac{\partial^2 J}{\partial y^2}\right|_0 . \tag{13.10}$$

Hence, by substitution in eqn. (13.8),

$$\frac{1}{h^2}(J_1 + J_2 + J_3 + J_4 - 4J_0) = \sigma\mu_0\mu_r \left.\frac{\partial J}{\partial t}\right|_0 . \tag{13.11}$$

Comparison with eqn. (13.7) shows that the RC network can be used to solve problems involving the diffusion equation.

Similarly, the wave equation in a loss-free medium can be studied by means of an LC network as illustrated in Fig. 13.4. Such networks have

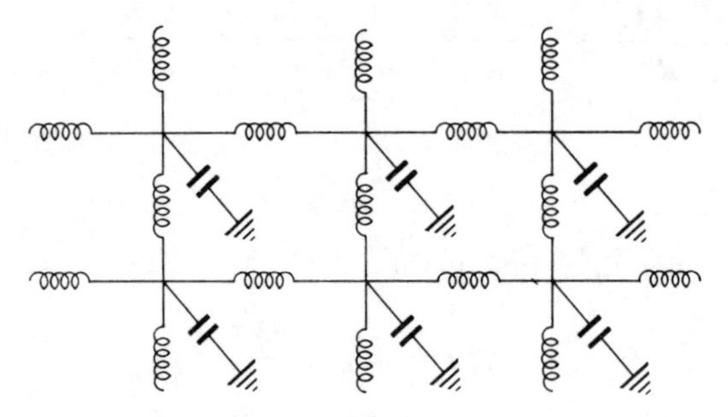

FIG. 13.4. An LC network for the wave equation.

been used in the study of waveguides. The introduction of mutual inductances makes possible the study of higher order equations which occur in other branches of engineering.

13.3. Numerical Computation

Interesting and useful as these various experimental methods are, they do not really begin to meet the needs of engineers seeking for the solution of electromagnetic problems. Before the advent of the large digital computer they often had to be employed for want of a better method. But today numerical calculation by means of high-speed digital computers is by far the most powerful and widely used method. The advent of these machines has gone a long way towards removing the geometrical limitations discussed at the beginning of this chapter. It is now possible to concentrate attention on the physical problems rather than on the shape of the apparatus. This represents a tremendous step forward. Another great advantage is that non-linear equations can be studied numerically even if they cannot be solved analytically.

It is, of course, impossible to do justice to this topic in a few pages. There is a considerable literature on numerical methods in electromagnetism† and it is growing rapidly. The subject needs a book to itself merely for an introductory survey. All that we can do is take an example to illustrate the general approach and then encourage the reader to launch out on his own.

13.4. The Finite-difference Method

For our example let us choose the powerful *finite-difference* method‡ which is used in the numerical solution of differential equations. We have already seen that many electromagnetic problems can be stated in terms of differential equations and a method which can be used to solve these equations therefore has a wide usefulness.

13.4.1. A Scheme for the Solution of Laplace's Equation

To illustrate the technique consider Laplace's equation in two dimensions

$$\frac{\partial^2 V}{\partial x^2} + \frac{\partial^2 V}{\partial y^2} = 0. \tag{13.12}$$

Let us solve the problem of finding the potential V in a region closed by boundaries at which V is known. Consider for instance the problem illustrated in Fig. 13.5. Since this problem involves simple rectangular boundaries it can be solved analytically, and the reader may feel we should attempt a more ambitious problem with irregularly shaped boundaries.

† See, for instance, P. Silvester, *Modern Electromagnetic Fields*, Prentice-Hall, 1968.

‡ For a full treatment see G. D. Smith, *Numerical Solutions of Partial Differential Equations*, Oxford University Press, 1965.

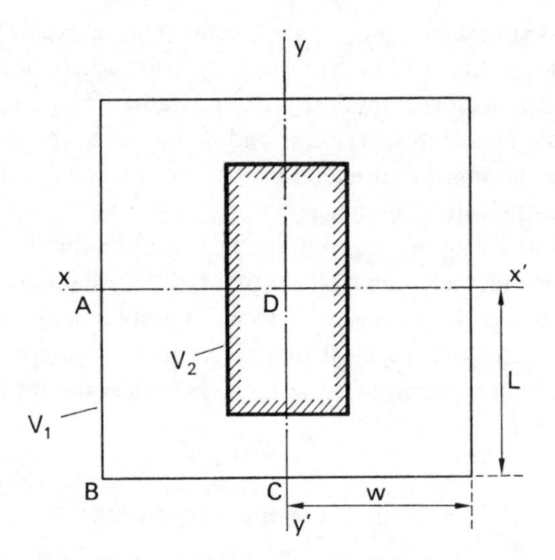

FIG. 13.5. A problem involving Laplace's equation.

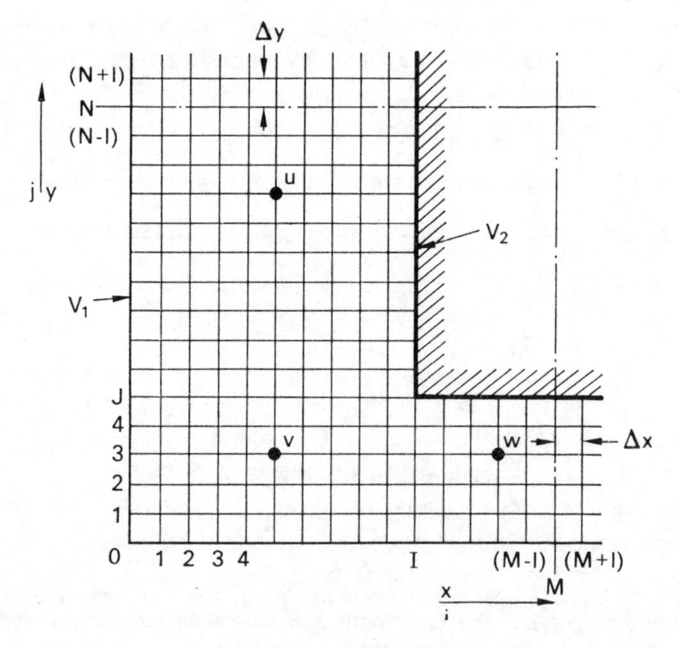

FIG. 13.6. A finite-difference grid.

We have chosen this simple problem, because it serves as a good introduction to the method. Because of symmetry about the axes XX' and YY' the solution needs to be obtained only for one quadrant as shown in Fig. 13.6.

The first step is to cover the area of interest with a computation grid as shown in Fig. 13.6. A rectangular grid is convenient, though not essential. With the origin of coordinates at the bottom left-hand corner, the grid lines are numbered in the x- and y-directions. A typical point (x, y) of the grid can then be identified by the coordinates (i, j) where $i = x/\Delta x$ and $j = y/\Delta y$. Δx and Δy are the grid dimensions. In the quadrant there are M steps in the x-direction and N in the y-direction. The potential at the node (i, j) is written $V_{(i,j)}$.

The second step is to replace the differential equation (13.12) by a finite difference equation relating the potential at any node (i, j) to the potential at the adjacent nodes. We use Taylor's series as in § 13.2.4.

$$V_{(i+1,j)} = V_{(i,j)} + \Delta x \left.\frac{\partial V}{\partial x}\right|_{i,j} + \frac{(\Delta x)^2}{2!} \left.\frac{\partial^2 V}{\partial x^2}\right|_{i,j} + \frac{(\Delta x)^3}{3!} \left.\frac{\partial^3 V}{\partial x^3}\right|_{i,j} + \cdots \tag{13.13}$$

In the same manner,

$$V_{(i-1,j)} = V_{(i,j)} - \Delta x \left.\frac{\partial V}{\partial x}\right|_{i,j} + \frac{(\Delta x)^2}{2!} \left.\frac{\partial^2 V}{\partial x^2}\right|_{i,j} - \frac{(\Delta x)^3}{3!} \left.\frac{\partial^3 V}{\partial x^3}\right|_{i,j} + \cdots \tag{13.14}$$

Adding eqn. (13.13) to eqn. (13.14),

$$V_{(i+1,i)} + V_{(i-1,j)} = 2V_{(i,j)} + (\Delta x)^2 \left.\frac{\partial^2 V}{\partial x^2}\right|_{i,j} + F_1(\Delta x), \tag{13.15}$$

where $F_1(\Delta x)$ represents terms containing fourth and higher order powers of Δx. Neglecting these terms we have

$$\left.\frac{\partial^2 V}{\partial x^2}\right|_{i,j} = \frac{V_{(i+1,j)} - 2V_{(i,j)} + V_{(i-1,j)}}{(\Delta x)^2}. \tag{13.16}$$

Similarly, under the same assumption,

$$\left.\frac{\partial^2 V}{\partial y^2}\right|_{i,j} = \frac{V_{(i,j+1)} - 2V_{(i,j)} + V_{(i,j-1)}}{(\Delta y)^2}. \tag{13.17}$$

If for convenience we choose a square mesh, so that $\Delta x = \Delta y$,

$$V_{(i+1,j)} + V_{(i-1,j)} + V_{(i,j+1)} + V_{(i,j-1)} - 4V_{(i,j)} = 0 \qquad (13.18)$$

is the finite-difference equation corresponding to Laplace's equation. This five point formula is illustrated in Fig. 13.7.

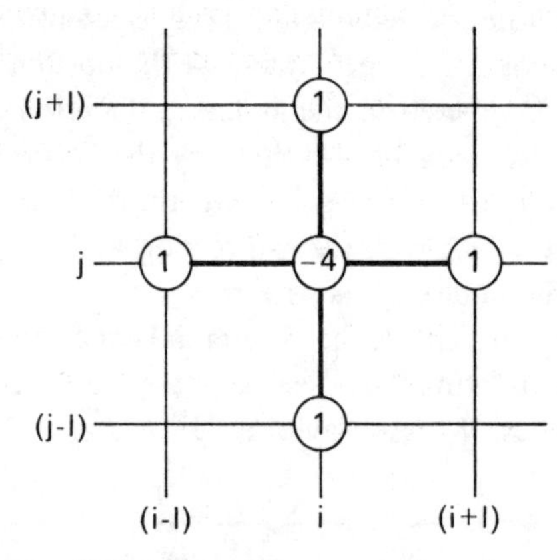

FIG. 13.7. A five-point computation scheme.

The third step is to express the boundary conditions in a suitable form to fit eqn. (13.18). In our example, Fig. 13.6, we have

$$\left.\begin{aligned} V_{(i,1)} &= V_1 \quad 0 < i < M, \\ V_{(1,j)} &= V_1 \quad 0 < j < N, \\ V_{(i,J)} &= V_2 \quad I < i < M, \\ V_{(I,j)} &= V_2 \quad J < j < N. \end{aligned}\right\} \qquad (13.19)$$

Also, symmetry about XX' and YY' requires that

$$\left.\begin{aligned} V_{(i,N+1)} &= V_{(i,N-1)} \quad j = N \\ V_{(M+1,j)} &= V_{(M-1,j)} \quad i = M, \end{aligned}\right\} \qquad (13.20)$$

so that at $j = N$ eqn. (13.18) becomes

$$V_{(i+1,N)} + V_{(i-1,N)} + 2V_{(i,N-1)} - 4V_{(i,N)} = 0 \qquad (13.21)$$

and at $i = M$

$$2V_{(M-1,j)} + V_{(M,j+1)} + V_{(M,j-1)} - 4V_{(M,j)} = 0. \qquad (13.22)$$

13.4.2. Methods of Solution for Problems Involving Laplace's Equation

There are two widely used methods. In the first, eqn. (13.18) is written at every node. Thus for the row $j = 1$ in our example,

$$\left.\begin{aligned} 2V_1 + V_{21} + V_{12} - 4V_{11} &= 0, \\ V_{11} + V_1 + V_{31} + V_{22} - 4V_{21} &= 0, \\ V_{21} + V_1 + V_{41} + V_{32} - 4V_{31} &= 0, \\ \cdots\cdots\cdots\cdots\cdots\cdots \\ V_{M-2,1} + V_1 + V_{M,1} + V_{M-1,2} - 4V_{M-1,1} &= 0, \\ 2V_{M-1,1} + V_1 + V_{M,2} - 4V_{M,1} &= 0. \end{aligned}\right\} \tag{13.23}$$

Similar equations can be written for the remaining rows. The complete set of equations relates the node potentials to the boundary conditions. These equations are then rearranged in matrix form and solved by Gaussian elimination.

An alternative method is to evaluate the potentials at each node by successive approximations or iterations. Several iterative schemes are available, the simplest of which rearranges eqn. (13.18) in the form

$$V_{(i,j)} = \tfrac{1}{4}\,|V_{(i+1,j)} + V_{(i-1,j)} + V_{(i,j+1)} + V_{i,j-1}|. \tag{13.24}$$

In Fig. 13.6 suppose that except at the boundaries, which are at fixed potentials, all other node potentials are assumed initially to be zero. Applying eqn. (13.24) to each node of row $j = 1$ we obtain

$$\left.\begin{aligned} V_{11} &= \tfrac{1}{4}\,|0 + V_1 + 0 + V_1|, \\ V_{21} &= \tfrac{1}{4}\,|0 + 0 + 0 + V_1|, \\ \cdots&\cdots\cdots\cdots\cdots\cdots \\ V_{M-1,1} &= \tfrac{1}{4}\,|0 + 0 + 0 + V_1|, \\ V_{M,1} &= \tfrac{1}{4}\,|2(0) + 0 + V_1|. \end{aligned}\right\} \tag{13.25}$$

Thus the node potentials are obtained from the initial assumed potentials. Repeated use of eqn. (13.24) obtains improved values until no further change results. The method relies on the estimation of $V_{(i,j)}$ in the $(n+1)$th scan from values obtained in the nth scan.

$$V_{i,j}^{n+1} = \tfrac{1}{4}\,|V_{i+1,j}^{n} + V_{i-1,j}^{n} + V_{i,j+1}^{n} + V_{i,j-1}^{n}|. \tag{13.26}$$

This is known as the Jacobi iterative scheme for eqn. (13.18).

The convergence of the process can be accelerated by using an iterative scheme which makes use of the latest estimated potentials as soon as these become available. Scanning from bottom to top by columns, we can use the formula

$$V_{i,j}^{n+1} = \tfrac{1}{4}\,|V_{i+1,j}^{n} + V_{i-1,j}^{n+1} + V_{i,j+1}^{n} + V_{i,j-1}^{n+1}|. \tag{13.27}$$

This is known as the Gauss–Seidel scheme. For a large number of iterations the Gauss–Seidel scheme can be shown to converge approximately twice as fast as the Jacobi scheme.

A still faster convergence can be obtained by using the successive over-relaxation scheme (SOR). The residual at the node (i, j) at the $(n+1)$th scan is defined as

$$R_{i,j} = V_{i+1,j}^{n} + V_{i-1,j}^{n+1} + V_{i,j+1}^{n} + V_{i,j-1}^{n+1} - 4V_{i,j}^{n}. \tag{13.28}$$

The SOR iteration scheme is given by

$$V_{i,j}^{n+1} = V_{i,j}^{n} + \tfrac{1}{4}KR_{i,j}, \tag{13.29}$$

where $1 < K < 2$. It will be seen that $K = 1$ reduces eqn. (13.29) to the Gauss–Seidel scheme. The best value of K to ensure the most rapid convergence has to be estimated.

In the SOR and Gauss–Seidel schemes the iteration formula must be applied systematically from one node to the next. This is not necessary in the Jacobi scheme, because only old potential values are used in the calculation of the new estimate.

13.4.3. Finite-difference Approximations to Other Differential Equations

The finite-difference method can readily be applied to the other differential equations met in electromagnetism.

Poisson's equation is of the form

$$\frac{\partial^2 V}{\partial x^2} + \frac{\partial^2 V}{\partial y^2} = f(x, y). \tag{13.30}$$

In finite-difference form we can write

$$V_{i+1,j} + V_{i-1\ j} + V_{i,j+1} + V_{i,j-1} - 4V_{i,j} = f_{i,j}. \tag{13.31}$$

The methods of solution used for Laplace's equation are also applicable to Poisson's equation.

The diffusion equation is of the form

$$\frac{\partial^2 V}{\partial x^2} = \alpha \frac{\partial V}{\partial t} \tag{13.32}$$

and the wave equation

$$\frac{\partial^2 V}{\partial x^2} = \beta \frac{\partial^2 V}{\partial t^2}. \tag{13.33}$$

Let the suffix i denote the space variable and j the time variable in these two equations. Consider first the diffusion equation:

$$\frac{\partial V}{\partial t} = \frac{1}{\Delta t} |V_{i,j+1} - V_{i,j}|. \tag{13.34}$$

Hence it follows that

$$\frac{V_{i+1,j} - 2V_{i,j} + V_{i-1,j}}{(\Delta x)^2} = \alpha \frac{V_{i,j+1} - V_{i,j}}{\Delta t}$$

or

$$V_{i,j+1} = V_{i,j} + \frac{\Delta t}{\alpha (\Delta x)^2} |V_{i+1,j} - 2V_{i,j} + V_{i-1,j}|. \tag{13.35}$$

For the wave equation:

$$\frac{V_{i+1,j} - 2V_{i,j} + V_{i-1,j}}{(\Delta x)^2} = \beta \frac{V_{i,j+1} - 2V_{i,j} + V_{i,j-1}}{(\Delta t)^2}$$

or

$$V_{i,j+1} = \gamma V_{i+1,j} + |2(1-\gamma) V_{i,j} + \gamma V_{i-1,j} - V_{i,j-1}|, \tag{13.36}$$

where

$$\gamma = \frac{1}{\beta}\left(\frac{\Delta t}{\Delta x}\right)^2. \tag{13.37}$$

The solution of eqns. (13.35) and (13.36) is somewhat different from that described for Laplace's equation. For the diffusion equation two space conditions and one time condition need to be specified, because the space coordinate appears as a derivative of the second order and the time coordinate as a derivative of the first order. Suppose that the boundary conditions are that $V(x, t)$ is given for all t at $x = 0$ and $x = L$, and that the initial condition $V(x, 0)$ is specified for all x. Figure 13.8 shows

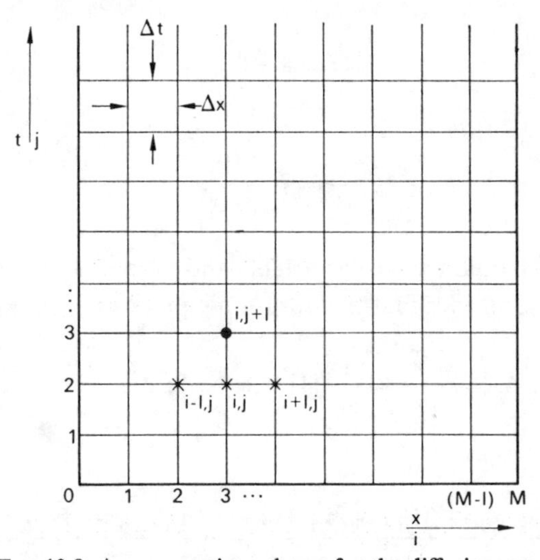

FIG. 13.8. A computation scheme for the diffusion equation.

a suitable computation grid. Examination of eqn. (13.35) shows that we can obtain one point ahead in time from the values of the previous time row. Thus the solution is directly evaluated for one time row and an iterative scheme is not required. The same is true for eqn. (13.36), where the information on two previous time rows is used for the solution of the unknown time row. The solutions therefore move forward in time, row by row. This is known as an *explicit* finite difference scheme.

It is possible by using alternative approximations to derive other finite-difference equations representing the diffusion and wave equations, in which the solution at each time row is not calculated explicitly from the previous time rows. For example, several unknown values can be related to several known values in one equation. Such *implicit* schemes are usually solved by iteration.

13.4.4. Difficulties Encountered in the Use of Finite-difference Equations

It is inevitable that in the representation of differential equations by finite-difference equations there should be some error due to approximation. Thus in eqn. (13.15) we decided to neglect terms containing fourth or higher orders of Δx. The accuracy of the representation will therefore depend on the mesh size. Since a smaller mesh requires more computation a compromise has to be achieved, which gives sufficient accuracy without being wasteful of computer time. The error due to the finite mesh size is called the *discretization error*. In addition there are *round-off errors* due to the fact that numerical methods have to be carried out with a limited number of significant figures.

In order to be useful a finite-difference scheme must be *compatible* with the differential equation and it must have *stability*. Compatibility ensures that the numerical solution converges to the solution of the original equation. This is not always easy to ensure, because a particular finite-difference scheme may represent several differential equations to different approximations. This may occur in the study of the diffusion and wave equations. It then becomes important to examine the discretization error to ensure that the representation favours the particular equation which is being solved.

Stability may be lost if the various errors accumulate without limit during the computation. Stability conditions can be derived for some equations. Thus for eqn. (13.35) stability is obtained only if

$$\frac{\Delta t}{\alpha\,(\Delta x)^2} \leqslant \frac{1}{2} \tag{13.38}$$

and for eqn. (13.36) if

$$\frac{1}{\beta}\left(\frac{\Delta t}{\Delta x}\right)^2 \leqslant 1. \tag{13.39}$$

13.4.5. A Typical Computer Program

Numerical solutions are usually obtained with the aid of a digital computer. Such computers are ideally suited to performing repetitive arithmetical operations. They possess the following advantages: speed, accuracy, the ability to process both very large and very small numbers and the

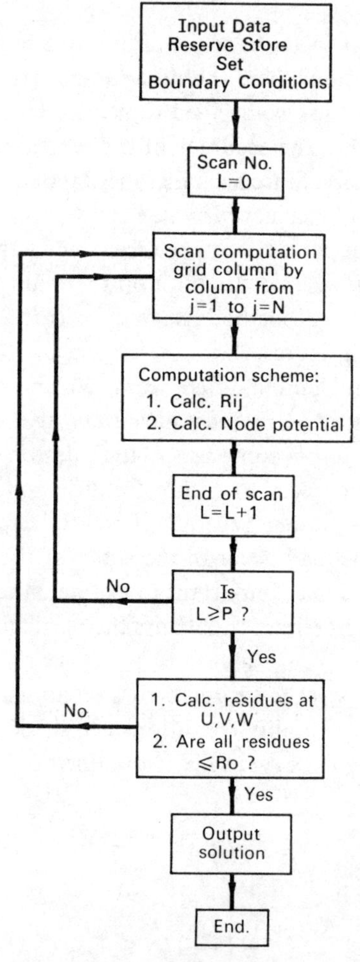

FIG. 13.9. A computer program.

ability to retain numerical data in store for further mathematical processing as required. Typical figures for a medium-size computer at the present time are: cycle time 2·25 μs, calculations to 11 significant figures, numbers in the range 5×10^{76} to -5×10^{76} can be processed, 32,000 numbers can be retained in the high-speed memory. This is an impressive specification, but there is no doubt that in the near future computers will be capable of much improved performance.

However large the computer is, it is important to make the best use of it. As in all engineering calculations, care has to be taken at the beginning to decide on the amount of information which is essential. The instructions given to the computer are called the computer program, and it is of great importance to devise an efficient program.

To illustrate the form a program may take, consider the solution of the problem shown in Fig. 13.6. A possible program structure based on the SOR scheme is shown in Fig. 13.9. The input data consist of Δx, Δy, V_1, V_2, K, I, J, M, and N. The core array S_1 of the computer is reserved for the computation grid and the boundary conditions are set. The computer is then instructed to apply eqn. (13.29) systematically, scanning the computation grid from $j = 1$ to $j = N$, beginning at column $i = 1$ and ending at column $i = M$. In principle the solution is obtained when the residual at every node in the computation grid is zero. In practice an upper limit is imposed on the residual so that there is sufficient accuracy. Thus the residuals at several points, e.g. U, V, W in Fig. 13.6, can be monitored continuously after P scans and then compared at the end of each following scan to a limit of say R_0. When the moduli of the residuals at the test points are less than R_0, the computation is terminated. Alternatively the maximum residual is examined. The choice of R_0 is important, otherwise computation time is wasted.

Summary

An analytical study of electromagnetism is essential in order to give insight into the subject. Analytical solutions are, however, severely limited because they require a simple geometry.

In this chapter we have discussed two alternative methods of finding solutions. The first is an experimental method, which employs various analogues to the electromagnetic field. The second is a numerical method and is generally carried out by means of high-speed electronic digital computers. The development of such computers has had a profound effect on electromagnetic investigations and numerical computation provides by far the most important and useful method.

Exercises

13.1. Explain why it is possible to determine an electrostatic field in non-conducting dielectric by measurements on a conducting sheet or electrolytic tank. What is the analogue for electric charge?

13.2. Explain why it is possible to represent a magnetostatic field without current sources by means of a conducting analogue. Show that there are two possible methods, in one of which a constant electric potential represents a highly permeable surface and in the other a flux line.

13.3. Explain why it is possible to represent a magnetostatic field with current sources by means of a conducting analogue. Show that a flux line becomes an equipotential and flux is proportional to current flow.

13.4. Explain how the distribution of flux in a transformer stamping can be investigated by means of a network analogue. How would you represent the condition (a) that the surface magnetic field strength is given, or (b) that the total flux is given? [Ans.: RC network; voltage at boundary; total current.]

13.5. A magnetic field strength $H = 100 \sin(100\pi t)$ is applied at time $t = 0$ parallel to the surfaces of an infinite copper sheet of half-thickness $d = 1{\cdot}0$ cm and electrical conductivity $\sigma = 5{\cdot}8 \times 10^7$ siemens. Obtain, using a digital computer, a numerical solution of the magnetic field in the sheet using the ordinary explicit method of eqn. (13.35) and the computation grid of Fig. 13.8.

[Ans.: In the interior of the sheet

$$\frac{\partial^2 H}{\partial x^2} = \mu_r \mu_0 \frac{\partial H}{\partial t} = \mu_0 \frac{\partial H}{\partial t}.$$

It is convenient to normalize the equation by putting

$$\bar{x} = x/d, \qquad \tau = t/T_0,$$

where T_0 is the periodic time 1/50 s. We then have

$$\frac{\partial^2 H}{\partial \bar{x}^2} = \alpha \frac{\partial H}{\partial \tau},$$

where

$$\alpha = \frac{1}{\pi}\left(\frac{d}{\delta}\right)^2.$$

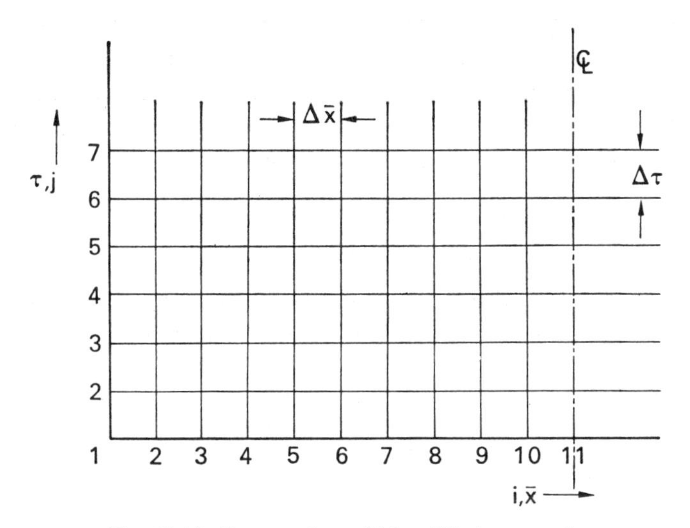

FIG. 13.10. Computation grid for diffusion equation.

Refer to Fig. 13.10. Then $x = (i - 1)\Delta x$ and $\tau = (j - 1)\Delta\tau$. Let the centre line of the sheet be at $i = 11$ and the surfaces be at $i = 1$ and $i = 21$. At the surface $i = 1$ we have

$$H(1, j) = 100 \sin [2\pi (j - 1)\Delta\tau]$$

and at the centre-line $i = 11$

$$\frac{\partial H}{\partial x} = 0$$

or

$$\frac{H(12, j) - H(10, j)}{2\Delta\bar{x}} = 0.$$

Using eqn. (13.35) we have for $1 < i < 11$,

$$H(i, j + 1) = H(i, j) + r[H(i - 1, j) - 2H(i, j) + H(i + 1, j)]$$

and at $i = 11$

$$H(11, j + 1) = H(11, j) + 2r[H(10, j) - H(11, j)],$$

where

$$r = \frac{\Delta\tau}{\alpha(\Delta\bar{x})^2} \leqslant \frac{1}{2}.$$

Since there are 10 spaces for the half-thickness, we have

$$\Delta\bar{x} = 0.1.$$

Hence, for stability of the computational process [eqn. (13.38)],

$$(\Delta\tau)_{\max} = \tfrac{1}{2}\alpha(\Delta\bar{x})^2 = 0{\cdot}00182.$$

It is convenient to have an integral number of time intervals per cycle and $\Delta\tau$ should be chosen on this basis.

Computation should scan the grid starting from $i = 2, j = 2$ to $i = 11, j = 2$, progressing row by row. It will be found that the numerical solution proceeds through the transient to the steady state, as is shown in Fig. 13.11. (It is intended that the computation should be carried out on a digital computer.)]

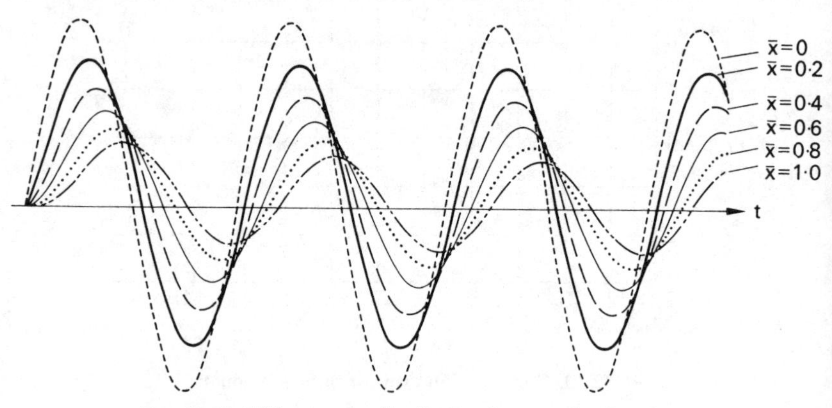

FIG. 13.11. Eddy current distribution in a conducting sheet.

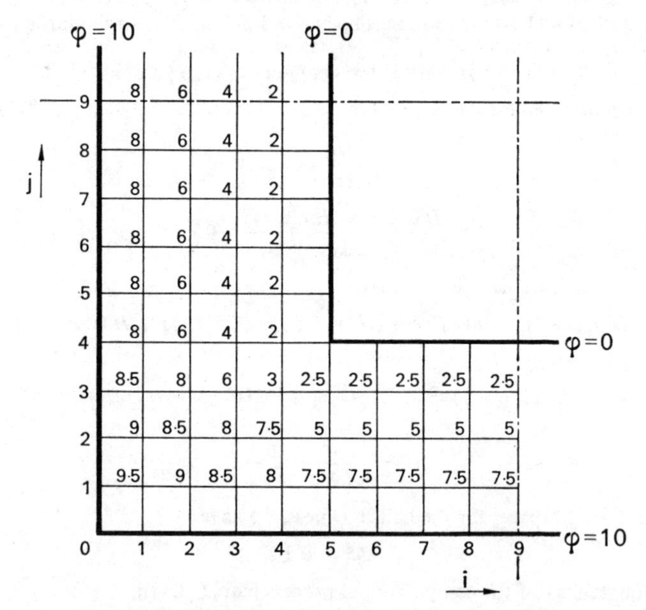

FIG. 13.12. Computation grid for Laplace's equation.

13.6. Figure 13.12 illustrates a potential problem governed by Laplace's equation. The potentials on the boundaries are known and those in the interior of the region have been guessed. The correct interior potentials are required. This problem can be solved with the help of a desk calculator or, if need be, by hand.

[*Procedure*: (1) Draw a computation grid of convenient size—say 2 cm by 2 cm—and insert by pencil the assumed potentials.

(2) Calculate the residuals defined by

$$R(i,j) = \phi(i+1,j) + \phi(i-1,j) + \phi(i,j+1) + \phi(i,j-1) - 4\phi(i,j).$$

For example $R(3, 2) = 7{\cdot}5 + 8{\cdot}5 + 6 + 85 - 4\times 8 = -1{\cdot}5$. Record the residuals at the nodes by pencil.

(3) Proceed with the relaxation process, starting with the largest residual. Suppose $R(3, 2)$ is the largest. Set it to zero by reducing the node potential from 8 to $8 - \frac{1}{2}(1{\cdot}5) = 7{\cdot}625$. This alteration will affect the residuals $R(4, 2)$, $R(2, 2)$, $R(3, 3)$, and $R(3, 1)$ and all these must be reduced by 0·375.

(4) Continue this process, noting that the boundary potentials are not changed, until the modulus of all residuals $\leqslant 0{\cdot}1$.

(5) Compare the solution with an estimate obtained by visual inspection. The problem has been chosen so that linear interpolation in the region $1 < i < 5$, $1 < j < 4$ will give an almost exact result.

(6) Try refining the solution in the region $1 < i < 5$, $1 < j < 4$ by halving the mesh size.

13.7. Obtain a solution for the problem of Fig. 13.12 using the SOR scheme with $K = 1{\cdot}5$ [see eqn. (13.29)].

[*Procedure*: Set the initial node potentials $(i,j) = 0$. Then apply eqns. (13.28) and (13.29) in the manner described for the Gauss–Seidel scheme, eqn. (13.27).]

13.8. Discuss the application of Fig. 13.12 to problems involving capacitance, inductance and resistance.

APPENDIX

A1. Three Common Coordinate Systems

A1.1. Rectangular coordinates x, y, z; Fig. A1.
A1.2. Cylindrical coordinates r, θ, z; Fig. A2.
A1.3. Spherical coordinates r, θ, ϕ; Fig. A3.

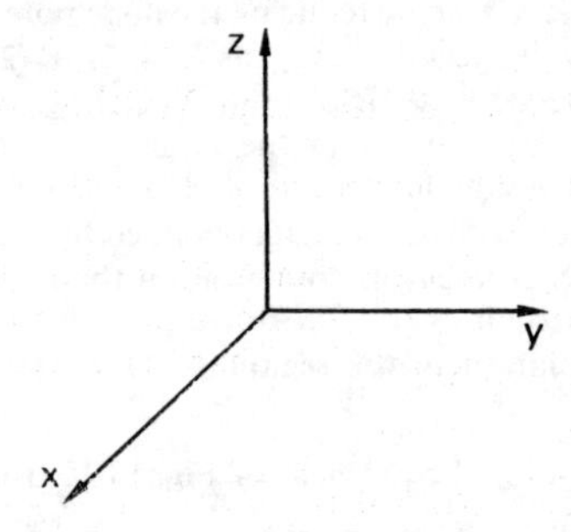

FIG. A1. Rectangular coordinates.

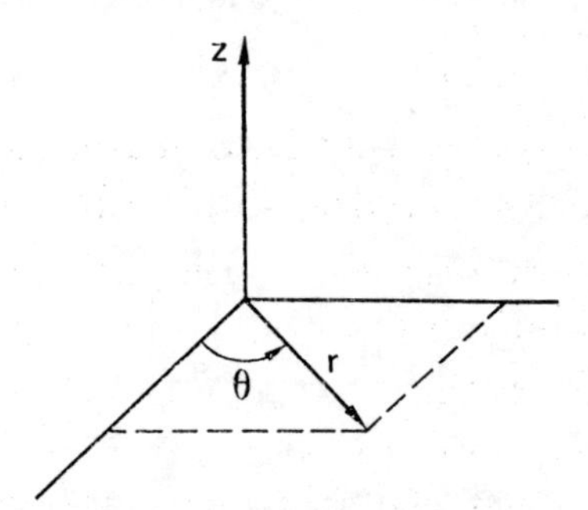

FIG. A2. Cylindrical coordinates.

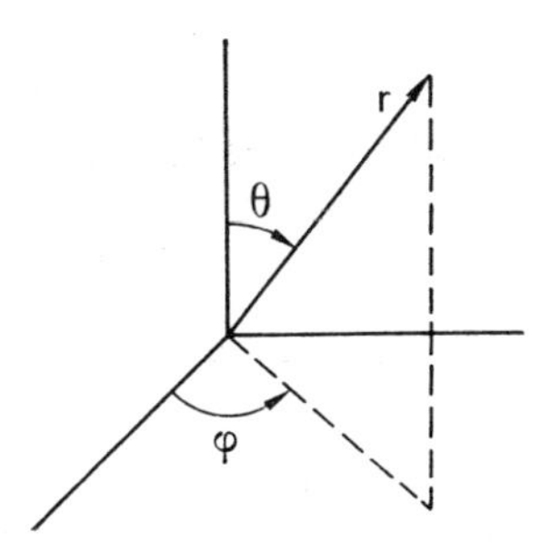

FIG. A3. Spherical coordinates.

A 2. Vector Addition and Multiplication

A2.1. $\mathbf{F} + \mathbf{G}$ is obtained by finding the diagonal of the parallelogram formed by the vectors.
$\mathbf{F} \cdot \mathbf{G} = FG \cos \alpha$, where α is the angle between the vectors.
$\mathbf{F} \times \mathbf{G} = \mathbf{u}_n FG \sin \alpha$, where $\mathbf{u}_n$ is the unit vector perpendicular to $\mathbf{F}$ and $\mathbf{G}$ in the direction of motion of a right-handed screw.

A2.2. In rectangular coordinates

$$\mathbf{F} + \mathbf{G} = \mathbf{u}_x (F_x + G_x) + \mathbf{u}_y (F_y + G_y) + \mathbf{u}_z (F_z + G_z)$$

$$\mathbf{F} \cdot \mathbf{G} = F_x G_x + F_y G_y + F_z G_z$$

$$\mathbf{F} \times \mathbf{G} = \begin{vmatrix} \mathbf{u}_x & \mathbf{u}_y & \mathbf{u}_z \\ F_x & F_y & F_z \\ G_x & G_y & G_z \end{vmatrix}$$

A2.3. In cylindrical coordinates (if $\mathbf{F}$ and $\mathbf{G}$ are at the same point)

$$\mathbf{F} + \mathbf{G} = \mathbf{u}_r (F_r + G_r) + \mathbf{u}_\theta (F_\theta + G_\theta) + \mathbf{u}_z (F_z + G_z)$$

$$\mathbf{F} \cdot \mathbf{G} = F_r G_r + F_\theta G_\theta + F_z G_z$$

$$\mathbf{F} \times \mathbf{G} = \begin{vmatrix} \mathbf{u}_r & \mathbf{u}_\theta & \mathbf{u}_z \\ F_r & F_\theta & F_z \\ G_r & G_\theta & G_z \end{vmatrix}$$

A2.4. In spherical coordinates ($\mathbf{F}$ and $\mathbf{G}$ at the same point)

$$\mathbf{F} + \mathbf{G} = \mathbf{u}_r (F_r + G_r) + \mathbf{u}_\theta (F_\theta + G_\theta) + \mathbf{u}_\phi (F_\phi + G_\phi)$$

$$\mathbf{F} \cdot \mathbf{G} = F_r G_r + F_\theta G_\theta + F_\phi G_\phi$$

$$\mathbf{F} \times \mathbf{G} = \begin{vmatrix} \mathbf{u}_r & \mathbf{u}_\theta & \mathbf{u}_\phi \\ F_r & F_\theta & F_\phi \\ G_r & G_\theta & G_\phi \end{vmatrix}$$

A2.5.
$$F^2 = \mathbf{F}\cdot\mathbf{F}$$
$$\mathbf{F}+\mathbf{G} = \mathbf{G}+\mathbf{F}$$
$$\mathbf{F}\cdot\mathbf{G} = \mathbf{G}\cdot\mathbf{F}$$
$$\mathbf{F}\times\mathbf{G} = -\mathbf{G}\times\mathbf{F}$$
$$(\mathbf{F}+\mathbf{G})\cdot\mathbf{H} = \mathbf{F}\cdot\mathbf{H}+\mathbf{G}\cdot\mathbf{H}$$
$$(\mathbf{F}+\mathbf{G})\times\mathbf{H} = \mathbf{F}\times\mathbf{H}+\mathbf{G}\times\mathbf{H}$$
$$\mathbf{F}\cdot\mathbf{G}\times\mathbf{H} = \mathbf{G}\cdot\mathbf{H}\times\mathbf{F} = \mathbf{H}\cdot\mathbf{F}\times\mathbf{G}$$
$$\mathbf{F}\times(\mathbf{G}\times\mathbf{H}) = (\mathbf{F}\cdot\mathbf{H})\,\mathbf{G} - (\mathbf{F}\cdot\mathbf{G})\,\mathbf{H}$$

A 3. Line, Surface, and Volume Integrals

A3.1. $\int \mathbf{F}\cdot d\mathbf{l}$ (work done by **F**)

$\int F_x\,dx + \int F_y\,dy + \int F_z\,dz$

$\int F_r\,dr + \int F_\theta r\,d\theta + \int F_z\,dz$

$\int F_r\,dr + \int F_\theta r\,d\theta + \int F_\phi\,r\sin\theta\,d\phi$

A3.2. $\iint \mathbf{F}\cdot d\mathbf{s}$ (flow of **F**)

$\iint F_x\,dy\,dz + \iint F_y\,dz\,dx + \iint F_z\,dx\,dy$

$\iint F_r r\,d\theta\,dz + \iint F_\theta\,dz\,dr + \iint F_z\,dr\,r\,d\theta$

$\iint F_r r^2\sin\theta\,d\theta\,d\phi + \iint F_\theta r\sin\theta\,dr\,d\phi + \iint F_\phi r\,dr\,d\theta$

A3.3. $\iiint U\,dv$

$\iiint U\,dx\,dy\,dz$

$\iiint Ur\,dr\,d\theta\,dz$

$\iiint Ur^2\sin\theta\,dr\,d\theta\,d\phi$

A 4. Gradient, Divergence, Curl, and Laplacian

A4.1. $\text{grad}\,V = \nabla V = \mathbf{u}_l\,\dfrac{dV}{dl}$ (maximum slope)

$\text{div}\,\mathbf{F} = \nabla\cdot\mathbf{F} = \lim\limits_{v\to 0}\dfrac{1}{v}\oiint \mathbf{F}\cdot d\mathbf{s}$ (outflow per unit volume)

$\text{curl}\,\mathbf{F} = \nabla\times\mathbf{F} = \lim\limits_{s\to 0}\dfrac{\mathbf{u}_n}{s}\oint \mathbf{F}\cdot d\mathbf{l}$ (circulation per unit area)

$\nabla^2 V = \text{div}\,(\text{grad}\,V) = \nabla\cdot\nabla V$

$\nabla^2\mathbf{F} = \text{grad}\,(\text{div}\,\mathbf{F}) - \text{curl}\,(\text{curl}\,\mathbf{F})$

A4.2. In rectangular coordinates it is possible to define a single differential operator

$$\nabla = \mathbf{u}_x \frac{\partial}{\partial x} + \mathbf{u}_y \frac{\partial}{\partial y} + \mathbf{u}_z \frac{\partial}{\partial z}$$

Hence

$$\nabla V = \mathbf{u}_x \frac{\partial V}{\partial x} + \mathbf{u}_y \frac{\partial V}{\partial y} + \mathbf{u}_z \frac{\partial V}{\partial z}$$

$$\nabla . \mathbf{F} = \frac{\partial F_x}{\partial x} + \frac{\partial F_y}{\partial y} + \frac{\partial F_z}{\partial z}$$

$$\nabla \times \mathbf{F} = \begin{vmatrix} \mathbf{u}_x & \mathbf{u}_y & \mathbf{u}_z \\ \dfrac{\partial}{\partial x} & \dfrac{\partial}{\partial y} & \dfrac{\partial}{\partial z} \\ F_x & F_y & F_z \end{vmatrix}$$

$$\nabla^2 V = \frac{\partial^2 V}{\partial x^2} + \frac{\partial^2 V}{\partial y^2} + \frac{\partial^2 V}{\partial z^2}$$

$$\nabla^2 \mathbf{F} = \mathbf{u}_x \, \nabla^2 F_x + \mathbf{u}_y \, \nabla^2 F_y + \mathbf{u}_z \, \nabla^2 F_z$$

A4.3. In cylindrical coordinates (and all curvilinear coordinates) no single differential operator can be defined. It is necessary to use the definitions in § A 4.1.

$$\nabla V = \mathbf{u}_r \frac{\partial V}{\partial r} + \mathbf{u}_\theta \frac{1}{r} \frac{\partial V}{\partial \theta} + \mathbf{u}_z \frac{\partial V}{\partial z}$$

$$\nabla . \mathbf{F} = \frac{1}{r} \frac{\partial (rF_r)}{\partial r} + \frac{1}{r} \frac{\partial F_\theta}{\partial \theta} + \frac{\partial F_z}{\partial z}$$

$$\nabla \times \mathbf{F} = \begin{vmatrix} \dfrac{\mathbf{u}_r}{r} & \mathbf{u}_\theta & \dfrac{\mathbf{u}_z}{r} \\ \dfrac{\partial}{\partial r} & \dfrac{\partial}{\partial \theta} & \dfrac{\partial}{\partial z} \\ F_r & rF_\theta & F_z \end{vmatrix}$$

$$\nabla^2 V = \frac{\partial^2 V}{\partial r^2} + \frac{1}{r} \frac{\partial V}{\partial r} + \frac{1}{r^2} \frac{\partial^2 V}{\partial \theta^2} + \frac{\partial^2 V}{\partial z^2}$$

$$\nabla^2 \mathbf{F} = \mathbf{u}_r \left(\nabla^2 F_r - \frac{2}{r^2} \frac{\partial F_\theta}{\partial \theta} - \frac{F_r}{r^2} \right)$$

$$+ \mathbf{u}_\theta \left(\nabla^2 F_0 + \frac{2}{r^2} \frac{\partial F_r}{\partial \theta} - \frac{F_\theta}{r^2} \right) + \mathbf{u}_z \, (\nabla^2 F_z)$$

A4.4. In spherical coordinates

$$\nabla V = \mathbf{u}_r \frac{\partial V}{\partial r} + \mathbf{u}_\theta \frac{1}{r}\frac{\partial V}{\partial \theta} + \mathbf{u}_\phi \frac{1}{r \sin\theta}\frac{\partial V}{\partial \phi}$$

$$\nabla . \mathbf{F} = \frac{1}{r^2}\frac{\partial (r^2 F_r)}{\partial r} + \frac{1}{r\sin\theta}\frac{\partial}{\partial\theta}(F_0 \sin\theta) + \frac{1}{r\sin\theta}\frac{\partial F_\phi}{\partial\phi}$$

$$\nabla \times \mathbf{F} = \begin{vmatrix} \dfrac{\mathbf{u}_r}{r^2 \sin\theta} & \dfrac{\mathbf{u}_\theta}{r\sin\theta} & \dfrac{\mathbf{u}_\phi}{r} \\ \dfrac{\partial}{\partial r} & \dfrac{\partial}{\partial\theta} & \dfrac{\partial}{\partial\phi} \\ F_r & rF_\theta & r\sin\theta F_\phi \end{vmatrix}$$

$$\nabla^2 V = \frac{1}{r^2}\frac{\partial}{\partial r}\left(r^2 \frac{\partial V}{\partial r}\right) + \frac{1}{r^2 \sin\theta}\frac{\partial}{\partial\theta}\left(\sin\theta \frac{\partial V}{\partial\theta}\right) + \frac{1}{r^2 \sin^2\theta}\frac{\partial^2 V}{\partial\phi^2}$$

A5. Differentiation of Vectors

$$\nabla (U + V) = \nabla U + \nabla V$$

$$\nabla . (\mathbf{F} + \mathbf{G}) = \nabla . \mathbf{F} + \nabla . \mathbf{G}$$

$$\nabla \times (\mathbf{F} + \mathbf{G}) = \nabla \times \mathbf{F} + \nabla \times \mathbf{G}$$

$$\nabla (UV) = U \nabla V + V \nabla U$$

$$\nabla . (V\mathbf{F}) = V\nabla . \mathbf{F} + \mathbf{F} . \nabla V$$

$$\nabla \times (V\mathbf{F}) = V\nabla \times \mathbf{F} + (\nabla V)\times \mathbf{F}$$

$$\nabla . (\mathbf{F}\times\mathbf{G}) = \mathbf{G} . \nabla\times\mathbf{F} - \mathbf{F} . \nabla\times\mathbf{G}$$

$$\nabla \times (\mathbf{F}\times\mathbf{G}) = (\mathbf{G} . \nabla)\mathbf{F} - (\mathbf{F} . \nabla)\mathbf{G} + \mathbf{F}\nabla . \mathbf{G} - \mathbf{G}\nabla . \mathbf{F}$$

$$\nabla (\mathbf{F} . \mathbf{G}) = (\mathbf{G} . \nabla)\mathbf{F} + (\mathbf{F} . \nabla)\mathbf{G} + \mathbf{G}\times\nabla\times\mathbf{F} + \mathbf{F}\times\nabla\times\mathbf{G}$$

$$\nabla \times \nabla V = 0$$

$$\nabla . \nabla \times \mathbf{F} = 0$$

A6. Integral Relationships of Vectors

$$\iiint \nabla . \mathbf{F}\, dv = \oiint \mathbf{F} . d\mathbf{s}$$ (Gauss's or divergence theorem)

$$\iint \nabla\times\mathbf{F} . d\mathbf{s} = \oint \mathbf{F} . d\mathbf{l}$$ (Stokes's or circulation theorem)

$$\iiint \nabla V\, dv = \oiint V\, d\mathbf{s}$$

$$\iiint \nabla\times\mathbf{F}\, dv = -\oiint \mathbf{F}\times d\mathbf{s}$$

$$\iiint (\nabla U . \nabla V + U\nabla^2 V)\, dv = \oiint U\, \nabla V . d\mathbf{s}$$

$$\iiint (U\,\nabla^2 V - V\,\nabla^2 U)\, dv = \oiint (U\nabla V - V\nabla U) . d\mathbf{s}$$

These relations are often called Green's theorem

A 7. Electromagnetic Relationships

$$\left.\begin{aligned}\nabla\times\mathbf{E} &= -\frac{\partial\mathbf{B}}{\partial t}\\ \nabla\times\mathbf{H} &= \mathbf{J}+\frac{\partial\mathbf{D}}{\partial t}\\ \nabla\cdot\mathbf{D} &= \varrho \text{ free}\\ \nabla\cdot\mathbf{B} &= 0\\ \mathbf{D} &= \varepsilon_0\varepsilon_r\mathbf{E}\\ \mathbf{B} &= \mu_0\mu_r\mathbf{H}\end{aligned}\right\}\text{ Maxwell's equations}$$

$$\mathrm{J} = \sigma\mathbf{E} \qquad \text{(Ohm's law)}$$

In electrostatics

$$\mathbf{E} = -\nabla V$$

$$\nabla^2 V = -\varrho/\varepsilon \qquad \text{(Poisson's equation)}$$

$$\nabla^2 V = 0 \qquad \text{(Laplace's equation)}$$

$$V = \frac{1}{4\pi\varepsilon}\iiint\frac{\varrho(\mathbf{r}')}{|\mathbf{r}-\mathbf{r}'|}\,dv'$$

($\mathbf{r}'$ is the source coordinate and $\mathbf{r}$ the field coordinate)

In magnetostatics

$$\mathbf{B} = \nabla\times\mathbf{A}$$

$$\nabla\cdot\mathbf{A} = 0$$

$$\nabla^2\mathbf{A} = -\mu\mathbf{J}$$

$$\mathbf{A} = \frac{\mu}{4\pi}\iiint\frac{\mathrm{J}(\mathbf{r}')}{|\mathbf{r}-\mathbf{r}'|}\,dv'$$

or if the sources are treated as poles

$$\mathbf{H} = -\nabla V^*$$

$$\nabla^2 V^* = -\varrho^*/\mu$$

In time-varying fields

$$\mathbf{B} = \nabla\times\mathbf{A}$$

$$\nabla\cdot\mathbf{A} = -\frac{1}{c^2}\frac{\partial V}{\partial t}$$

$$\mathbf{E} = -\frac{\partial\mathbf{A}}{\partial t} - \nabla V$$

$$\nabla^2 V - \frac{1}{c^2}\frac{\partial^2 V}{\partial t^2} = -\varrho/\varepsilon$$

$$\nabla^2 \mathbf{A} - \frac{1}{c^2}\frac{\partial^2 \mathbf{A}}{\partial t^2} = -\mu \mathbf{J}$$

$$V = \frac{1}{4\pi\varepsilon}\iiint \frac{\varrho\left(\mathbf{r}', t - \frac{|\mathbf{r}-\mathbf{r}'|}{c}\right)}{|\mathbf{r}-\mathbf{r}'|}\,dv'$$

$$\mathbf{A} = \frac{\mu}{4\pi}\iiint \frac{\mathbf{J}\left(\mathbf{r}', t - \frac{|\mathbf{r}-\mathbf{r}'|}{c}\right)}{|\mathbf{r}-\mathbf{r}'|}\,dv'$$

A 8. Some Useful Constants

Velocity of light	$c = 2{\cdot}998 \times 10^8$ m/s
Magnetic constant	$\mu_0 = 4\pi \times 10^{-7}$ H/m
Electric constant	$\varepsilon_0 = 8{\cdot}854 \times 10^{-12}$ F/m
Electronic charge	$e = 1{\cdot}602 \times 10^{-19}$ C
Electronic mass	$m = 9{\cdot}108 \times 10^{-31}$ kg
	$e/m = 1{\cdot}759 \times 10^{11}$ C/kg
Electron volt	eV $= 1{\cdot}602 \times 10^{-19}$ J
Gravitational constant	$g = 9{\cdot}81$ m/s^2
Electrical energy	kWh $= 3{\cdot}6 \times 10^6$ J

INDEX